RISC/CISC Development and Test Support

RISC/CISC Development and Test Support

Marvin Hobbs

PRENTICE HALL
Englewood Cliffs, New Jersey 07632

Library of Congress Cataloging-in-Publiction Data

Hobbs, Marvin.
RISC/CISC development and test support / Marvin Hobbs.
p. cm.
Includes bibliographical references and index.
ISBN 0-13-388414-7
1. RISC microprocessors. 2. CISC microprocessors. I. Title.
TK7895.M5H62 1992
621.39'16—dc20 90-26110
CIP

Acquisition Editor: Karen Gettman
Editorial supervision: Cyndy Lyle Rymer
Cover Design: Karen Stephens
Prepress buyer: Kelly Behr
Manufacturing buyer: Susan Brunke

Printed in the United States of America
10 9 8 7 6 5 4 3 2 1

ISBN 0-13-388414-7

Prentice-Hall International (UK) Limited, *London*
Prentice-Hall of Australia Pty. Limited, *Sydney*
Prentice-Hall Canada Inc., *Toronto*
Prentice-Hall Hispanoamericana, S.A., *Mexico*
Prentice-Hall of India Private Limited, *New Delhi*
Prentice-Hall of Japan, Inc., *Tokyo*
Simon & Schuster Asia Pte. Ltd., *Singapore*
Editora Prentice-Hall do Brasil, Ltda., *Rio de Janeiro*

Contents

Chapter 2
RISC Microprocessors / 26

Chapter 3
32-Bit CISC Microprocessors / 56

Chapter 4
A Comprehensive Development System for 32-Bit CISC Microprocessors / 81

Chapter 5
Logic Analyzers and Monitors / 106

Chapter 6
RISC and Simulation / 132

Chapter 7
Evaluation and Development Boards / 159

Chapter 8
In-Circuit Emulators / 195

Preface

This book is about the development and test support tools and techniques for RISC and CISC 32-bit microprocessors. During the past few years the quest for higher-speed microprocessors has resulted in the development of several versions of RISC architecture. The CISC architecture has also seen considerable development as its advocates competed with the RISC processors to protect their software base. While RISC microprocessors are capable of higher-speed performance than CISC microprocessors, the latter are competing very well in some applications. There is little doubt that both types will be around for the foreseeable future. The development of both types has placed new demands on development and test support tools. In addition, built-in development and self-testing within the chips will impact development and test tools in the future. This book covers the role of in-circuit emulators, logic analyzers, and software simulators in the development phase and the role of in-circuit testers and emulators in the test phase. The approaches taken to built-in development support and built-in self-testing are covered.

Chapter 1 reviews the role of in-circuit emulation (ICE) in dedicated and universal development systems. In the universal role ICE can be used with a wide range of computer hosts, including personal types. The role of logic analyzers is seen to be rising for both RISC and CISC applications and to be serving as an important supplement to either emulation or simulation tools. Design automation as realized by the combination of CASE, electronic CAE/CAD, and CAD/CAM is covered as one approach. It is seen that built-in development support can play a significant role in simplifying external tools. In-circuit and functional testing is shown to be on the threshold of change due to built-in chip test features, such as boundary scan.

Chapters 2 and 3 cover RISC and CISC microprocessors. In Chapter 2 the evolution of RISC architecture from the universities to industry is reviewed. RISC principles are detailed. The SPARC and RISC architectures are compared. The types comprising the Motorola 88000 family are covered. The features of Intel's 80860, AMD's 29000, and VLSI Technology's VL86C010 and VL86C020 are detailed. In Chapter 3 the evolution of Intel's 80X86 and Motorola's 68000 families is reviewed. The general characteristics of the Intel and Motorola 32-bit CISC microprocessors are described. These include Intel's 80386 and Motorola's 68020 and 68030. Early applications of higher clock frequencies are described. It is seen how Intel studied the RISC technology and applied what they learned to CISC designs, such as the 80486, to compete with RISC. The features of Motorola's 32-bit CISC microcontroller are covered in detail.

In Chapter 4 the development system used by Bell Laboratories to support AT&T's 32100 and 32200 CISC microprocessors is described in detail. Although these microprocessors have been used almost exclusively in AT&T systems, this development system is a good example of one that can be used to support the development of microprocessor-based products and systems to be implemented by CISC microprocessors operating at clocks up to at least 25 MHz. The use of in-circuit emulation and the hardware of the analysis pod are covered. The four different environments in which this development system can be used are described in detail.

In Chapter 5 logic analyzer operation and its use in hardware and software debugging are covered. The use of a logic analyzer with a device events monitor to support applications of AT&T microprocessors operating up to clock frequencies of 18 MHz is described. Those who used it say that it operated satisfactorily up to 25 MHz. The in-circuit debugger for Intel's 80486 is shown to be closely associated with high-speed logic analyzers. A C language source code debugger is described. Analyzers have distinct advantages over emulators in the development of high-speed systems. A logic analyzer application to Motorola's 88000 RISC microprocessor is described.

Chapter 6 shows how software simulation has been used to support the application of SPARC, MIPS, and Motorola's 88000 RISC microprocessors. It is demonstrated that the hardware development relies on logic analyzers.

Chapter 7 covers evaluation and development boards. A typical evaluation board used by AT&T in the application of its 32-bit CISC microprocessors is described. Detailed coverage is given to a SPARC evaluation board used by Cypress Semiconductor. A special feature, called stretched clock logic, is a part of this board. An evaluation board for the MIPS R3000 RISC microprocessor is also described in similar detail. The relationship between the CPU and memory is emphasized.

Chapter 8 deals with in-circuit emulators in detail. The Hewlett-Packard 647000 series is covered with a discussion of the problems posed by the operation of microprocessors at higher clock frequencies. An emulation system for ASIC verification is detailed. The methods of using in-circuit emulation in functional board testing are covered. They are CPU emulation, memory emulation, and bus cycle emulation.

Chapter 9 covers the software tools associated with the development of both RISC- and CISC-based products and systems. The development cycle environment for SPARC RISC microprocessors is covered first. Optimizing compilers for both SPARC and MIPS microprocessors are covered in some detail. The software tools used in the application of Intel's 80960 RISC microcontroller are described. The requirements and main features of debuggers follow. This chapter concludes with a listing of the support software tools for a 32-bit CISC microprocessor.

Chapter 10 reviews the built-in development support features of Intel's 80386 and 80486. The extensive trace features of their 80960KB microcontroller are detailed. The evolution of development support in Motorola's 68000 family is shown to lead to the rather extensive built-in support in their 68332 microcontroller. The built-in debugging support of Intel's RISC chip, the 80860, is discussed briefly, and it is indicated that built-in development support is not a feature of most RISC chips at this time.

Chapter 11 covers boundary scan, a design for testability (DFT) technique proposed by the Joint Test Action Group (JTAG). It has been adopted as IEEE Standard P1149.1. The material in this chapter is largely based on the JTAG proposal.

In Chapter 12 Texas Instruments is cited as the first manufacturer to offer a range of products compatible with boundary scan testing. More detail is given about the built-in self-test (BIST) features of Intel's 80486 microprocessor and Motorola's 68332 microcontroller. It is shown that they are not compatible with boundary scan but nevertheless represent rather extensive implementations of BIST techniques. The boundary scan mode of Intel's 80860 RISC microprocessor is described in detail.

This book is for the electronics engineer, whose responsibility is to combine the software and hardware to realize a working system or product. The engineer does not necessarily need to be trained in computer science, but an understanding of its basics would be helpful. Training in the theory and operation of microprocessors is necessary, and some experience would be helpful. This book should be of particular interest to electronics engineers who have decided to specialize in microprocessor applications at all levels, whether in college or industry.

Acknowledgments

This book has been made possible by the support of leading companies in the field of microprocessor development and test technology. These include the Intel Corporation, Hewlett-Packard Company, American Telephone and Telegraph, Cypress Semiconductor, Ross Technology, Motorola, Inc., Integrated Device Technology, Inc., MIPS Computer Systems, Sun Microsystems, Inc., Fujitsu, Talon Instruments, Advanced Micro Devices, Tektronics, Inc., Quickturn Systems, Inc., and John Fluke Mfg. Co.

Societies and publications in the field have also been very helpful. These include the Institute of Electrical and Electronics Engineers, BUSCON, MICRO, the IEEE *Spectrum, Personal Engineering and Instrumentation News, Electronic Engineering Times*, and Dempa Publications.

Chapter 1

Introduction

For the next decade the development of tools to support the design and testing of microprocessor-based products and systems will be driven mainly by two trends—one, the rise in clock frequencies, and two, the increase in the number of transistors on a chip. A projection of the increase in clock frequencies over the next decade from 33 MHz to 250 MHz is shown in Figure 1-1, and the projected increase of transistors per chip from about 1 million to possibly 50 million is shown in Figure 1-2. Such projections are not being made only by dreamers but by the R&D management of leading semiconductor manufacturers.

The rise in clock frequencies has already had a significant impact on the design of in-circuit devices such as emulators, debuggers, and interface pods. Simulation with logic analyzers has been favored by some as an alternate in many high-speed applications.

The increase in transistors per die will require the implementation of scan techniques, which will undoubtedly be necessary to successfully test so many transistors on a single chip. However, the placement of so many transistors on a chip will make it possible to have the equivalent of a "board-on-a-chip." Consequently, it will become much more difficult to access points within a chip and thus complicate development techniques as they are known today.

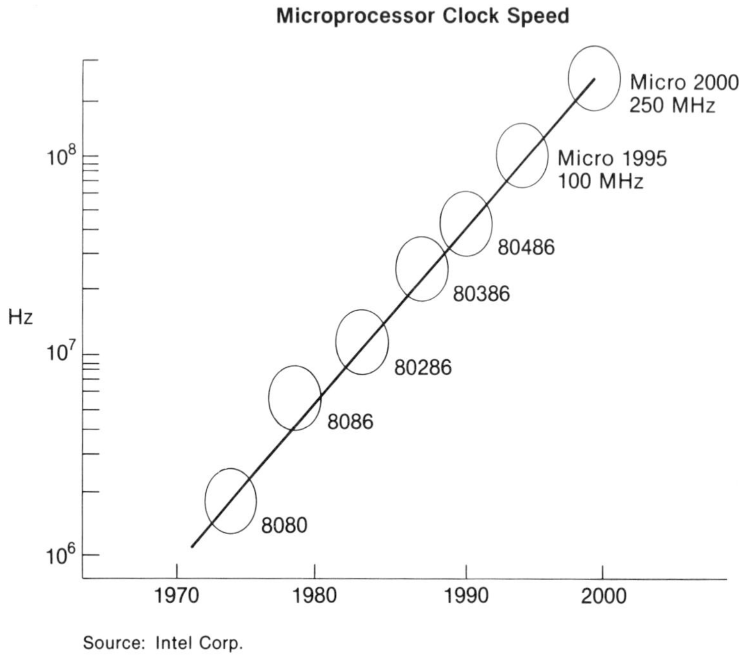

FIGURE 1-1
The projected rise of microprocessor clock speed. Reprinted courtesy of Intel Corporation.

In the meantime, tools and techniques that have been employed during the past decade will continue to be used in various forms as they evolve to keep pace with microprocessor development. These tools and techniques are covered in this book.

In its application, no microprocessor stands alone. It is either assembled on a circuit board or embedded into a product or system with other components essential to its function. Software and hardware are required to bring it into a viable entity. Software development may require the production of many thousands of words of code. Software and hardware development must be integrated. After the design has been finalized and put into production, it must be suitable for adequate testing to ensure the quality of the product in a competitive world.

Software development using high-level languages, such as C or Pascal, requires the basic tools shown in Figure 1-3. The editor is loaded, files are set up, and code is typed in. A compiler or cross-compiler delivers an object

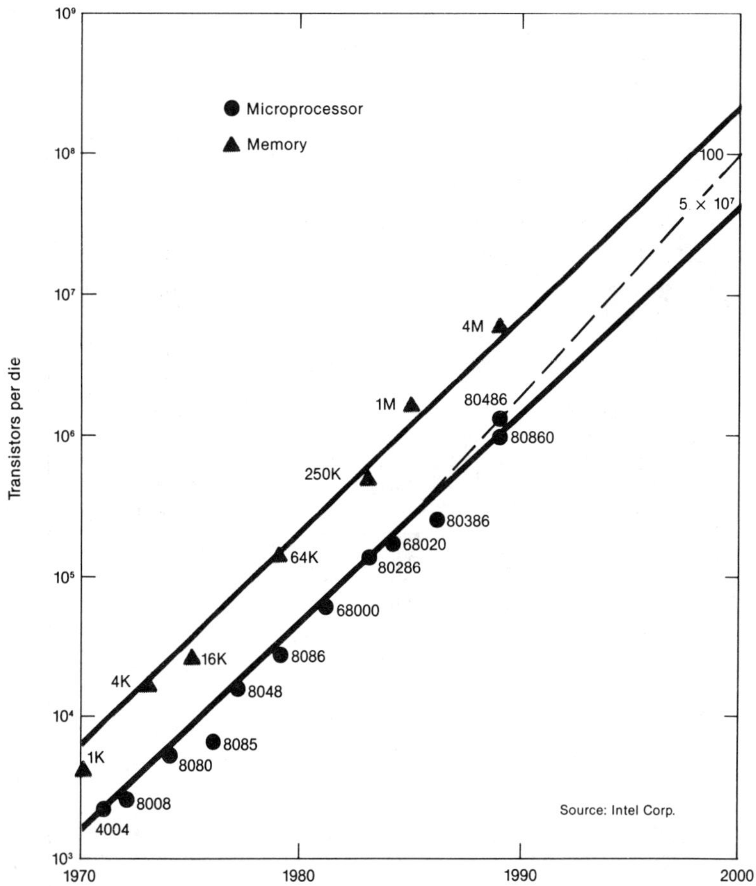

FIGURE 1-2

A projected increase of transistor density per chip. Patrick P. Gelsinger, Paolo A. Gargini, Gerard H. Parker and Albert Y.C. Yu, "Microprocessors circa 2000," *IEEE Spectrum*, Oct. 1989. © 1989 by IEEE.

file to a linker, which combines it with library files and the assembly code produced by an assembler, if any assembly language is used. Finally, the program is loaded into memory and run or downloaded into the target hardware. Such a program is not likely to run properly the first several times it is tried, and some form of debugger will be required to make it workable. All of these steps may be taken on a host computer (PC, workstation, or the like) before the hardware of the microprocessor-based product or system is developed or built.

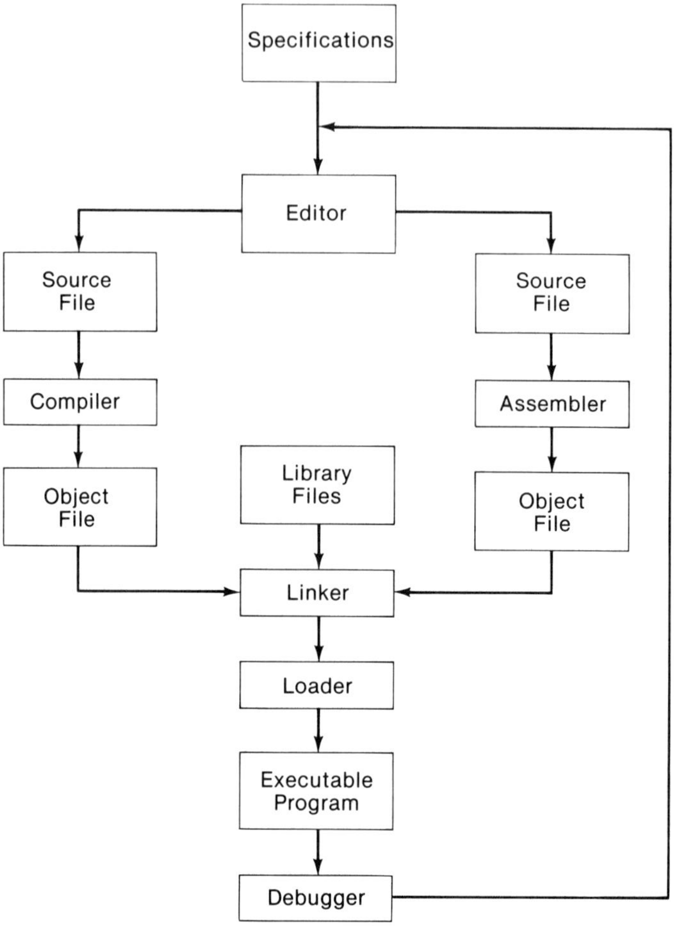

FIGURE 1-3
The basic software development process

The hardware design will almost always proceed in parallel with the system software development. It will be done using development tools such as CAE (computer-aided engineering) workstations, schematic capture packages, logic and circuit simulators, and PCB (printed circuit board) layout software. Once the hardware has been built, the first prototypes may be checked out with instruments such as waveform generators, scopes, and logic analyzers to ensure basic functionality. At this point hardware/software integration can take place. It is here that the designer has two basic approaches by which he or she can proceed. One of these might be called the hardware approach and the other the software approach.

Until RISC came into the picture and until clock speeds were increased to the current levels of many 32-bit microprocessors (33 MHz and higher), designers preferred the hardware approach for integration. Since about 1975 the tool for this method has been the microprocessor development system (it should be clear that such a system is not to develop microprocessors, but rather to complete the development of microprocessor-based products or systems). Some of the principal RISC vendors have used the software approach. Here a software simulator not only allows for the execution and debugging of the user's target hardware, but the simulator is also capable of debugging applications programs residing on top of an operating system. Even the most sophisticated programs can be constructed and debugged before the target hardware is finished.

Software Testing and Debugging

Over the past decade the most widely accepted arrangement of hardware development tools for microprocessor-based and microcontroller-based products and systems has been one in which a device, called an emulator, is used as a substitute for the microprocessor or microcontroller in the target area. Certain steps in test and debugging procedures can be performed with the emulator in a stand-alone mode, but its capabilities are greatly enhanced by association with a host computer. Such an arrangement permits the developer to patch in changes in the program after the need for them has been determined by displaying portions of the program. The original concept of emulator application, as introduced by Intel in 1975, was implemented by an arrangement shown in Figure 1-4. Here the host computer and the emulator were housed in a single package (called a microprocessor development system, or MDS).

Software development tools (on floppy disks or hard disks) were used as required with the host computer to produce a program and to test and debug it after it was downloaded through the emulator into the target area. Such systems were classed as "dedicated," because the MDS was tailored to the microprocessor that it replaced in the target area. This requirement for such an arrangement meant that a different MDS had to be provided for each microprocessor family or architecture.

Although this concept was applied almost exclusively for a few years after its introduction, the cost and inconvenience of it resulted in separation of the host computer and emulator, as shown in Figure 1-5. This arrangement, called an "open" system, permitted the developer to use any general-purpose computer with enough power that might be available. It was only necessary to procure an emulator designed for the microprocessor or microcontroller architecture under development in the target area. This approach offered an open invitation for the developer to try a personal com-

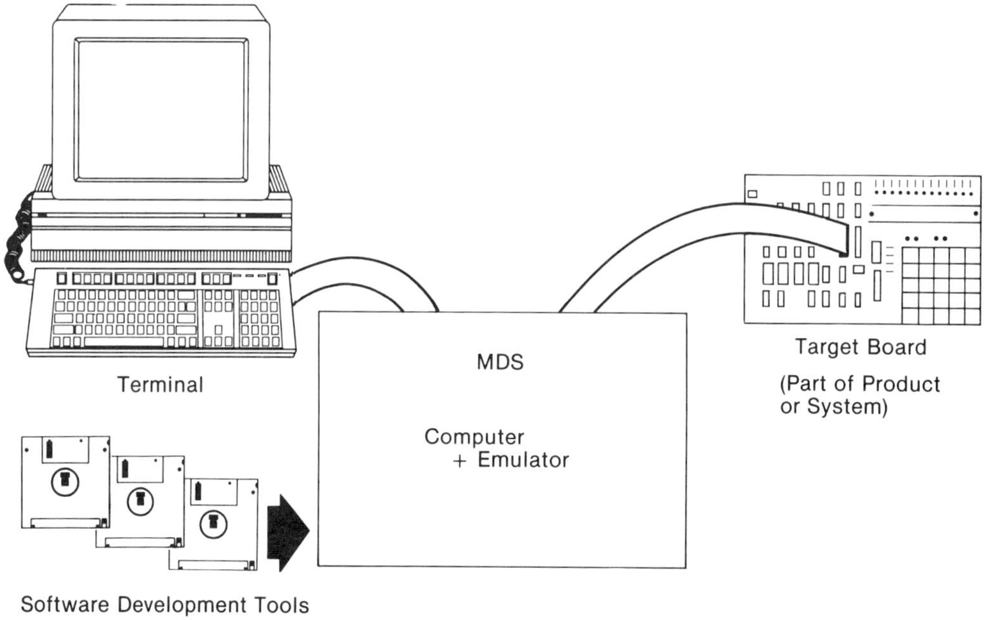

FIGURE 1-4
The original MDS concept

puter as the host. The result has been that personal computers (particularly the IBM PC AT or a suitable clone) are now widely used as host computers in microprocessor development systems.

Another approach, shown in Figure 1-6, utilizes a logic analyzer together with an emulator or analysis pod and a host computer. With this arrangement, state analysis of the program and timing analysis of the hardware performance of the product or system are provided to facilitate the testing and debugging process. Of course, the logic analyzer can be used without the emulator, but in that role it may not realize all of the benefits to be derived from association with a host computer and the in-circuit features of the emulator.

In still another approach, a host computer can download directly into the target area as shown in Figure 1-7 and, with the aid of specialized software development tools, perform a degree of testing and debugging.

In all of the arrangements described in Figures 1-5, 1-6, and 1-7, the software development tools shown to the left of the host computer consist of editors, compilers and/or assemblers, linkers, hex converters, and the like, to enable the developer to write the program. These tools are combined with special software packages to enable the developer to test and debug the program and integrate it with the product or system hardware into a

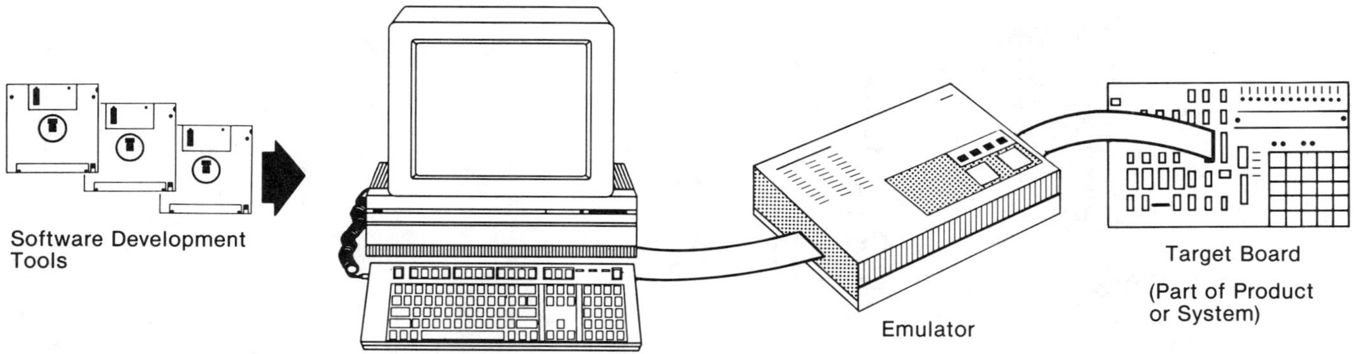

FIGURE 1-5 The widely used arrangement of host computer and emulator

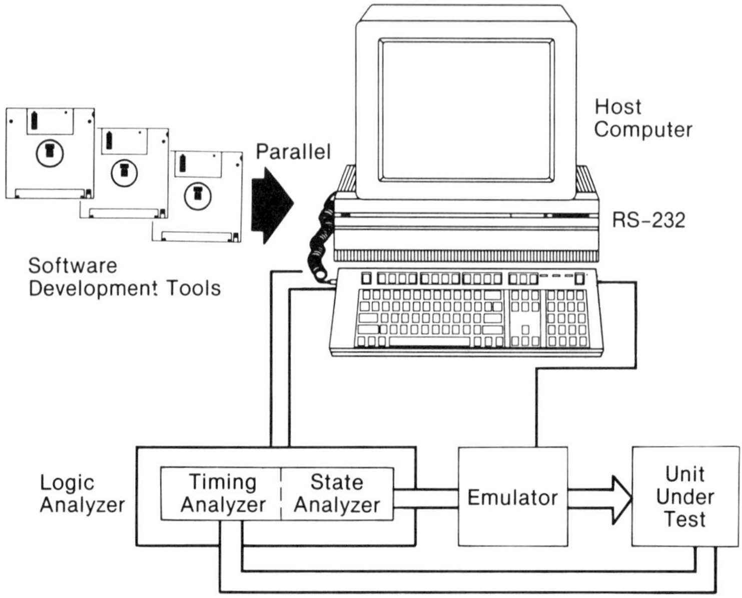

FIGURE 1-6
Logic analyzer combined with emulator in development system

FIGURE 1-7
Some software tools make possible development in the target board

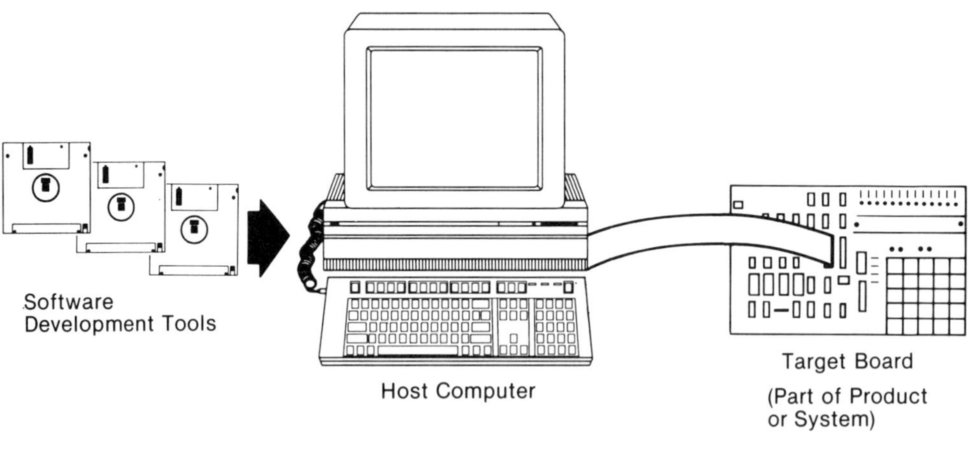

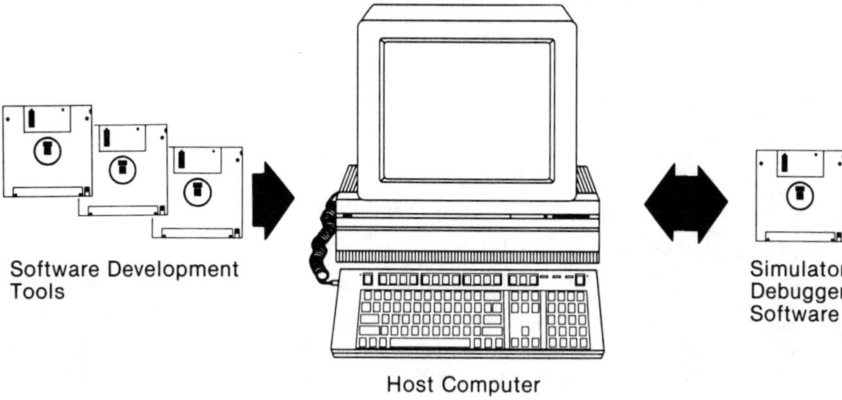

FIGURE 1-8
Simulation makes possible software testing without target hardware

workable result. In the case of the simulator/debugger (Figure 1-8), the simulation program is in additional software packages.

Dedicated Microprocessor Development Systems (MDS)

In 1975 the Intel Corporation introduced the in-circuit emulator (trademarked as ICE), which allowed the system developer to run 8-bit programming code in a microprocessor-based test circuit. When the emulator was combined with a small microcomputer in a desktop unit, the first microprocessor development system was born. It was referred to as an MDS, an acronym that was used later in some instances to describe a broader range of products. An MDS was originally a microprocessor-based computer with dual floppy disks, 64 kilobytes (KB) of memory, and ports for a printer and CRT. It could support one or more in-circuit emulation modules. An in-circuit emulator plugged into the user's system under test (the target) in place of the microprocessor and could intercept signals between the microprocessor and the rest of the target circuit. The emulator could select which memory, either in the MDS system or in the target, would respond to any particular memory request cycle from the microprocessor.

An MDS normally had at least two breakpoints (i.e., it could recognize a particular microprocessor cycle, such as a fetch at a particular address) and 500 to 1000 words of trace memory (a saved record of executed microprocessor cycles). It could start and stop the processor and display the trace in several formats, including disassembly (a reverse-assembled code listing).

As microprocessor-based projects grew larger, the original MDS became useful for only smaller projects because of several limitations:

1. Disk storage capacity was too small and 8-inch floppy disks were too slow.
2. Memory size was too small.
3. The operating system had limited capabilities.
4. The microprocessor was 8-bit.

By 1977 Intel introduced its Series II Development Systems Workstation, also a dedicated system whose purpose was to develop software programs and debug them for Intel's 8086 (16-bit) microprocessor. Although the dedicated system approach fit with a given microprocessor because its program could be developed on the same microprocessor as that which was to be used in the product or system, the expense of having a different MDS for each new microprocessor was not viewed favorably by the user industry. However, during this period of time the semiconductor manufacturers continued to offer MDS equipment to support each new microprocessor as it reached the marketplace.

Universal Microprocessor Development Systems

As soon as test equipment manufacturers became interested in providing development equipment for microprocessors, they chose instruments that could develop and debug a much wider range of microprocessor types. Although they were limited by the necessity to provide different emulators for each microprocessor type, the remainder of the test equipment was unchanged for the number of microprocessors they chose to support. Tektronix entered the field with a universal system, made by Millenium Systems, in 1977. By 1979 they were supporting 18 different microprocessors, many of them from Intel. This led Intel to broaden the application of its Series II System to support 10 different microprocessors of its own manufacture. In 1979 Intel held about 50 percent of the market for development systems. Other semiconductor manufacturers combined held about 40 percent of the market, and Tektronix held only about 10 percent despite its universal approach. Apparently, at least part of the reason for this was that most users were designing only one microprocessor type into their products or systems and preferred to obtain the test support equipment from the semiconductor maker because of the company's specialized knowledge of the microprocessor.

At the Wescon in the autumn of 1979, Hewlett-Packard introduced its model 64000 universal microprocessor development system, which was designed to serve several users at separate stations. The master station of the system included a 16-bit microcomputer, a 20-megabyte (MB) hard disk drive, a tape cassette for program entry and file backup, a PROM programmer, an emulator, and a high-speed line printer. Up to five additional stations could be added, each of which contained its own host processing system and memory, plus 10 card slots that could hold the controllers for such options as cassette tape, emulators, up to 128 kilobytes of 100-nanosecond emulation memory, real-time transparent analysis, and a PROM (programmable read-only memory) programmer. In this first version of the HP64000, four microprocessors (Intel's 8080 and 8085, Motorola's 68000, and Zilog's Z80) were supported with relocation macroassemblers, real-time emulation at full processor speed, and real-time transparent analysis. This first version also used assembly language: higher-level languages soon followed. Also, additional microprocessor types were soon accommodated. The HP64000 has survived to this day and now accommodates at least eight families of microprocessors and more than a dozen other types of all of the leading semiconductor manufacturers. In 1986 Hewlett-Packard introduced a new version designated HP64000-UX. It offered a complete measurement set to test, exercise, and debug microprocessor software during all phases of development. Emulators for over forty microprocessors are available to display target system activity, beginning with the first code module. Analyzers for state analysis, timing analysis, software performance analysis, and high-level language debugging provide real-time tests of program execution.

It is said that the heart of this analysis and emulation system is the instrumentation card cage. It accepts standard HP64000 feature cards (such as the current MC68000 emulation system or a 120-channel state analyzer) and connects to an HP Model 320 computer via an instrumentation bus. Even greater capabilities can be realized by the use of multiple card cages shared by a single workstation or in a time-share configuration. Different emulation and analysis subsystems in one or more card cages can be used simultaneously by multiple users or can be accessed by one person using multiple windows.

The HP64000-UX environment can be implemented through one of three basic systems: the Entry System, the Workstation System, and the Instrumentation System. The Entry System, shown in Figure 1-9, is a multiuser configuration with the HP9000 Model 320 computer and the instrumentation card cage. Up to eight users are supported and up to four card cages can be controlled by one Model 320 computer. The system may be networked via LAN IEEE Standard 802.3 to the systems. The Workstation System, shown in Figure 1-10, can be either single-user or multiuser with the Model 320 computer networked via a local area network (IEEE 802.3 LAN) to other single-user or multiuser Model 320 computers. In this system, one Model 320 computer acts as a "file-server" node. In addition to supporting one or more users, the file server manages the disk storage for the entire design team. Other "client" nodes have additional disks for local

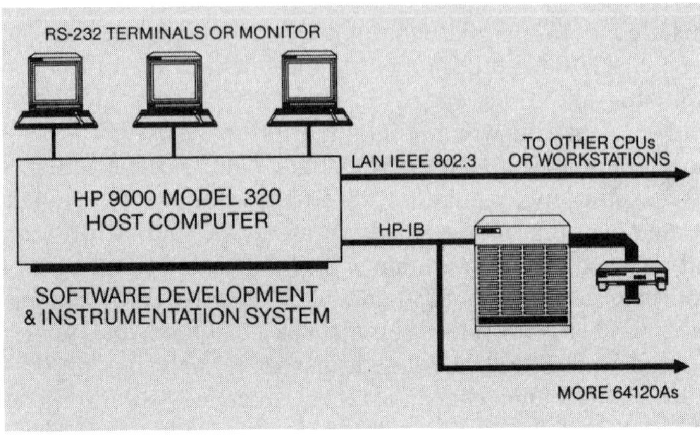

FIGURE 1-9
Hewlett-Packard's entry system

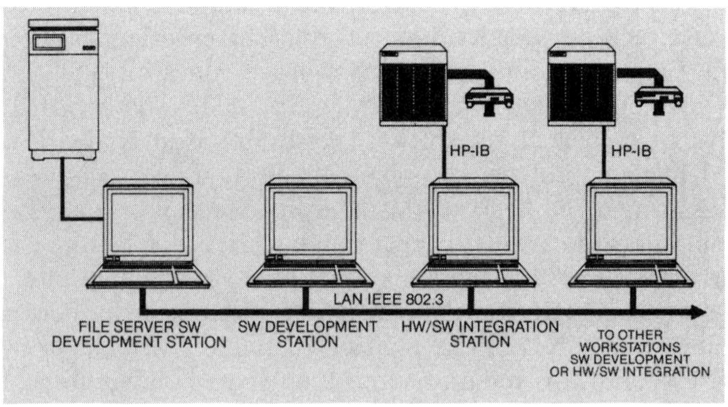

FIGURE 1-10
Hewlett-Packard's workstation system

storage. This configuration is arranged so that regardless of which node a user logs into, private or public files can be accessed in the same manner.

The instrumentation system is a networked connection of a time-shared host computer to an HP64000-UX system dedicated to instrument control as shown in Figure 1-11. The host computer can be the DEC VAX, an IBM PC, or an HP9000 Model 320 computer. This configuration allows for a continuity of software development tasks currently underway or previously developed on host computers. It provides time-shared users on the computer access to powerful emulation and analysis tools via the IEEE 802.2 LAN and ARPA/Berkeley series.

The principal advantage of the HP64000-UX Microprocessor Development Environment is the ability to connect easily to a broad variety of

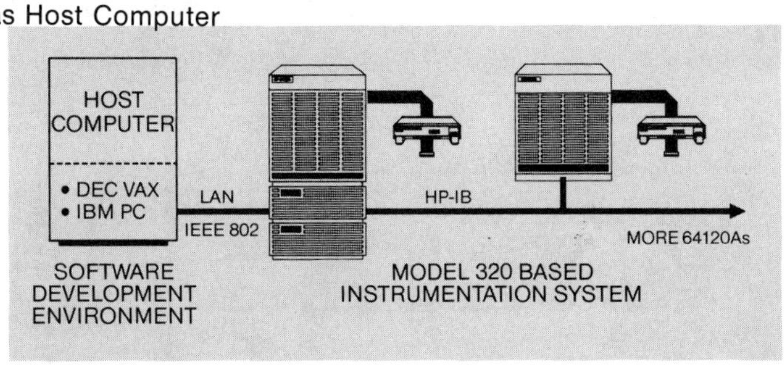

FIGURE 1-11
Hewlett-Packard's instrumentation system

other design clusters, host computers, or system resources, as shown in Figure 1-12. Design teams can be located in other rooms or buildings and still share project data efficiently. Compute nodes can be accessed rapidly to allow task unloading for specific applications. Previous investments in time-shared computers or the HP64000 Logic Development Systems are said to be increased in value due to the expanded functionality available through the 64000-UX microprocessor.

The PC as Host Computer

By the mid-1980s it became apparent that the design engineering community had adopted as de facto computing standards the IBM PC AT and the Microsoft MS-DOS operating system for work such as microprocessor-based product or system development. A number of companies brought out emulators to work with PC hosts on the market. In 1987 Hewlett-Packard introduced their HP64700 series of emulators, which they said represented the latest stage in the evolution of the HP64000 Microprocessor Development System. The earlier versions of the HP64000 had been pointed toward larger host computers. It appeared that the best approach was to make any microprocessor development system host independent, so that it could be used with PCs as well as workstations and other microcomputers as the hosts.

At the lowest level this approach implied a totally self-contained microprocessor emulation vehicle. It would contain its own local controller, which had sufficient operating software on board to relieve the remote controller of all base-level control functions. All commands to the emulators would be constructed from simple ASCII character strings. The emulator would communicate with its controller over a standard-protocol serial network, and at its RS-232-D/RS-422 port the world would appear as an 80 character-by-24 line "dumb" terminal. File transfers using industry standard formats would be accepted. Thus any host computer (or terminal) could control the HP64000 series emulators, making them useful for a wide range

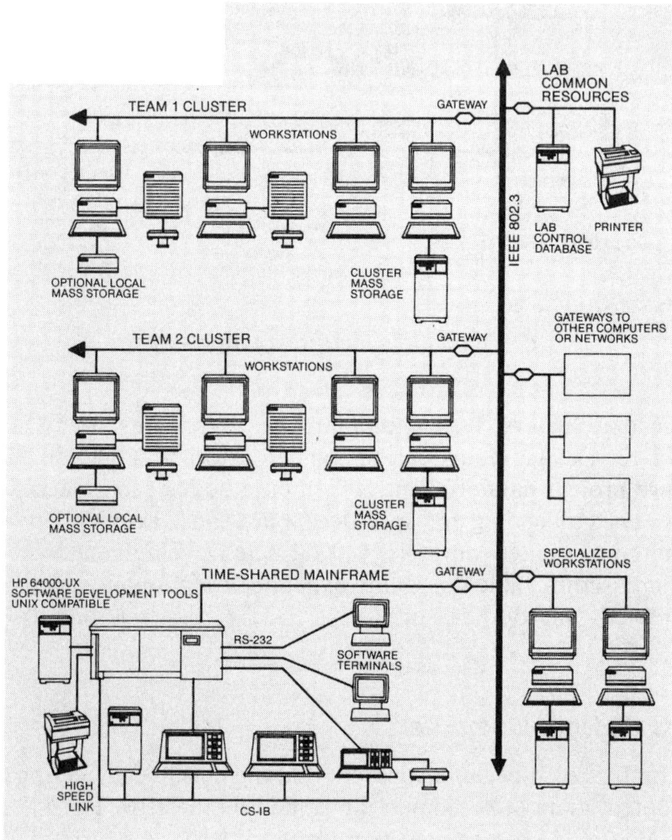

FIGURE 1-12
The networking environment of the HP-64000-UX

of tasks, from code development to field service applications. One of the strong features of the 64700 series was that it could operate from the same environment as a companion HP64000-UX emulator and could provide for state and time analysis.

Logic Analyzers

Emulators set breakpoints, memory and register display, memory overlay, and real-time trace. Logic analyzers were first used with emulators to provide state and timing analysis. A typical example of the development system for 32-bit CISC microprocessors is described in Chapter 4. There has also been a tendency to eliminate the emulator and to perform the hardware/software integration with a monitor and logic analyzer. An example of this approach is described in Chapter 5. However, it is only recently that logic analyzers capable of providing limited emulation have appeared. There is a demand for such instruments because designers prefer nonintrusive tools for debugging 32-bit microprocessor-based products and systems operating at clock frequencies of 33 MHz or higher.

New analyzers from Gould, Arium, and Tektronix provide different degrees of limited emulation. The Gould CLAS4000 and the Arium ML4400 offer symbolic debugging. The Arium ML4400 also provides for ROM emulation. The Tektronix Prism 3000 (Figure 1-13) features a prototype debugging tool (PDT), which provides for emulatorlike control of the system under test without the intrusiveness of conventional emulators.

The PDT allows the target microprocessor to run at full speed with no wait states. Registers and memory can be examined and patched. Software breakpoints can be set and the system can be stopped and restarted. The PDT's 8K samples of memory can be used as a trace buffer due to tight coupling with the state section. The full triggering capabilities of all modules installed in the Prism 3000 are used to set complex hardware breakpoints. Mnemonic disassembly and performance analysis modes round out the integrated state, timing, and PDT capabilities of this logic analyzer. Disassembly software automatically loads and configures upon power-up if the memory probe adapted is in place.

RISC microprocessors have challenged the capabilities of logic analyzers as well as emulators, and only the most sophisticated models of these analyzers have been recommended to designers. These challengers arise for a number of reasons. One of these is the operating speed of RISC processors, where 33 MHz is common and 80 to 100 MHz is required for ECL (emitter coupled logic) versions. In most previous CISC microprocessor applications, bus execution speeds have been somewhat lower than the processing speed. With some CISC chips the bus speed has been less than half of the processing speed. Because the intent of the RISC design is to execute at least a single instruction during each bus cycle, the bus execution speed matches that of the processor. Many RISC processors have more buses with wider bandwidth than some CISC types. Often separate buses for data and addresses are used and both of them have a 32-bit width. The result is demand for more channels with higher synchronous clock speeds.

On the hardware side RISC processors make extensive use of caching and pipelining, which aggravates problems that were already present with CISC types but on a lesser scale. Some RISC chips employ more than one cache. As an example, the Motorola 88000 employs as many as four separate caches. They can be turned off to observe data, but this may seriously alter the development environment. One alternative is to deactivate the cache and to match the observed data to possible problems and then observe the effect when the cache is reactivated. Longer instruction words and a larger

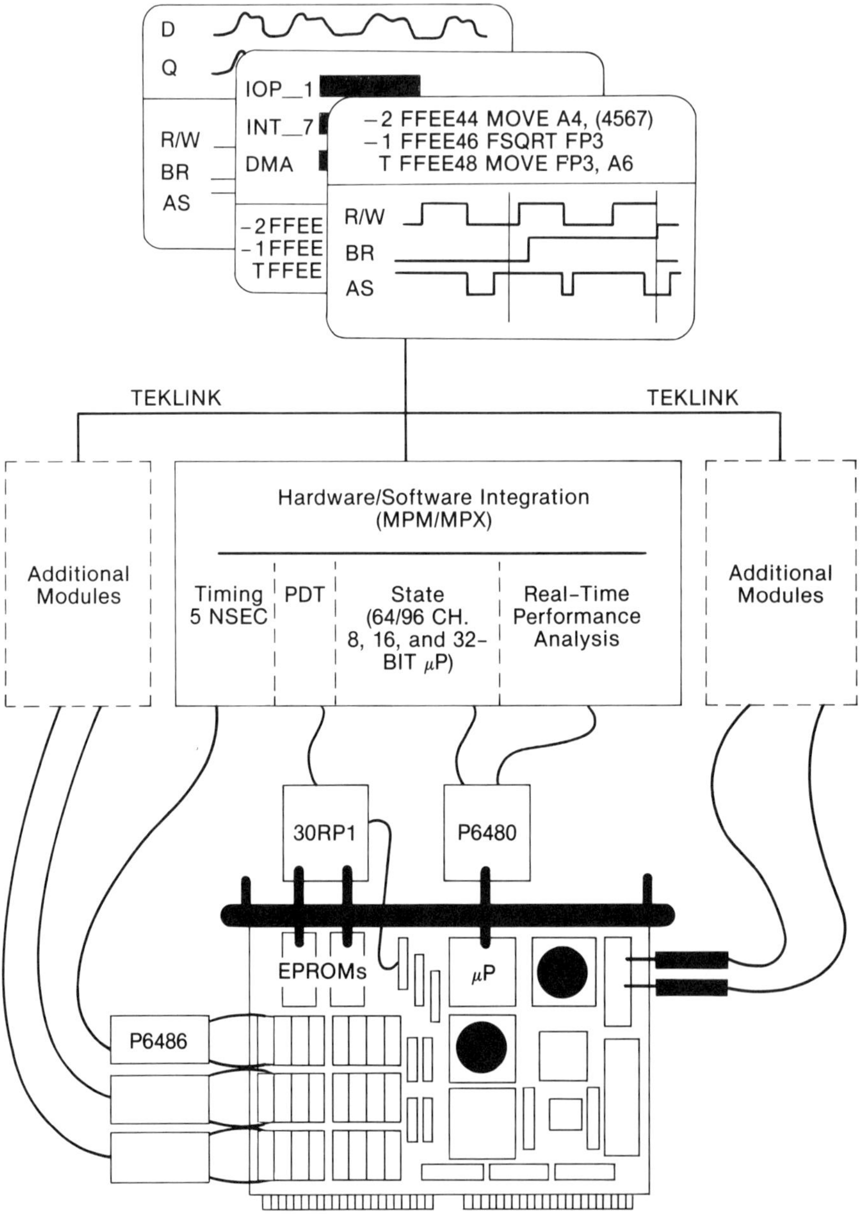

FIGURE 1-13
Tektronix Prism 3000 links the logic analyzer to the target system without a conventional emulator. Reprinted courtesy of Tektronix, Inc.

register format are additional factors. Floating-point coprocessors that perform their math operations internally add to diagnosis difficulties. On the software side, the code size is at least 30 percent greater than CISC types and there is little or no assembly language.

Despite all of these challenges, logic analyzers appear to be doing a better job of keeping up with RISC than emulators. The models mentioned above as well as the Hewlett-Packard 165000 and others are providing important development support to RISC-based development.

Simulation

Leading developers of RISC-based products and systems have chosen to use simulation without emulation. MIPS Computer Systems provides an example of this approach. They use a system software development tool, called a system programmers package (SPP), which combines an instruction simulator, called Sable, and a cache-memory simulator with a software monitor and development tools that support the MIPS R2000 and R3000 RISC chips. They use this system to ready the software before the hardware is completed. The instruction simulator allows the developers to run system software and key applications with a source-code debugger. These developers haven't found a need for an in-circuit emulator. When the hardware is available, they check it with a 100-MHz logic analyzer.

Other RISC suppliers, such as Sun and Motorola, have used similar approaches that depend on software and do not require a hardware emulator. In all cases an instruction simulator is used to debug the key code and a logic analyzer is used to check the hardware. Tektronix has a simulator which delivers 100,000 instruction executives/second using a 68020. Running on the Motorola 88000 it can deliver 600,000 instructions/second. Here again an emulator is not required. The simulator and the DAS92000, a sophisticated Tektronix logic analyzer, are used for checking hardware/software integration.

Design Automation

When Intel introduced the in-circuit emulator (ICE) in 1975, they spoke of design automation. However, there was a long way to go to arrive at what is called design automation today. For microprocessor-based product and system design in 1990, Hewlett-Packard links CASE, electronic CAE/CAD, and mechanical CAD/CAM. Figure 1-14 shows a flowchart of the software, electrical/electronic, and mechanical aspects of a product design, which is implemented by a combination of CASE/CAE/CAD solutions. By putting these technologies together, the designer can implement the product design with the latest advances in the three fields. Under CASE it includes the

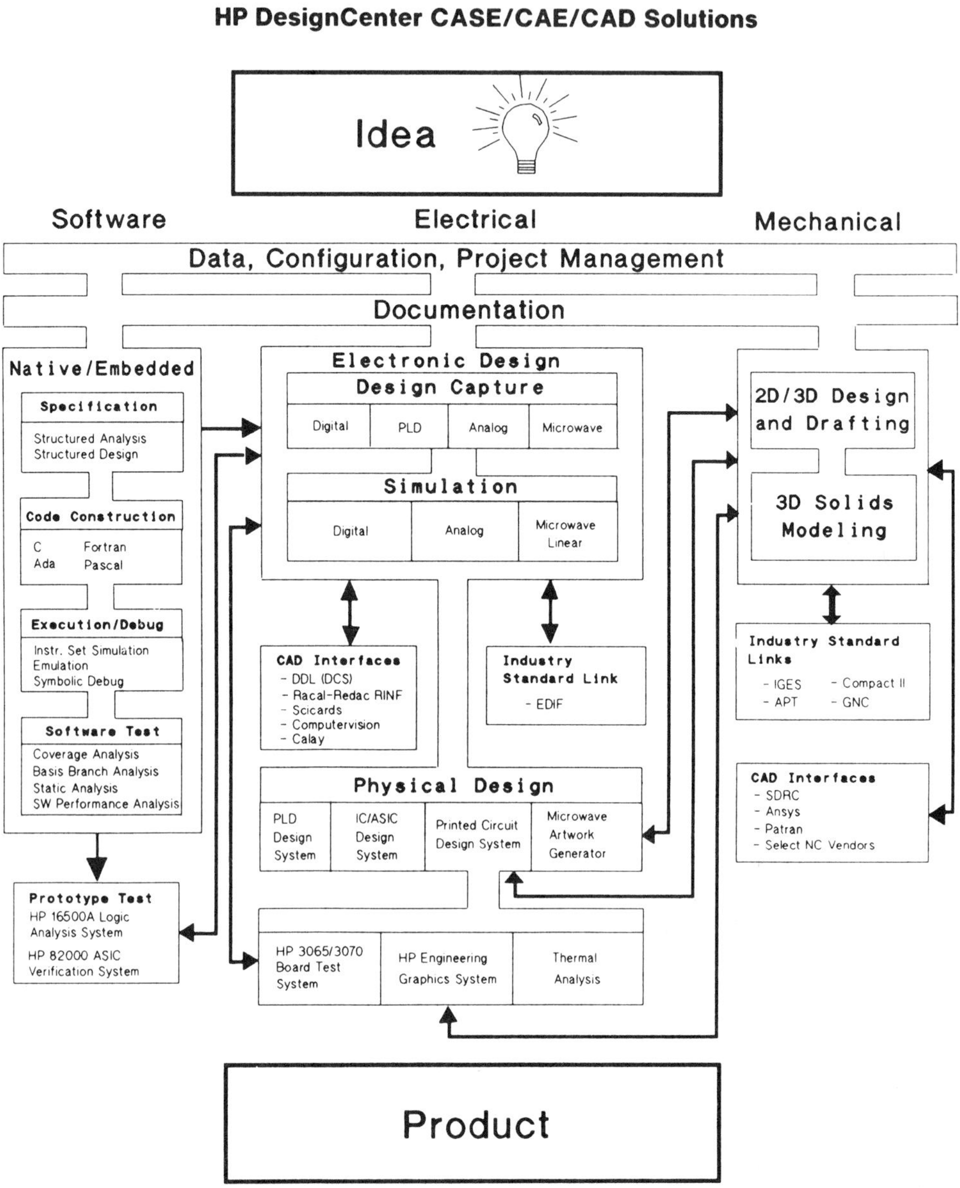

FIGURE 1-14
Hewlett-Packard's flowchart of design automation. Reprinted courtesy of Hewlett Packard Company.

HP64000 microprocessor development environment, which may include any combination from single-user, PC-based, real-time emulation to large team installations with a network of several HP9000 Series 300 workstations to develop the software and hardware necessary for multiprocessor systems.

In this combination emulators provide an essential link between the software development environment and the target system. Programs developed on the HP64000 development environment are run on the emulation subsystem for real-time debugging and analysis. An emulation bus analyzer provides the displays and triggering conditions for the emulator and serves as the access point for interactive emulation analysis. Processor run controls in the emulator allow one to single-step, display, and modify memory. Modifications and improvements to the software can be made quickly and easily in the early design phases with emulation giving the flexibility to experiment before committing a product to firmware. Hewlett-Packard appears to be strongly committed to keeping emulators in the picture and has indicated that it is developing emulators for high-speed RISC and CISC microprocessors.

In addition to the software/hardware integration provided by the emulation features of the HP64000, very useful views of source code can be obtained during its execution with logic analyzers, software performance analyzers, and state analyzers. These analysis tools are installed in a card cage controlled by the HP9000 Series 300 workstation running on the HP-UX operating system. A software performance analyzer provides overview measurements to aid in evaluating total system effectiveness of programs operating in real time. Global measurements let software designers determine where resources are being used in terms of execution times, memory usage, and interaction traffic. Software performance measurements provide histogram displays for quick comparison of software activity, with tabular displays with continually updated means and standard deviations on current measurements. Measurement modes of memory and program activity, event duration, and intermodule linkages are included.

Comprehensive logic analysis includes emulation bus analysis, software coverage analysis, and state/timing analysis.

Built-in Development Support

While on the one hand we see a greater proliferation of expensive test equipment to achieve design automation, on the other hand we see more development support being built into chips to simplify and reduce the cost of external equipment. For some time, chips like the 80386 have contained debug registers to assist in their application. This feature has been carried over and enhanced in the 80486. Another recent example, with additional features, is the Motorola 68332 microcontrollers. This 32-bit CISC unit includes the equivalent of a software monitor in its microcode. It is said that this embedded solution will not replace an in-circuit emulator for all problems, but it is all that most developers will need for hardware/software checkout. A serial port links the controller to an external tester. Using this port, chip operation can be stopped, breakpoints can be set, and memory or registers can be interrogated for debug values.

Motorola has available a companion development/evaluation board that supports the microcontroller with 64K of 100-ns SRAM (static random access memory), 128K of on-board PROM, and an RS-232 serial interface. This board is the size of a standard business card—2 inches × 4 inches. A companion board contains the development interface. Using a 68HC11 as a controller, four channels of hardware breakpoints, BCC EPROM (business card computer erasable programmable read-only memory) programming and the on-chip debugger are supported. It has been indicated that this evaluation board will sell for about $500.

Intel's 80960 microcontroller also features extensive built-in debug capabilities. It provides facilities for monitoring the activity of the processor by trace events. Intel's 80860 microprocessor supports debugging with both data and instruction breakpoints. The built-in development support features of each of these chips are described in detail in Chapter 9.

Board Testers

Microprocessor-based boards are tested in production by in-circuit and functional testers. In-circuit testers check for shorts, opens, continuity, and defective components. The PCB to be tested is held pneumatically or mechanically on "elastic" nails, which are arranged in front of the test points of the PCB by a specific Plexiglas grid. The grid serves as the interconnecting device to the universal grid, which leads to the central electronics. Figure 1-15 shows the general arrangement of the PCB is a "bed-of-nails" test equipment console, which is shown in Figure 1-16. The console houses a computer controller and switching and measurement equipment, as well as the "bed-of-nails" assembly. Although it is very expensive, this type of equipment has been very popular and efficient for testing non-surface mount (SMT) boards in large-quantity production. However, it has limitations for SMT testing.

There are various approaches to SMT testing by methods involving physical contact with test points, but boundary scan is the most likely approach to provide a major improvement in testing this type of board. However, most boards will contain some conventional logic along with boundary scan components for the next several years. In-circuit testers will combine with boundary scan, functional, and "bed-of-nails" testing, depending upon which is applicable in a particular design.

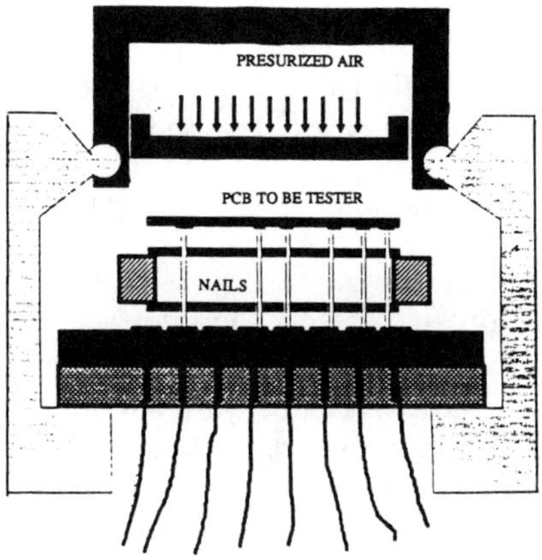

FIGURE 1-15
"Bed-of-nails" test arrangement for PC boards. Louis Balme and Anne Mignotte, "New Testing Equipment for SMT PC Boards," International Test Conference 1988 Proceedings, Sept. 12, 1988 © 1988 IEEE.

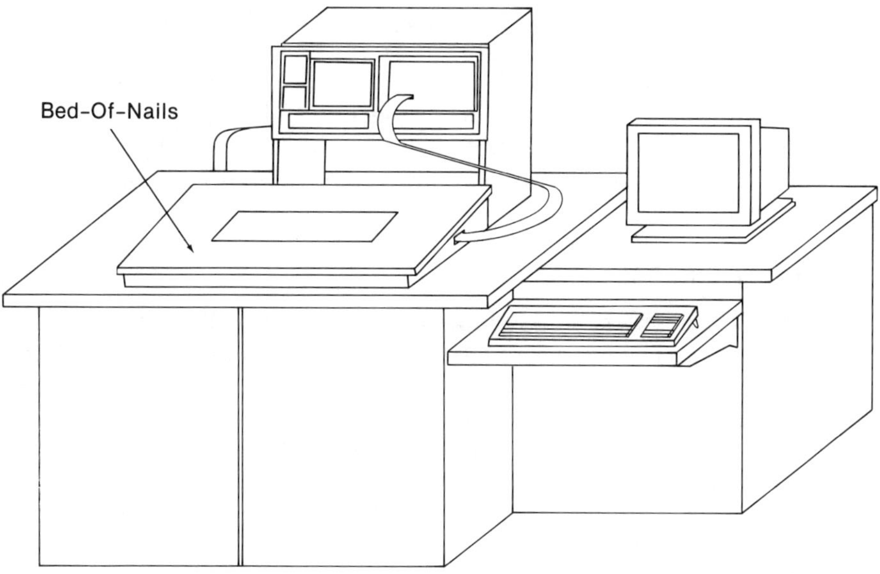

FIGURE 1-16
Automatic test equipment (ATE) console

Functional board testing takes various forms, one of which is emulative testing. Such testers are often installed as a subsystem of a "bed-of-nails" tester, as shown in Figure 1-16. Of course, they may be used alone, especially by board manufacturers producing less than a thousand units/month of a given design. Such testers are made by Fluke, Talon, and Temco. Emulative board test techniques are said to locate more faults quicker than any other approach. They troubleshoot the kernel circuitry first and thus provide a solid base for testing the remainder of the board. As an example, Figure 1-17 shows the configuration of the Fluke 9100 Digital Test System. Pods are available to support more than fifty microprocessor chips. The pods are used in the following applications:

Microprocessor Interface Pods

To emulate the board's microprocessor, one selects the interface pod that corresponds to the microprocessor on board. In addition to containing its own microprocessor, each pod has its own RAM (random access memory), ROM (read-only memory) and I/O (input/output), making it a complete kernel. Replacing the board's microprocessor, the pod controls all bus-related devices on the board. When the pod is plugged in, the clock circuit of the board under test is channeled to the pod's microprocessor, so that tests are performed with the board running at normal speed.

Memory Interface Pod

This pod features the capability to be reconfigured for additional microprocessors by replacing a plug-in personality module. It provides the test capabilities of READs, WRITEs, RAM test, and so on. The UUT (unit under test) connections are made with interfacing through the UUT ROM and via a sync module.

Built-In Bus, RAM, and ROM Tests

With the above pods the built-in bus, RAM, and ROM test routines can be quickly executed. These tests can be run without writing additional code.

The card edge fixture provides a fast way to interface to any digital circuit board with a card edge connector. Signals may be accessed at the card edge, and boards with an external bus may be tested.

For card edge stimulus, the I/O module provides real-time signals to the card edge, allowing limited bus emulation with handshake input, pattern looping, and dynamic trisate control to each bit. These modules are 40 bits wide and up to four modules can be used at one time.

The single-point probe may be used for higher-frequency signals up to 40 MHz instead of the I/O module. Also, it can be used for parts of the board which cannot be accessed with an I/O module. Like the I/O module, the probe can drive a node high or low, stimulate a node, and gather various types of response data. It can take signatures, sense logic levels, and count events or frequencies. An external clock module provides leads for testing asynchronous circuitry.

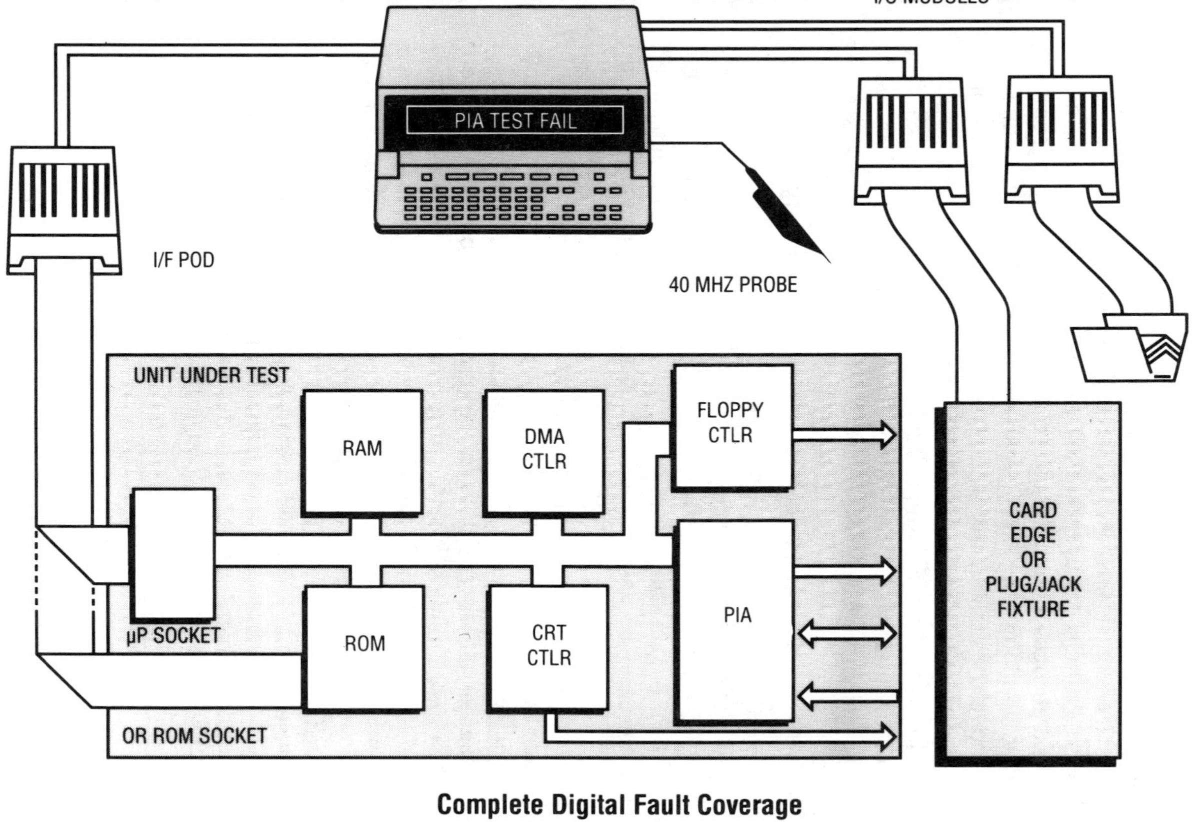

FIGURE 1-17 Block diagram of Fluke's Model 9100 functional test equipment. Reproduced with permission from the John Fluke Mfg. Co., Inc.

Built-in Self-Test

Since their inception it has been necessary to design microprocessor-based products and systems that could be tested in production if they were to be built in any quantity. If this had not been the case, there would not be so many functioning microprocessor-based products and systems in existence today. Increased emphasis has recently been placed on design for testability because of increased chip complexity, increased component density on circuit boards, and changes in board construction. Some microprocessor chips have more than a million transistors, and there is every indication that this number will move up significantly in the years ahead. Higher levels of integration have resulted in an increase in the complexity and number of integrated circuits to make the testing of circuit boards more difficult. The combination of functions within chips has isolated many critical points within them from accessibility. The use of surface mount technology has made many points on circuit boards less accessible.

In-circuit testers employing the "bed-of-nails" technique require extensive access to the circuits under test. With less accessibility to the internal structure of the chips and less accessibility to the circuits on boards, this test technique is no longer as applicable as previously. Functional tests are often applied to the edge connectors of circuit boards, but chip and circuit density may increase the cost of such testing to the point where it is unacceptable.

Fortunately, there is one solution that appears to have considerable potential. Since the number of transistors on the chips has risen a hundredfold or more in the last decade, it is no longer uneconomical to utilize a small percentage of these transistors for built-in test purposes. Thus there is great interest in providing self-test features on the chip, especially if they can be accessed on the circuit board without excessive complexity. Such a solution is provided by boundary scan, which is covered in Chapter 10.

FIGURE 1-18
Simple boundary scan cell design. A. Hassen, J. Rajski, and V.K. Agarwal, "Testing and Diagnosis of Interconnects Using Boundary Scan Architecture." International Test Conference Proceedings, Sept. 12–14, 1988 © 1988 IEEE.

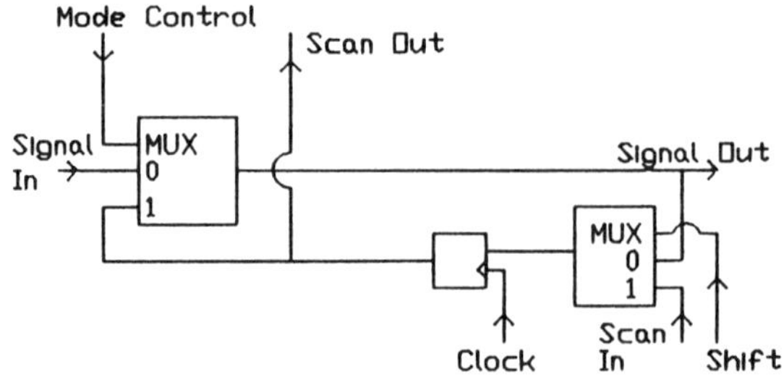

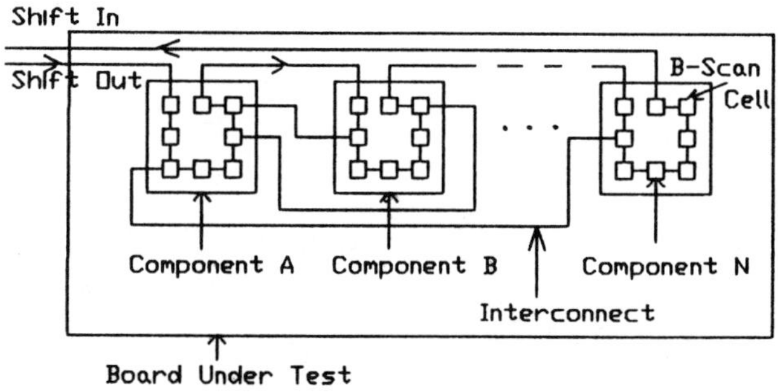

FIGURE 1-19
Boundary scan on a PC board. A. Hassen, J. Rajski, and V.K. Agarwal, "Testing and Diagnosis of Interconnects Using Boundary Scan Architecture." International Test Conference Proceedings, Sept. 12–14, 1988 © 1988 IEEE.

What is now referred to as structured design for testability, or DFT, involves some form of scan. The boundary scan concept allows one to access and control all of the primary input and output pins on the printed circuit board from outside the board. This is accomplished by connecting all of the primary inputs and outputs of the chip into a shift register which has a boundary scan input and output. A simple boundary scan cell is shown in Figure 1-18. The shift registers on all of the integrated circuits on a board can be connected to form a larger shift register with a single scan "in" edge and a single scan "out" edge, as shown in Figure 1-19.

Using this concept, it is possible at the board level to confirm that each integrated circuit is performing its required function, that they are connected in the proper manner, and that they interact correctly so that the complete board performs its intended function. Due to the complexity of many circuit boards, the automation of such testing is not a simple matter. Structured techniques are required to accomplish this feat in a cost-effective way.

Chapter 2

RISC Microprocessors

Whether RISC or CISC microprocessors dominate the microcomputer market in the future, there is little doubt that the development of RISC technology has stimulated more extensive CISC developments as well as spawning a considerable array of its own developments. Before covering specific RISC microprocessor implementations currently available, it is well to review the philosophy of RISC theory and development.

In a study conducted in the mid-1970s, an IBM research team found that about 80 percent of the computation for a typical program required only about 20 percent of the instructions in a processor's instruction set. The most frequently used instructions were very simple operations such as load, store, and add. Thus the computer industry's concentration on complexity was actually burdening processor designs with instructions that would rarely be used in practice. The IBM researchers concluded that the two-level implementation being used to implement the complex instructions was causing a more serious problem. It was slowing execution of the more frequently used simple instructions by requiring them to use the same microcode mechanism.

IBM went on to build a minicomputer, their Model 801, which eliminated the conventional two-level microcode approach. Instead, the Model

801 emphasized the cooperation between a well-chosen set of simple instructions implemented directly in hardware and an optimized compiler. Such a compiler was able to increase program execution speed by rearranging the detailed steps of the program to utilize the CPU resources more efficiently. Because this approach abandoned microcode implementations of complex instructions in favor of a few simple instructions implemented with hardwired control, it came to be called a reduced instruction set computer (RISC) architecture.

As IBM pursued development of the first RISC-based processor, researchers at the University of California at Berkeley (under Dr. David A. Patterson) and at Stanford University (under Dr. John L. Hennessey) began parallel efforts in RISC technology. The Berkeley researchers concentrated on understanding the most effective use of the area on a VLSI (very large-scale integration) chip for building a computer. Because the limited number of instructions of a RISC design require only a relatively small amount of on-chip control logic, more chip space can be dedicated to system functions that enhance the performance and versatility of the processor. The Berkeley group chose to allocate most of the available silicon area to on-chip registers in its RISC I and II 32-bit VLSI microprocessors. The Berkeley team used these registers to hold information that was passed from one program to another. In conventional computers, such information is usually stored in system memory before the subroutine is called. This operation is very time-consuming. The Berkeley method of using registers to pass parameter information significantly reduced the time required to execute subroutine calls.

The Stanford RISC research project, called Microprocessor without Interlocking Pipeline Stages (MIPS), combined optimizing compilers with Berkeley's development of a VLSI RISC. The pipeline technique keeps a number of instructions active at once by overlapping fetching, decoding, and execution of instructions in pipelines. To improve the efficiency of processor pipelines, the Stanford MIPS project introduced the concept of using compilers to optimize the use of their microprocessor's pipeline at compile time.

As time went on, the Scalable Processor Architecture (SPARC) grew out of the work at Berkeley and the MIPS architecture grew out of the work at Stanford. Both approaches share a load/store architecture, but one has been called the register-window approach and the other has been called the single-register-set approach. The former relies on a large number of registers to store variables that would otherwise have to be fetched from a slower main memory, and the latter pares hardware down to a minimum and relies on a smart compiler to allocate variables more efficiently to registers.

RISC Principles

As RISC research and products have progressed, a common definition and set of design principles have begun to emerge. The fundamental theme of

RISC architectural design is that of maximizing the effective speed of a design by performing most functions in software, except those features whose inclusion in hardware yields a net performance gain, as shown by detailed studies of substantial high-level language programs.

To implement this theme, RISC designs emphasize "single-cycle execution." The speed of a machine cycle is not dependent on the architecture, but rather on the type of technology in which the architecture is implemented, such as CMOS (complementary metal-oxide semiconductor). Thus a conventional processor and a RISC implemented in identical technology would have essentially the same lower limit on basic cycle time. The difference between the two computers with respect to cycles would be the number of cycles required to execute a single instruction. For example, a DEC VAX 11/780 superminicomputer requires an average of five machine cycles for each instruction. A Motorola 68020 microprocessor requires an average of eight machine cycles for each instruction. In contrast, RISC processors strive to execute one instruction per machine cycle, and single-cycle execution rate is the basis for the performance gains possible in RISC. Single-cycle execution is achieved in RISC designs through the combination of some or all of the following design principles:

Hardwired control, with little or no microcode

Simple instructions and few addressing modes

Load-store, register-to-register design

Deep pipelining

Reliance on optimizing compilers

High-performance memory hierarchy

Little or No Microcode

RISC designs eliminate microcode ROM and implement the instruction set directly in hardware. This means there is no intermediate translation step from machine instructions to primitive microcoded operations to increase the number of cycles required to execute an instruction. In the case of a RISC implemented as a microprocessor, this also frees up chip space for performance-enhancing functions.

Simple Instructions and Addressing Modes

RISC designs are able to eliminate microcode by implementing a small and simple set of instructions and addressing modes. In addition to the complex instruction set of conventional computers, conventional architectures include an extensive set of ways to address system memory, many of which require the processor to calculate effective memory addresses during program execution based on several different parameters. Complex instructions and addressing modes mandate microcode or multicycle execution. In addition, they complicate the design of the processor hardware and the compilers.

Load-Store Design

Whenever an instruction requires accessing information that is not resident in the processor, such as data stored in system memory, performance is decreased. RISC designs limit references to system memory to Load and Store instructions. All other instructions operate on data in registers. A Load instruction moves data from system memory to registers, where the data can be rapidly processed and temporarily held for further access. At the appropriate time, a Store instruction returns the data to its place in system memory. Conventional computers support instructions that process information directly in system memory. These instructions require multiple machine cycles to complete. RISC designs include a larger number of registers that are typically present in conventional designs. This limits even the need to reference system memory through Load and Store instructions by allowing the data to remain in registers for a longer period of time, where it can be reused without intervening Loads and Stores. A load-store, register-to-register design is the key to achieving single-cycle execution.

Pipelining

Modern processor design generally includes a multistage pipeline to increase the rate of instruction execution. Because each stage is responsible for an individual execution phase, such as instruction decoding or operand fetching, a pipelined processor is actually working on several instructions at once. Conventional processor pipelines are subject to various kinds of inefficiencies that slow execution or even temporarily stop it altogether. For example, when a processor encounters an instruction that requires longer to complete than expected, it must delay, or "stall," the normal movement of one or more other instructions through its pipeline, until execution is completed. Besides degrading performance, the existence of pipeline stalls forces designers to further complicate processor microarchitecture with hardware to account for these stalls. The predictability of the time required to perform operations in a RISC design allows pipelines to operate at maximum efficiency.

Optimizing Compilers

Even in a RISC design, references to system memory and conditional branches can cause problems for pipelining. However, instead of using complex expensive hardware to solve these problems, RISC designs rely on optimizing compilers.

For example, delays caused by references to system memory can be minimized through the judicious use of registers. Optimizing compilers can recognize when the contents of a register can be reused in subsequent operations without needing to reload the data from system memory. In those instances when a memory reference cannot be avoided, the compiler can rearrange instructions so that useful work not dependent on the referenced data can be performed while the processor is waiting for the data to be loaded into a register.

Similarly, an optimizing compiler can deal with the unpredictability of a conditional branch through use of a "delayed branch." This technique rearranges the sequence of code to allow the instruction following the branch to be executed while the processor is deciding whether or not to take the branch.

The use of such techniques as pipelining and optimizing compilers is not unique to RISC design; however, they are more effective with RISC processors than with conventional computers. For example, compiler writers for conventional computers must develop complicated structures to ensure that their compilers choose the most efficient addressing mode and instructions at different points in the program being compiled. Nor do they have access to the relatively large number of registers normally associated with RISC designs.

The large number of registers, the small, simple instruction set, and the single addressing mode of RISC designs make it easier for an optimizing compiler to limit references to system memory, recognize computations that can be streamlined, and reorganize instructions to ensure maximum pipeline efficiency. An optimizing compiler for RISC machines can calculate the savings of an optimization more simply, because all instructions require only a single cycle for execution. On the other hand, optimizing compilers for CISC machines are forced to consider a number of alternative instruction sequences, each accounting for a variable number of processor cycles.

Memory Hierarchy

As RISC design increases the performance of the CPU, it becomes important to provide a fast, efficient memory hierarchy to keep pace with the processor. In a RISC, the memory hierarchy typically consists of numerous on-chip registers, fast static RAMs for split data and instruction caches, write buffers, and on-chip memory management. Recent advances in the cost, density, and speed of semiconductor memory have contributed to the ability to design high-performance memory hierarchies to support the speed of a RISC processor.

Summary

The trade-off of complexity is one of the leading principles of RISC technology, and it extends from details of processor microarchitecture and compiler technology up to the system level itself. Using RISC principles, designers can leverage the price and performance advantages of available technologies in software, semiconductor memories, logic, and packaging. Rather than focus narrowly on microprocessor design, this approach emphasizes the total system equation. In the case of moving complexity from the processor to the compiler, RISC technology simplifies the job of the processor in favor of a heavier demand on the compiler.

At the system level, the greater dependence on software means that memory is often required to hold a few more individual instructions than would be required in a conventional computer. One might reasonably ask the question, "How can a RISC execute a program faster than a conventional computer if it must execute more instructions?" The length of time to run a given program can be computed as the length of a single clock cycle times the instruction count times the average cycles per instruction. If a conventional computer and a RISC are implemented in identical technology, for example, CMOS of equivalent speed, density, and size, the length of the single clock cycle will be essentially the same. The instruction count of a program will be approximately 20 percent greater for a RISC than for a conventional computer. The important difference in this equation is the dramatic decrease in the average cycles per instruction from a conventional computer to a RISC. Whereas the typical range for a conventional computer is five to ten cycles per instruction, the typical range for a RISC is one and one-half to two cycles per instruction. Multiplying these factors demonstrates that a RISC may have an architectural speed advantage of two to five times that of a conventional computer. In summary, RISC designs try to gain performance by reducing the cycles per instruction much faster than they lose performance by executing more instructions.

While the decrease in the average cycles per instruction from a CISC to a RISC microprocessor has been considered to be one of the main advantages of RISC in high-speed processing, recent designs of CISC microprocessors have reduced cycles per instruction considerably by taking advantage of some aspects of RISC technology without sacrificing binary code compatibility (see Chapter 3).

SPARC

The Scalable Processor Architecture (SPARC) is a product of Sun Microsystems, although today the microprocessors and chip sets for it are made by other manufacturers, who are licensed to do so by Sun. The scalability of SPARC refers to the number of CPU registers that can be implemented in its various versions. A distinguishing feature of SPARC from other RISC implementations is the use of overlapping register windows. They reduce the number of loads and stores as well as simplifying the compiler register allocation algorithms. In this scheme, results placed in registers need not be explicitly saved or stored during procedure calls.

SPARC supports from 2 to 32 register windows. Each window has 24 working registers plus 8 global registers. The 24 registers are divided into three groups: eight "in" registers, eight local registers, and eight "out" registers. The out registers from one window become the in registers for the next window. A "current window pointer" keeps track of which window is currently active. An overlapping register window implementation is shown in Figure 2-1. In addition to register windows, SPARC provides two other sets of registers visible to the user program at any time. Global integer registers can be used for global variables and pointers, either visible to the programmer or maintained as part of the program's execution environment. Global floating-point registers are also managed by software. The current compilers use them as caller-saved register variables and temporaries.

The first commercial implementation of SPARC was by Fujitsu with its MB86900 CMOS 32-bit processor. At that time it produced the integer unit MB86900, and a floating-point controller MB86910, but Fujitsu did not offer a complete chip set for those who needed a floating-point subsystem. For a complete chip set, it was necessary to add two Weitek chips—their 1164 multiplier and 1165 ALU. This combination operated at a clock speed of 16.67 MHz and had a sustained performance of 10 VAX Mips. Toward

FIGURE 2-1
SPARC window implementation. Reprinted courtesy of Cypress Semiconductor.

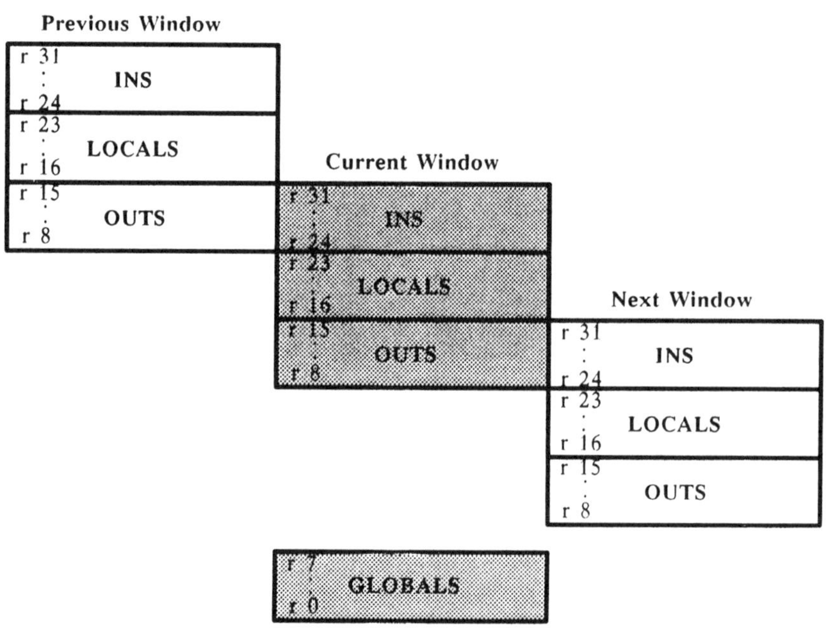

the end of 1988, Fujitsu offered a more complete SPARC chip set consisting of their MB86901 integer unit (Figure 2-2), a MB86911 floating-point controller, and a MB86920 memory management unit. However, a Texas Instrument SN74ACT8847 floating-point unit was required if an optional floating-point module was to be implemented. Such a combination is shown in the block diagram of Figure 2-3, where a complete evaluation board with VMEbus and memory components is illustrated. This chip set operates at 25 MHz with a sustained performance of 15 VAX Mips and 3.3 Mflops with the optional floating-point module.

Other licensees of the SPARC architecture are Cypress Semiconductor, LSI Logic, and Bipolar Integrated Technology. LSI Logic produced its L64801 integer unit, which was somewhat comparable with the Fujitsu MB86901. This was followed by enhanced versions by both Cypress and LSI Logic. Cypress produced the CY7C601, and LSI Logic produced their

FIGURE 2-2
Block diagram of Fujitsu's MB86901 SPARC RISC processor.
Reprinted courtesy of Fujitsu Microelectronics, Inc.

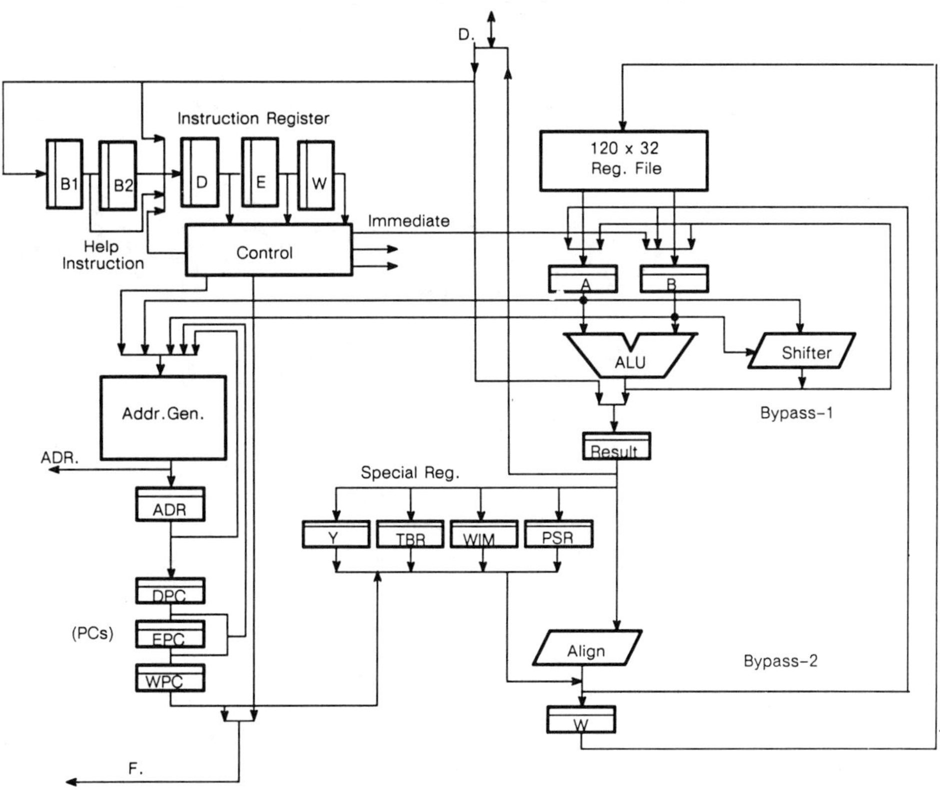

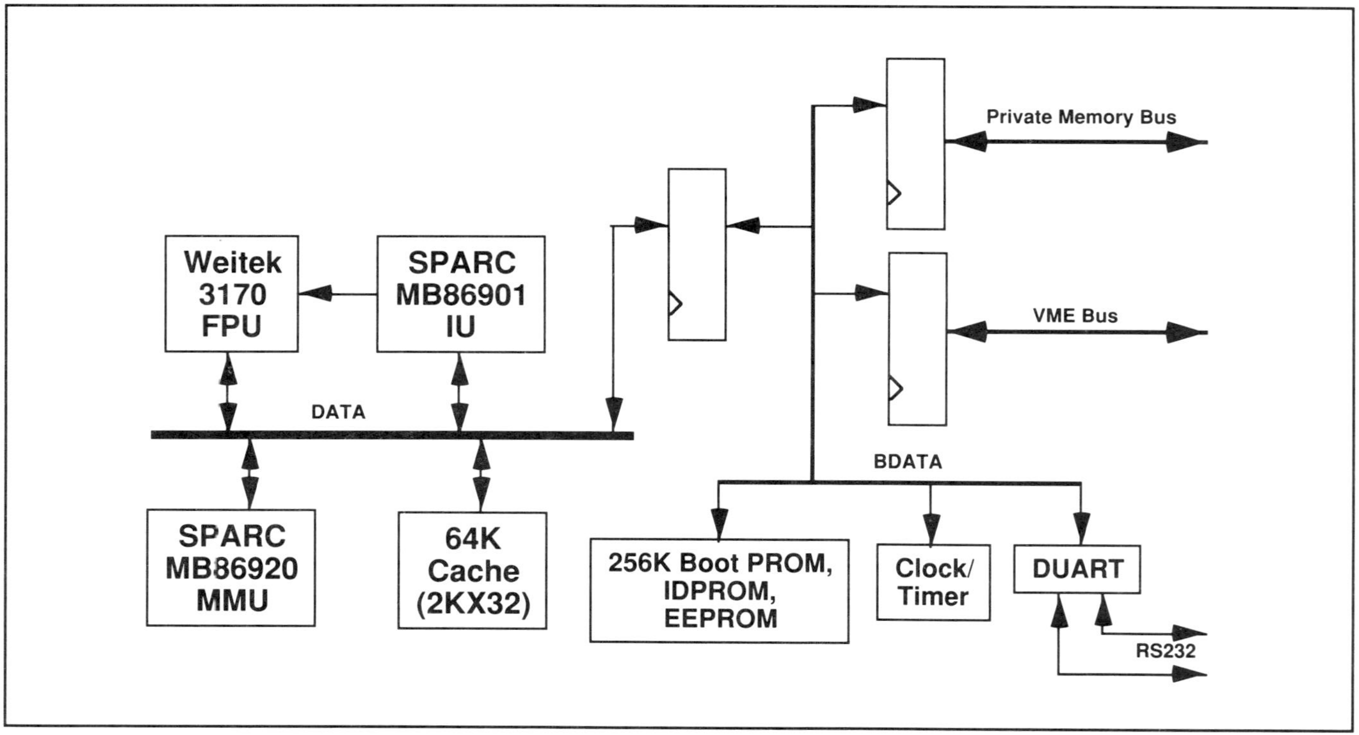

FIGURE 2-3 Block diagram of MB86900 evaluation board. Reprinted courtesy of Fujitsu Microelectronics, Inc.

L64811, both of which operated at 33 MHz and delivered sustained performance of 24 VAX Mips for the Cypress unit and 20 VAX Mips for the LSI Logic unit.

MIPS

The Microprocessor without Interlocking Pipeline Stages (MIPS) architecture was developed by the MIPS Computer Company. It contains thirty-two 32-bit registers, and all instructions and addresses are 32 bits. The CPU has a five-stage pipeline design, which assists in obtaining an execution rate approaching one instruction per cycle. A high-bandwidth memory interface handles separate instruction and data caches ranging in size from 4 to 64 KB each. Both caches are accessed during a single CPU cycle. All cache control logic is on the chip. There is an on-chip memory management unit with a fully associative, 64-entry translation lookaside buffer (TLB) that provides fast address translation for virtual-to-physical memory mapping of the virtual address space. The RISC CPU generates all addresses and handles memory interface control for up to three additional tightly coupled external coprocessors.

Like Sun with its SPARC architecture, MIPS Computer licenses out the chip manufacturing to other companies—in this case, LSI Logic, Integrated Development Technology, and Performance Semiconductor. The first commercial implementation of this architecture was the 2000 series. LSI Logic called it the R2000 and Performance Semiconductor called it the R2000A. The R2000 was a 32-bit CMOS RISC computer with thirty-two 32-bit registers, an on-chip TLB, a memory management unit, and cache control circuitry. The processor consisted of two tightly coupled processors implemented on a single chip. In addition to the RISC CPU there was a system control processor, which contained the TLB buffer and control registers to support a virtual memory subsystem.

The R3000 series implementation is an enhanced version of the R2000. LSI Logic calls it the LR3000; Integrated Development Technology calls it the 79R3000; Performance Semiconductors calls it the R3000. The LR3000 chip set is composed of three members: The LR3000 CPU, the LR3010 floating-point accelerator, and the LR3020 write buffer. The LR3000 CPU is a 32-bit RISC with an on-chip memory management unit, a cache controller, and an interface for up to three additional coprocessors. The cache controller interfaces directly with separate data and instruction caches composed of standard SRAM. Cache size can vary from 4 to 256 KB for both instructions and data. The memory management unit is a fully associative 64-entry TLB providing fast address translation for virtual-to-physical memory mapping of the 4-gigabyte (GB) virtual address space.

The performance of the CPU is due largely to a five-stage pipeline.

The compilers organize instructions so that the pipeline is kept full and operations are performed on data stored in an on-chip 32 × 32 register file. All instructions except for multiply and divide execute in a single cycle, allowing the program execution rate to approach one instruction per cycle.

An optional coprocessor is provided by the LR3010. It interfaces seamlessly with the CPU by cointerpreting instructions. The compilers have been designed to take advantage of this parallelism. Another chip, the LR3020, provides an interface between the cache and the main memory. When the CPU writes to the cache, the write buffer updates the slower main memory without stalling the CPU. This "write-through" capability ensures that data in the cache is consistent with the copy stored in the main memory. Each write buffer consists of a four-deep FIFO (first in, first out) storing 8 bits of address and 9 bits of data. Typically four write buffers are used per system. Hierarchical memory systems based around the write buffers can be built with standard 25-ns SRAMs.

IDT's 79R3000 processor consists of two tightly coupled processors on a single chip. The first of these is a full 32-bit CPU (central processor unit) incorporating RISC techniques. The second processor is a system control coprocessor containing a TLB and control registers to support a virtual

FIGURE 2-4
Functions of IDT's 79R3000 RISC processor. Reprinted with permission from Integrated Device Technology, Inc.

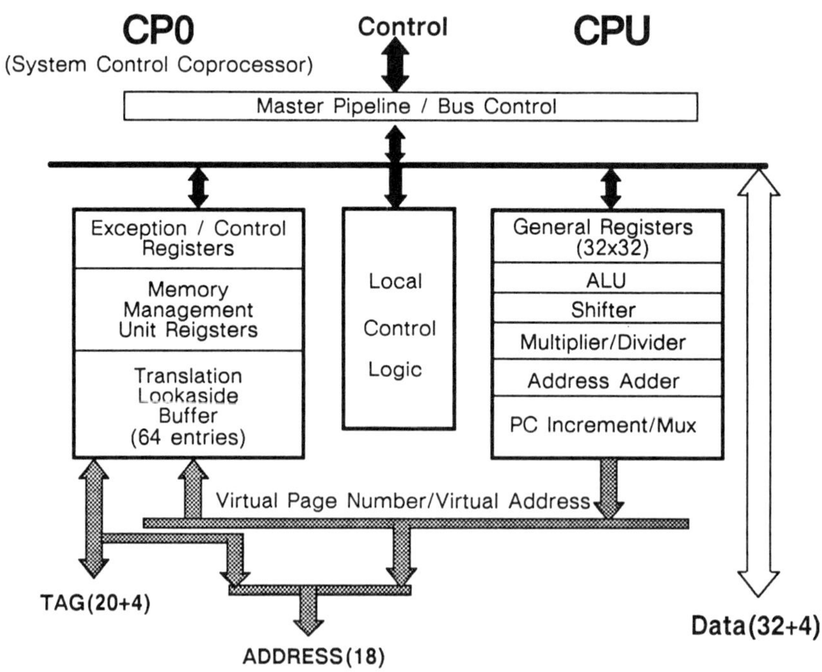

memory subsystem with a dual-cache bandwidth of up to 133 MB/second. The functions incorporated in the 79R3000 are shown in Figure 2-4.

The 79R3000 provides thirty-two general-purpose 32-bit registers and two 32-bit registers that hold the results of integer multiply and divide operations. All 79R3000 instructions are 32 bits long, and there are only three instruction formats. This approach simplifies instruction decoding. More complicated (and less frequently used) operations and addressing modes can be synthesized by the compiler using sequences of simple instructions. The 79R3000 instruction sets are as follows:

Load/Store instructions that move data between memory and general registers

Jump and Branch instructions that change the control flow of a program

Coprocessor instructions that perform operations in these sections of the chip

Special instructions that perform a variety of tasks, including moving data between special and general registers, making system calls, and executing breakpoints

A detailed block diagram of the 79R3000 is shown in Figure 2-5.

FIGURE 2-5
Block diagram of 79R3000 System

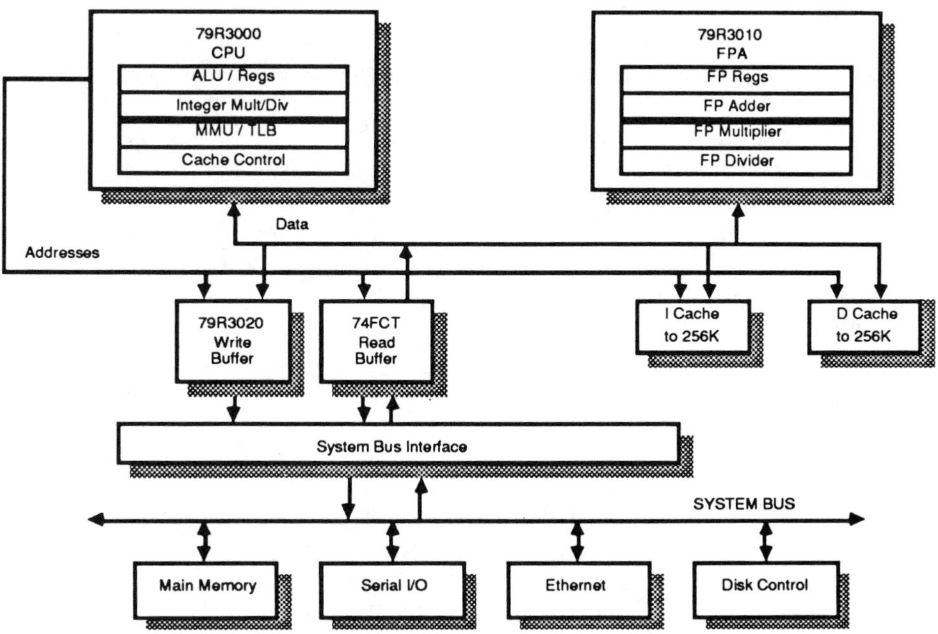

Motorola's 88000 Family

The initial architectural implementation of the 88000 comprised two VLSI chips—the 88100 CPU and the 88200 cache/memory management unit (CMMU). Both chips were available at 20 MHz with higher-speed versions to follow. A block diagram of the 88100 CPU architecture in Figure 2-6 shows the five major execution units. They include the instruction, data memory, and integer units, and the floating-point adder and multiplier pipelines. These execution units share a common 32-entry register file. A register scoreboard, for maintaining pipeline coherency among all five execution units, is associated with the register file. A Harvard bus architecture allows instructions to be fetched while operands are transferred between memories.

The triported register file contains 32 registers, each of which is 32 bits wide. As shown in Figure 2-7, the 88100 contains three types of registers that provide data and execution information to the execution units. Supervisor and user modes of operation are provided. Most control registers can be accessed only in the supervisor mode. Register R0 is always hardwired to zero. A register scoreboard provides a hardware interlock in order to maintain pipeline coherency among the various pipelined execution units with respect to the general-purpose register file. Every register has a scoreboard bit assigned to it except for R0, which cannot be modified.

The floating-point unit (FPU) executes the floating-point arithmetic, integer/floating-point conversions, and integer multiply and divide instructions. Single-precision, double-precision, and mixed-mode arithmetic operations can be performed for floating-point instructions. The FPU is implemented as two pipelines—a five-stage floating-point adder pipeline and a separate six-stage floating-point multiplier pipeline. Add, Subtract, Compare, Divide, and Convert instructions are implemented by one pipeline and Multiply instructions are implemented by the other. The Divide instruction is the only floating-point operation that is not completely pipelined. A Divide instruction iterates through one execution stage once for each bit of accuracy required in the result.

The FPU contains 11 control registers. Nine of these are shadow registers containing status and intermediate results are used for execution recovery; the other two registers are the floating-point status register and the floating-point control register. The FPU requires more than one clock cycle per instruction. Floating-point instructions prevent subsequent instructions from using the results prematurely by setting a bit in the scoreboard register during their execution.

The 88000 architecture provides for a degree of scalability through the concept of special function units (SFUs). More than twenty-six thousand opcodes, as well as vector table space, have been reserved for up to seven SFUs. The floating-point unit is designated to be SFU number one. As changes are made to future implementations of the architecture, such as new SFUs, the scoreboard guarantees upward code compatibility.

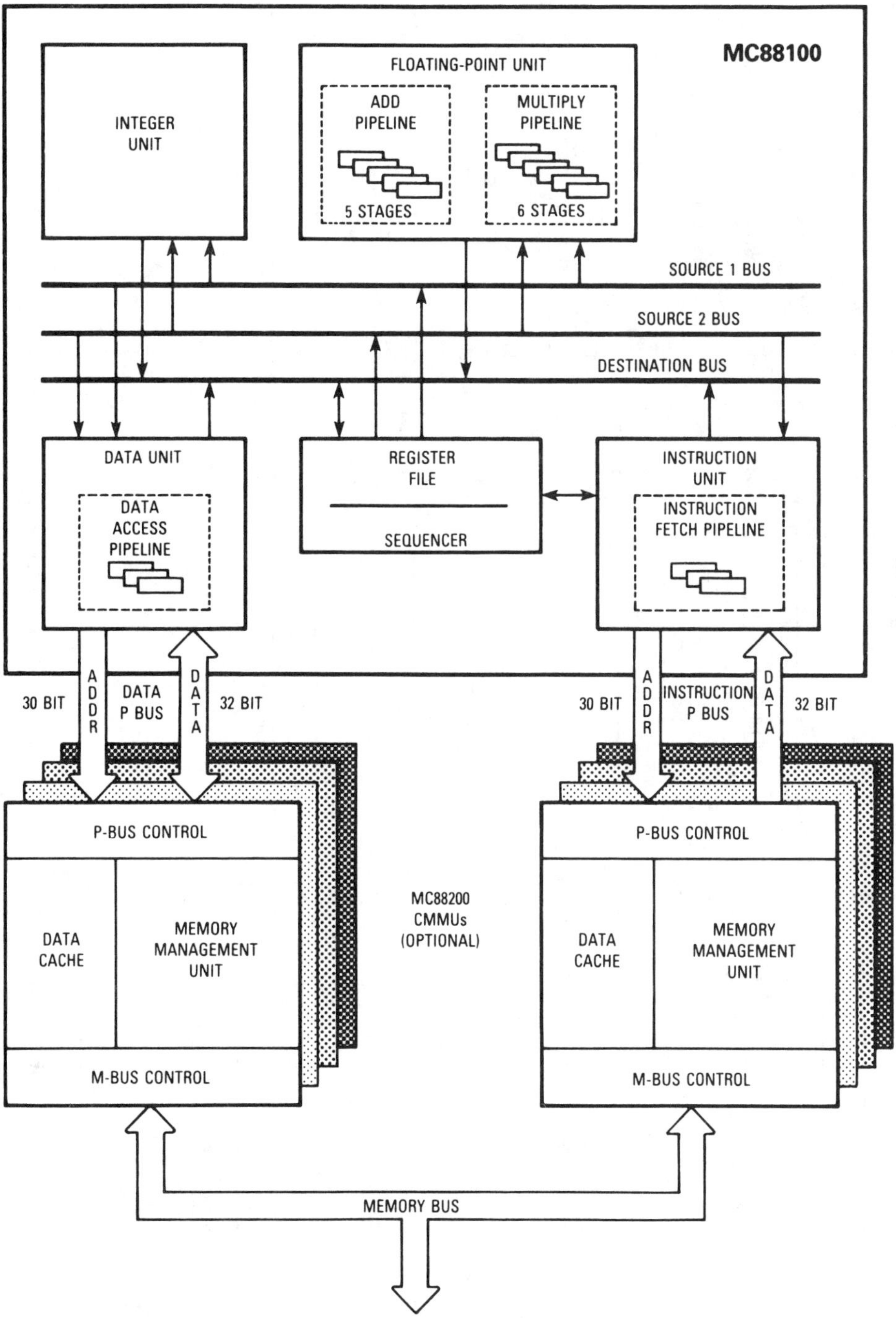

FIGURE 2-6
Architecture of Motorola's MC88100 RISC processor. Reprinted with permission of Motorola Inc.

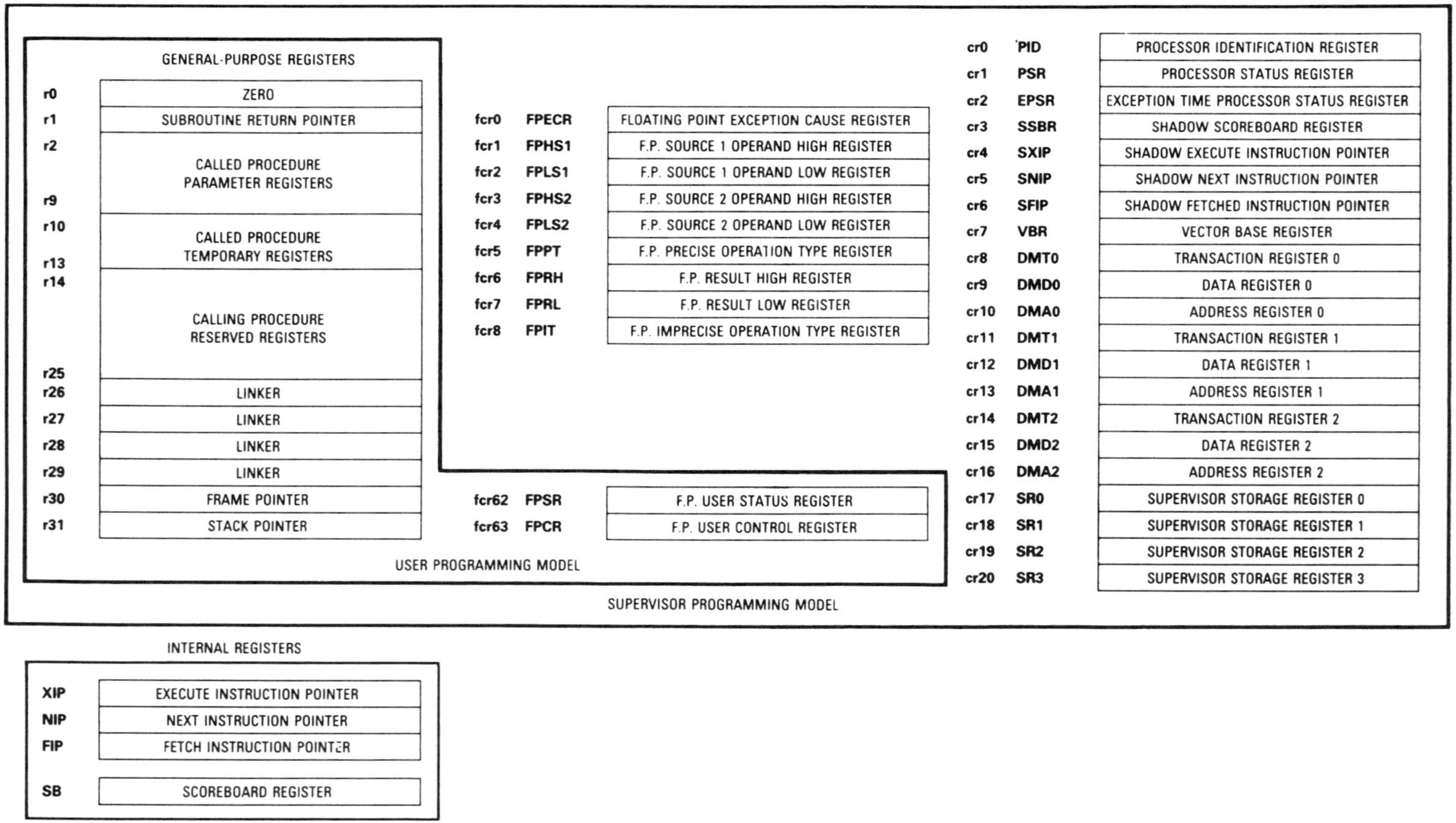

FIGURE 2-7 The registers of the MC88100. Reprinted with permission of Motorola Inc.

The 88200 C/MMU provides the interface between the 88100 CPU and the system memory and I/O. In order for the CPU to execute instructions in one clock cycle, a high-speed mechanism is required to supply the necessary instructions and operands demanded by the various execution units. The 88200 fills this function, with the following features:

A 16-KB, four-way associative write-through or copy-back cache with LRU (least recently used entry) replacement

Bus monitoring cache coherency logic (a bus snooper)

Fully associative page address translation cache

Fully associative block address translation cache

Memory bus (Mbus) interface to system memory

A usual arrangement is to connect a minimum of two CMMUs to an 88100, with one 88200 on the code Pbus (processor bus) and one 88200 on the data Pbus. A maximum of four 88200s may be connected to the code Pbus and four 88200s to the data Pbus.

With a 25-MHz clock the 88000 is rated at 20 VAX Mips. Also, it executes 46000 dhrystones, 20 single-precision Mwhetstones, and 3.6 double-precision Linpack Mflops.

Intel's 80860 (i860)

The i860 CPU incorporates the following units on a single chip (see Figure 2-8):

RISC Integer Core

Memory Management Unit with Paging

Bus Control Unit

Floating-point Control Unit

Floating-point Adder Unit

Floating-point Multiplier Unit

3-D Graphics Unit

Instructions Cache

Data Cache

The i860 CPU register model is divided into integer and floating-point sets to support parallel execution. There are 32 integer registers in the RISC core, supporting general integer operations. These registers may be loaded or stored from 8-, 16-, or 32-bit quantities in memory. A second set of thirty-two 32-bit registers is contained in the floating-point control unit, supporting

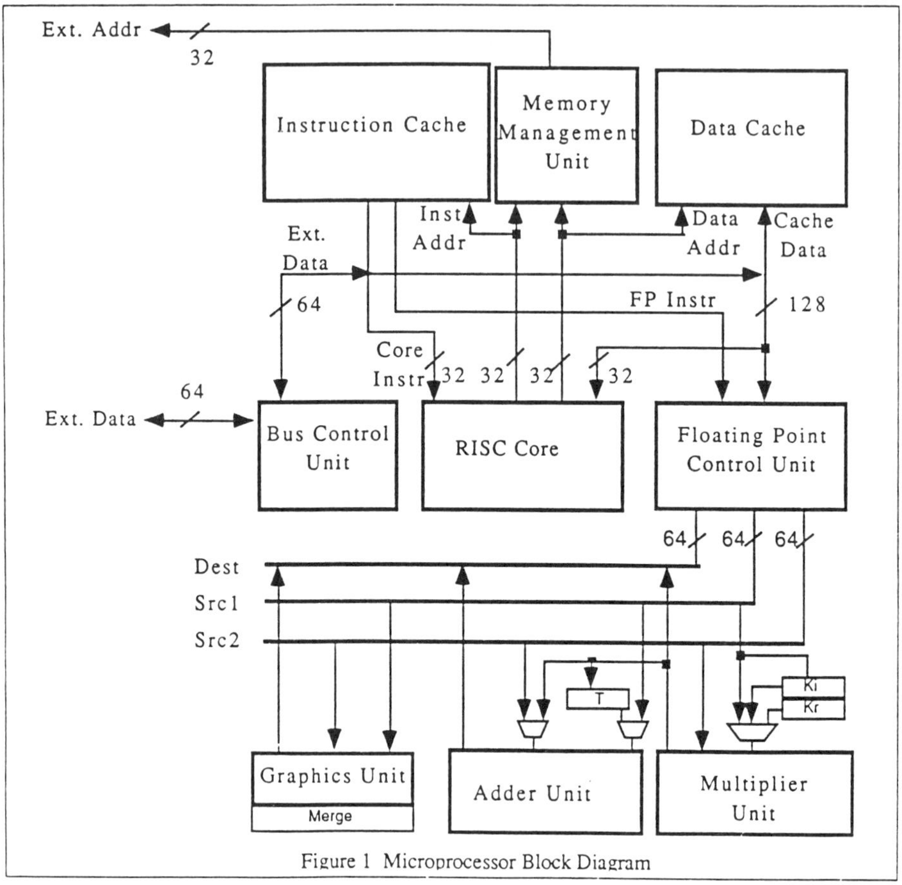

FIGURE 2-8
Block diagram of Intel's i860 RISC processor. Reprinted with permission of Intel Corporation.

floating-point and some integer operations. These registers may be loaded or stored individually, in pairs or in quads from 32-, 64-, or 128-bit quantities in memory. The 128-bit-wide data cache allows all loads and stores to be performed in one clock. Being able to perform 128-bit loads or stores in one clock is crucial to supplying data at the rate needed to keep the floating-point units busy. In addition, these instructions provide for fast block transfers of general data blocks. The two register files contain a total of eight ports to sustain the parallel operations.

The integer instruction set contains about thirty-five instructions, and follows the three operand, load/store style of RISC designs (see Figure 2-9). The RISC core fetches both integer and floating-point instructions. It executes all Load, Store, integer, bit transfer, and control transfer instructions. It uses a four-stage pipeline and extensive bypassing and scoreboarding to execute most instructions in one clock.

To combat creeping elegance, what is called the "5 percent/10 percent rule" was employed. Only operations that provided greater than five percent performance improvement were considered, and a feature with significant hardware complexity needed at least 10 percent performance improvement to be considered. Operations that lengthened the critical path timing or required multiple clock cycles were avoided. For example, integer divide and multiply are performed using the floating-point hardware.

Notable features of the RISC core include:

Auto increment addressing with an arbitrary increment for vector operations.

Conditional branches optimized for branch-taken and branch-not-taken cases.

FIGURE 2-9
RISC code instruction set of the i860. Reprinted with permission of Intel Corporation.

Load/Store

LOAD/STORE INTEGER: Relative Load/Store, Indexed Load
LOAD/PIPELINED LOAD/STORE/PIXEL STORE: Relative/Indexed/Auto Increment
IXFR
LOCK/UNLOCK
LOAD/STORE CONTROL REGISTER
FLUSH DATA CACHE BLOCK

Logical

AND
ANDNOT
OR
XOR

Shift

SHIFT LEFT/RIGHT: Logical/Arithmetic
SHIFT RIGHT DOUBLE

Arithmetic

ADD SIGNED/UNSIGNED
SUBTRACT SIGNED/UNSIGNED

Control

CALL/INDIRECT CALL
SUPERVISOR CALL/OVERFLOW TRAP
BRANCH INDIRECT/BRANCH RELATIVE/BRANCH CC SET/BRANCH CC CLEAR
LOOP
BRANCH EQUAL/NOT EQUAL

Single clock test/branch and loop instructions for efficient tight loops like vector routines.

A pixel store used in conjunction with the graphics unit for hidden cc (coprocessor condition) elimination.

Lock and Unlock instructions that allow a general sequence of up to 32 instructions to be interlocked for multiprocessor synchronization. This allows the traditional Test and Set operations as well as more sophisticated operations such as Compare and Swap.

A pipelined load optimized for accesses to data structures that are too big to fit into the on-board cache.

The i860 CPU implements the two-level paged virtual memory structure of the i386™ CPU with a 4-KB page size. This allows source-code-compatible OS (operating system) kernels to be developed for the i860 and the i386 CPUs and supports heterogeneous multiprocessor systems containing both types of processors. User/supervisor and read/write protection is provided for each page. The i860 CPU has a 64-entry, four-way set-associative TLB, with miss processing done in hardware.

The floating-point instruction set contains about thirty-five instructions and operates on single- and double-precision IEEE floating-point values and 32- and 64-bit integers (see Figure 2-10). The pipelined add and multiply units are each three stages. For single precision, each can produce one result per clock for a peak rate of 80 Mflops with a 40-MHz clock. For double precision, the multiplier can produce a result every other clock for a peak rate of 60 Mflops. Reducing the double-precision multiply rate saved half of

FIGURE 2-10
Floating-point instruction set of the i860. Reprinted with permission of Intel Corporation.

Multiply Unit	Add Unit		Dual Operations
FMULTIPLY	FADD	FIX	PFADDMULTIPLY
PFMULTIPLY	FSUB	PFIX	PFSUBTRACTMULTIPLY
FMULTIPLYLOW	PFSUB	FTRUNC	
FRECIPROCAL	PFGREATER	PFTRUNC	
FRECPRCIPROCALSQR	PFLESS	PFEQUAL	

Graphics Unit		
FINTADD	FZBUFCHECKS	FADDPIXEL
PFINTADD	PFZBUFHCHECKS	PFADDPIXEL
FINTSUB	FZBUFCHKL	FADDZ
PFINTSUB	PFZBUFCHL	PFADDZ

the multiplier tree and is consistent with the data bandwidth available for double-precision operations.

AMD's (Advanced Micro Devices, Inc.) 29000

AM29000 second-generation RISC architecture (Figure 2-11) features include the following:

192-word register file, organized as a stack cache, to speed calls and returns and minimize memory references

On-chip branch target cache and delayed branch instructions, to minimize pipeline delays caused by branches

FIGURE 2-11
AM29000 block diagram. Copyright © Advanced Micro Devices, Inc. Reprinted with permission of copyright owner. All rights reserved.

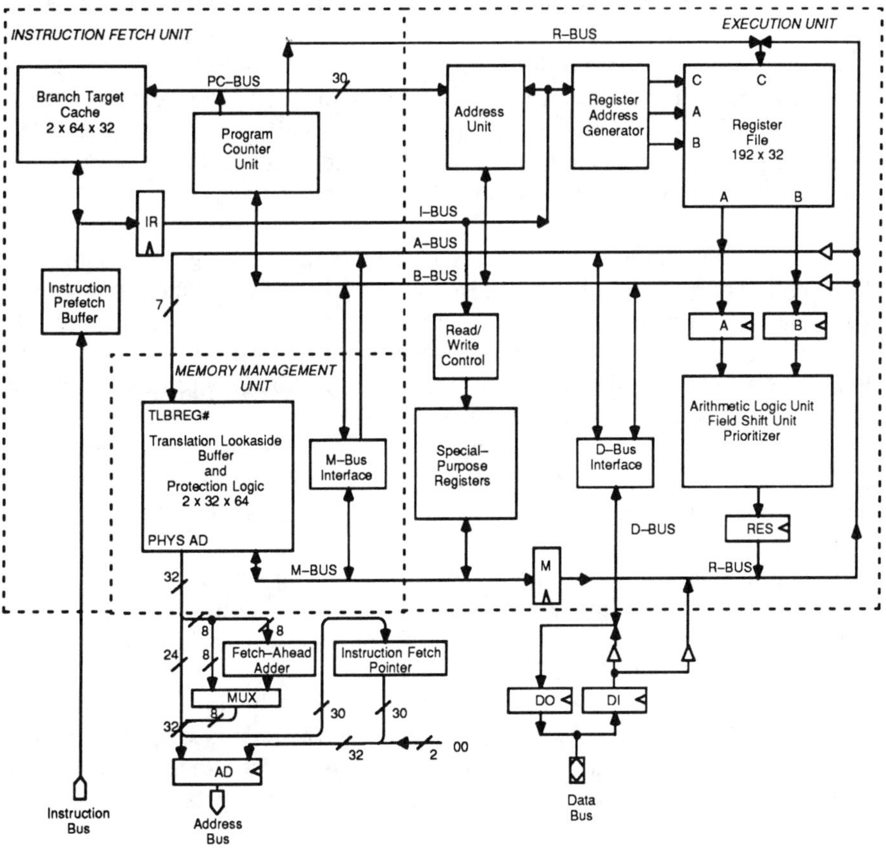

Three-bus memory interface, to eliminate contention for the bus and maximize bus bandwidth

Three-operand instructions, to reduce the number of instructions

No addressing modes, to maximize the ability of compilers to optimize address calculations

Overlapped loads and stores and delayed branch instructions, to maximize instruction throughput

Highly configurable, on-chip memory management unit, to minimize delays due to address translation

The AM29000 provides 192 general-purpose registers, more than any other commercially available microprocessor (see Figure 2-12). All 192 registers are capable of storing data or addresses, and all are accessible by any instruction. The resulting performance increase can be enormous, because the processor no longer must waste cycles waiting for external data. The register file is divided into 128 local registers and 64 global registers. The global registers have fixed addresses while the local registers are addressed relative to a stack pointer.

Stack pointer addressing for the local registers allows implementation of a run-time stack cache. In most cases, parameters can be passed to a function or procedure without communication with external memory. On average, procedure calls and returns can be executed with five to ten times fewer cycles compared to processors that implement stacks in external memory.

Register banking provides very fast context switching by dividing the local registers into eight banks of 16 registers. In multitasking applications, these banks can hold the processor status and variables for up to eight different tasks. Register bank protection prevents tasks from corrupting data belonging to other tasks. In this configuration, a task switch can occur in as few as 17 cycles.

Single-cycle instruction execution at 25 MHz is made possible by the AM29000's four-stage pipeline. Each stage—instruction fetch, instruction decode, instruction execution, and result write-back—completes in a single cycle.

Branches might cause a pipelined processor to wait while the pipeline is being refilled. To minimize this performance limitation, the AM29000 incorporates a branch target cache and implements delayed branches, which allow single-cycle branches.

Unlike a traditional instruction cache, the branch target cache contains only four instructions per branch target. These instructions keep the pipeline filled while a new instruction stream is established (see Figure 2-13). This technique lets the 128-word cache contain 32 branch targets, increasing the branch target cache hit ratio.

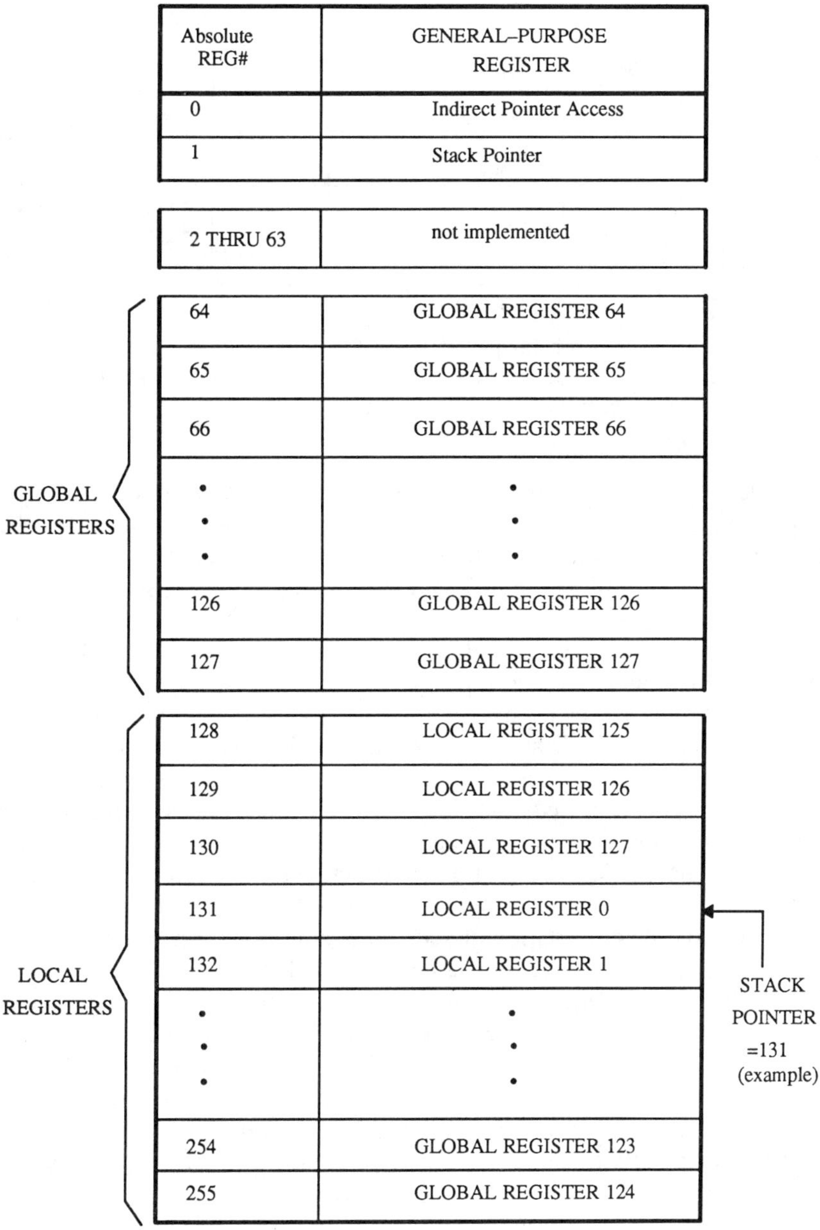

FIGURE 2-12
Register file of the AM29000.

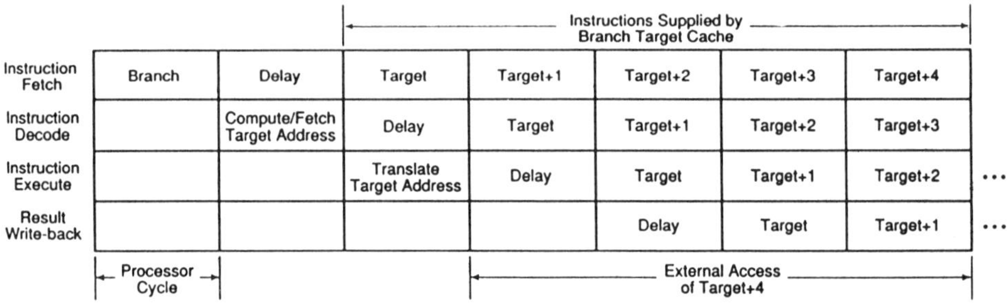

FIGURE 2-13
Effects of the branch target cache in the AM29000.

The AM29000 performs all external accesses over a three-bus channel. This channel includes an address bus, an instruction bus, and a data bus. The three-bus architecture provides the maximum data and instruction bandwidth without multiplexed buses, since the address bus is shared between instruction and data accesses in a way that does not induce contention. This is also designed for easier speed enhancements in future AM29000 processors, memories, and peripheral products.

The channel can be used with three access protocols: simple, pipelined, and burst mode. In simple accesses, the address is valid throughout the entire cycle. In pipelined accesses, the address of the next access is transmitted while a previous access is still in progress; pipelined access leads to higher performance than does simple access.

The highest possible transfer rate is achieved through the use of burst-mode accesses. After the address of the first access has been sent, subsequent requests for information at sequential addresses can occur at the rate of one per cycle without requiring additional address transfers.

With burst-mode accesses, the AM29000 can load or store data at 100 MB/s; thus it can be used to support memory-to-memory moves at up to 60 MB/s. This is 4 times the bandwidth of the FDDI (fiber distributed data interface) fiber optic network and 40 times the bandwidth of Ethernet. The AM29000's ability to handle high-speed data requirements, combined with its character handling capability, make it the ideal processor for high-performance communications and local area networking.

All AM29000 operations (except Loads and Stores) are register-to-register operations. Most instructions specify three general-purpose registers for operands (two source operands and one destination operand). The three-address architecture matches the notation used internally by compilers. This approach avoids the destruction of partial results, simplifies register allocation, and avoids the time-consuming operand copying often required in two-address processors.

The AM29000 has a load/store architecture for external data references; that is, all external data transfers occur to and from general-purpose registers as the result of explicit Load and Store instructions. This approach is used by all RISCs. In contrast to most processors (including first-generation RISCs), the AM29000 performs no address computation—there are no "addressing modes." All external addresses are register-indirect (that is, they are 32-bit quantities in a register).

Because of the large number of registers, most scalar data will be kept on the stack cache; external data references will be primarily to arrays or records. And compilers can efficiently construct accesses to these external data structures from simple register-indirect addressing. By thus eliminating the address calculation, the AM29000 saves one cycle in the pipeline—a cycle that is used for address translation in the MMU.

The AM29000 overlaps Load and Store operations with the execution of other instructions. Optimizing compilers are able to take advantage of this feature by scheduling Loads to occur while other instructions are being executed. Processor hardware detects data dependencies while overlapped Loads are being performed, so proper operation is ensured even if a Load cannot be properly scheduled.

In the event that the result of a Load operation is required immediately by an instruction, it is forwarded directly to the appropriate execution unit, bypassing the register file, thus reducing access latency by one full cycle.

The AM29000 on-chip memory management unit provides a demand-paged virtual memory environment. Key features of the MMU include the following:

- 64-entry, two-way set associative TLB
- 4 GB virtual memory per process for up to 256 processes
- Pipelined address translation to reduce latency
- Software TLB reload allowing user-defined memory-management architecture
- Least-recently-used hardware to assist reload
- Memory protection

Software TLB reload allows system designers to choose a memory-management scheme best matched to the particular environment. TLB reload performance is enhanced by least-recently-used hardware as well as the AM29000's low trap overhead.

All memory accesses on the AM29000 are fully restartable, allowing a demand-paged environment. In the event of a page fault during a Load or Store, the AM29000 can restart the Load or Store operation without reexecuting the Load or Store instruction.

VL86C010 and VL86020

If any RISC microprocessor fits into an inexpensive personal computer, it would be the VL86C010 Acorn RISC manufactured by VLSI Technology. In fact, it is used in a personal computer produced by Acorn Computer, Ltd., in England, where the chip was designed. Apparently, it has not caught on for PCs in the United States due to its lack of a software base. Because of the absence of this market, VLSI Technology has directed this chip toward embedded applications. The application of its second-generation version, the VL86C020, has also been pointed in the same direction.

The VL86C010 Acorn RISC Machine (ARM) processor family is designed to provide a superior cost/performance ratio, not just superior performance. This is not to say that the VL86C010 offers poor performance, for it performs better than the MC68030 and the iAPX80386. The VL86C010 processor block diagram is shown in Figure 2-14.

FIGURE 2-14
Block diagram of VLSI Technology's VL86C010 CPU. Ron Cates and James Farrell, VLSI Technology, Inc., "VL86C020 (ARM III)-RISC Enters the Second Generation," Midcon/89 Conference Record, 1989, pp. 60–63.

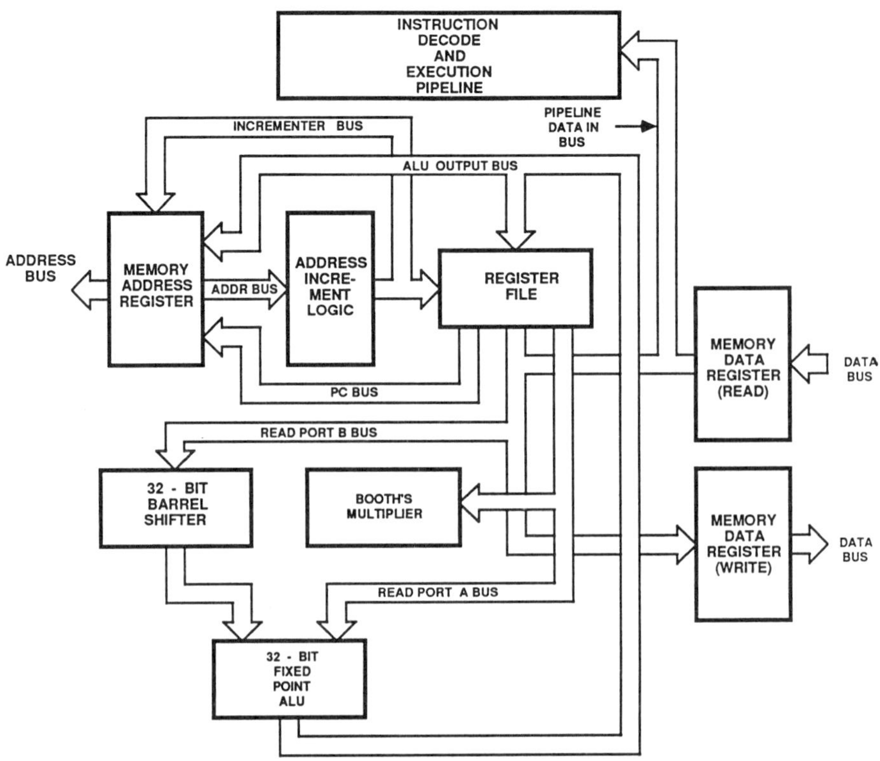

There are several architectural features worthy of note. The VL86C010 contains 27 partially overlapped registers with only 16 visible to the program at any one instant. Figure 2-15 shows the register file implementation. Many RISC microprocessors use overlapping register files across procedure call/return sequences. Whenever the processor performs a procedure operation, new registers are provided as temporary storage for the called procedure. In addition, part of the registers are shared with the calling procedure so that parameter passing can be performed in the registers, eliminating the need for stack-based linkage. The VL86C010 is different in that it overlaps the registers across the four operating modes of the processor instead of via procedure calls. The four modes are: User, Supervisor, Normal Interrupt, and Fast Interrupt. By overlapping the registers across modes, the additional registers reduce the overhead associated with context switching. For example, the seven registers in the Fast Interrupt (FIRQ) model can be used to hold status, DMA (direct memory access) pointers, and DMA word counts, eliminating the need to save state and reload these values during exception processing. By overlapping the registers across the processor modes, the VL86C010 guarantees a high utilization of the extra silicon. Uti-

FIGURE 2-15
Partially overlapped registers of the VL86C010. Ron Cates and James Farrell, VLSI Technology, Inc., "VL86C020 (ARM III)-RISC Enters the Second Generation," Midcon/89 Conference Record, 1989, pp. 60–63.

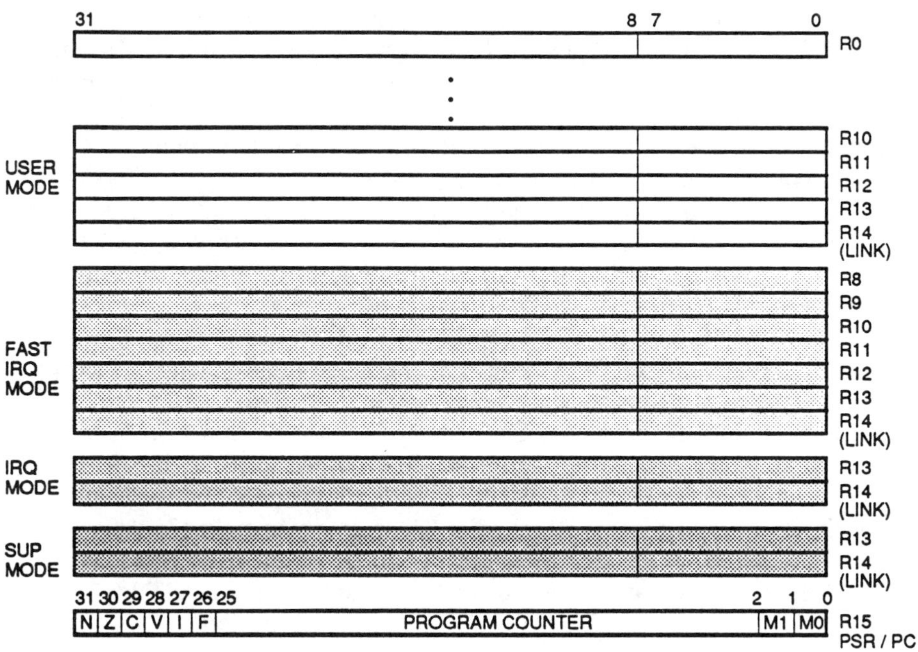

lization of registers when overlapped across procedure calls is much less predictable and highly variable, depending on program dynamics.

The next feature of interest is the pipeline structure. All RISC processors depend on pipelining to allow instruction prefetch and execution to be performed during the same clock cycle. Pipelining is needed to allow RISC processors to approach the goal of executing an instruction in a single clock cycle. The VL86C010 uses a very simple three-stage pipeline to achieve its average of 1.895 clocks per instruction. A complex pipeline is not necessary, as the VL86C010 is implemented using hardwired logic, eliminating the need for predecoding of instructions. The pipeline stages are prefetch, shift, and execution. A simple pipeline improves the branch execution time as the pipeline requires only three blocks to fill. Many other processors can require up to 15 clocks to execute a branch instruction due to their long pipeline architectures.

The VL86C010 contains a full 32-bit barrel shifter that is highly useful in many applications, particularly graphics. One of the two source operands passes through the shifter on each ALU (arithmetic/logic unit) cycle, enabling it to be shifted or rotated to the proper bit alignment. The shifting operation is performed essentially for free as a nonshifted value simply has a zero shift length and can be executed in a single clock, and the barrel shifter is not in the critical path of the processor. If the operand has a variable shift length (it is contained in a register), then the shift will require one additional clock.

The VL86C010 coprocessor interface is optimized towards high performance and uses a bus-watching protocol, whereby the VL86C010 and the coprocessor take a copy of each instruction as it is fetched from main memory. While the VL86C010 decodes the instruction and determines whether it is to be executed (remember that all VL86C010 instructions are conditionally executed), the coprocessor decodes it to determine whether it can commit immediately. The handshake to commit execution can take just one cycle required by VL86C010 to step its instruction pipeline, and this is the only overhead for executing the instruction. If the coprocessor is busy or requires time to prepare data items requested by the instruction, the handshake puts the VL86C010 into an interruptible busy-wait stage until the coprocessor is ready.

The VL86C020 is a second-generation RISC processor that is upwardly software-compatible to the VL86C010. The latest family member incorporates a 4-KB on-chip cache memory to increase performance and lower memory bandwidth demands. Since multiprocessor systems are now feasible, a new-semaphore instruction has been added to assist in shared resource allocation. Manufactured in VLSI Technology's standard 1.6-μm, double-level metal process, the VL86C020 is approximately 135,000 square mils and is still one of the smallest die of the 32-bit MPU devices now available. Operation at frequencies of 16 and 20 MHz are available. Cost is kept to a minimum by using a 160-pin quad flatpack package.

The VL86C020 processor possesses the same CPU characteristics as the VL86C010. A new instruction is supported that performs a Read-Modify-Write operation to the memory allowing for multiprocessor implementations. Figure 2-16 shows a block diagram of the VL86C020. The major areas of the device are: cache memory, clock generation, coprocessor interface, and CPU.

The problem for a coprocessor connected to a cached CPU is that the CPU no longer collects all instructions from main memory; most of them come from an invisible on-chip cache. The coprocessor bus watch mechanism therefore requires rethinking. The obvious solution is to attach the coprocessor to the bus between the CPU and the cache, down which all the

FIGURE 2-16

Block diagram of the VL86C020. Ron Cates and James Farrell, VLSI Technology, Inc., "VL86C020 (ARM III)-RISC Enters the Second Generation," Midcon/89 Conference Record, 1989, pp. 60–63.

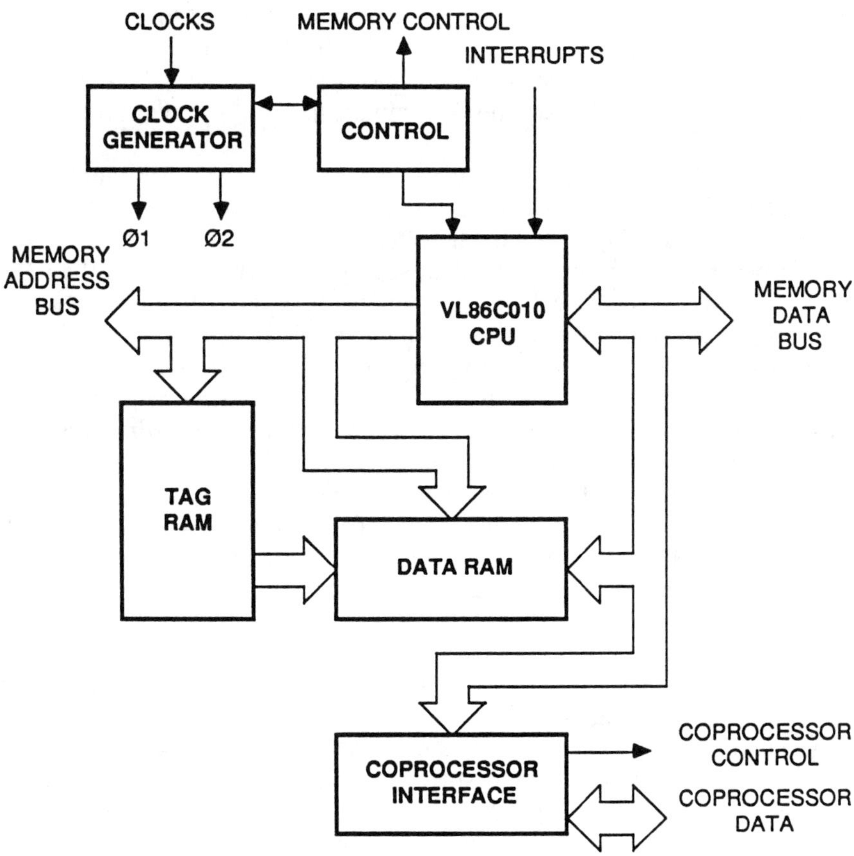

instructions flow. The problem is that this requires additional pins to bring this bus off-chip, and the CPU clock rate is likely to be so high that sending every instruction fetch off-chip at the CPU clock rate will require a lot of power. Additionally, the delay in getting the instruction off the chip and into the coprocessor will either compromise the on-chip timing or the coprocessor timing skew.

The solution adopted on VL86C010 accepts the cost of the additional bus but puts a one-clock pipeline delay into the route off-chip. This additional cycle is long enough for the coprocessor instructions to be identified, and the bus drivers can be disabled for all others, saving the unnecessary power dissipation. The coprocessor therefore runs one cycle behind the VL86C010, and the minimum handshake time increases to two (fast) clock ticks. Data transfers from main memory to the coprocessor spend one tick on VL86C020 to match this delay, and this keeps memory timing very safe. Data transfers from the coprocessor to main memory must be timed to allow for the delay, but again pipelining is used to ensure safe timing.

The VL86C020 interface uses extra cycles to ensure consistent memory timing, but the increased CPU clock rate means that the actual handshake time is reduced. As a result of this implementation, the coprocessor is also able to exploit the increased memory bandwidth available from the cache.

VL86C020 has two asynchronous clock signals; one is locked to external memory cycles, the second determines the rate of execution when all accesses are in the cache.

VL86C010 data transfers are synchronous to the CPU clocks. On VL86C020 a standard CPU core is used, but now it must accept transfers from either the on-chip cache or the off-chip main memory. Therefore the CPU is made synchronous to the data source for each cycle, and the CPU clocks switch between the external memory clock and the fast clock (for cache accesses). Furthermore, to maximize flexibility in the choice of the fast clock rate, it was decided to allow the two clocks to be completely unrelated. Every switch between the two clock frequencies requires resynchronization, with the unavoidable possibility of synchronization failure and metastability in the clock selection circuitry.

The pressure on the synchronizer design is high because the resynchronization time is mostly wasted time and should therefore be minimized, but the safety of a synchronizer increases exponentially with the synchronization time. Finding the optimal trade-off point for safe operation with minimal delay requires very careful engineering design.

The VL86C020 provides a two-and-a-half times performance increase over the previous generation VL86C010 while maintaining exactly the same memory bandwidth and software compatibility. By offering the CPU and cache memory on a single device, the processor offers a superior upward migration path, because the highest-frequency memory path is not required to cross a chip boundary. With the cache on-chip, two levels of buffers are removed from the most critical performance path. In addition, as process rules continue to shrink, the clock frequency can be increased for on-chip operations accordingly without making undue demands on memory system bandwidth.

Chapter 3

32-Bit CISC Microprocessors

There are about as many families of 32-bit CISC microprocessors as there are families of RISC microprocessors. However, only two of the CISC families dominate in the marketplace. These are the 80×86 family of Intel and the 680×0 family of Motorola. It is in these two families that a major effort has been made to enhance performance to compete with RISC microprocessors. The latter were covered first, in Chapter 2, because the improvements in the Intel and Motorola CISC families appear to have been propelled to some extent by the performance of the RISC devices.

In addition to the 80×86 and 680×0 families, there are the following groups:

The Z80 family of Zilog with the Z80000 32-bit microprocessor (μP)
The 32000 family of National Semiconductor with the 32532 32-bit μP
The 32000 family of AT&T with the 32100 and 32200 32-bit μP
The V family of NEC Electronics with the V60, V70, and V80 32-bit μPs
The 340×0 family of Texas Instruments with the 34010 and 34020 μPs
The TRON family of Japan with several participants

Of the above types, the Z80000 design efforts have been focused on embedded military applications, the 32100 and the 32200 on telephone and telecommunication applications, the 32532 and the V series on embedded control for commercial applications, and the 34010 and 34020 specialized for graphics.

All of these types have met strong resistance in the PC and workstation markets when being considered for the CPU application. The 34010 and the 34020 have a specialized niche and are not part of the PC or workstation mainstream. As a result the 80×86 and the 680×0 families dominate the PC and workstation market, where there is a large and growing demand for large quantities of units. The demand for 32-bit microprocessors in the embedded control market is expected to grow substantially in the 1990s, but it has not yet manifested itself.

In the 80×86 family and the 680×0 family, design efforts have been primarily to increase clock frequencies and to incorporate RISC-type cores in some designs without sacrificing backward binary program compatibility.

CISC Evolution at Intel

In 1974, Intel brought out the 6000-transistor 8080, and in 1978, the 8086, a 16-bit microprocessor with 29,000 transistors. The 8086—or more precisely, its 8-bit counterpart, the 8088—was a particularly important chip in that it became the brain of IBM's first personal computer. This machine created a new standard in the computer industry: the 86 architecture coupled with the DOS operating system. This led application software developers to begin writing the many DOS-compatible programs.

In 1982, Intel unveiled its second-generation 8086 chip—the 16-bit 80286. The 286 featured 130,000 transistors and the potential to multitask—albeit only with properly written multitasking software. Most important, it retained software compatibility with its predecessor, thus securing the software investment people had made in the 86 architecture. Guaranteeing compatibility between microprocessor generations became the company's second unwritten mission.

Despite its seemingly large transistor budget, the 286 still could not multitask existing DOS applications. Moreover, as time went on, new applications began to tax the 286 beyond its limits. If software compatibility could be maintained, changes in architecture could be accepted. The following suggestions were made:

- Use a 32-bit programming model, which would let programs handle 32-bit data better and facilitate the use of large data structures
- Develop a better way to multitask DOS applications

Incorporate page-based virtual memory to improve the chips' ability to deal with large programs

Obtain higher clock speeds

This time the Intel engineers emerged with the 275,000-transistor 386™ DX, which met all customer criteria. Most important, the 386 DX could run the same exact software as the 86 and 286—as well as software written for the UNIX™ operating environment. Due to the 386 CPU design enhancements, however, 86- and 286-based machines would not be able to run software written to take advantage of the new 386 capabilities.

The 386 CPU was well received because it offered more than two times the performance of then-high-end ATs and could run UNIX as effectively as workstations. Moreover, the 386 CPU could multitask. DOS multitasking software like DESQview™ and Windows™/386 soon become popular as they enabled applications to run in tandem, a real boon in the interrupt-driven office environment. Secretaries, accountants, and publishers began to migrate to the chip, particularly as it became available in lower-cost systems.

In 1988, Intel decided to reach out to those who might still be intimidated by the awesome power of the 32-bit microprocessor. The result was the 386 SX. The entry-level member of the 386 family, it features the same 32-bit internal architecture as the 386 DX, but it has a 16-bit data bus, resulting in a bit less performance and less price. It found a place in the mid-range PC market.

In April, 1989, Intel introduced the two newest 386 family members—one at either end: a new low-power 386 SX, ideal for battery-operated laptops, and a high-end 33-MHz 386 DX best suited for high-end PC, workstation, and file server applications. Also, simultaneously, Intel brought out the 386 family's heir-apparent—the i486™ microprocessor.

Featuring 1.2 million transistors and an initial performance benchmark of 15 to 20 VAX Mips, the i486 CPU is essentially a faster, more refined, and more integrated microprocessor. Previously peripheral functions and multiprocessing support were brought on-chip, yet there was still enough room for the compatibility logic. Just like the rest of the 386 family, the i486 CPU can run all 86 software written to date—$15 billion worth.

Intel foresaw that the software industry would need time to catch up with the 386 microprocessor, so the i486 CPU was designed to take advantage of today's—and tomorrow's—32-bit operating system. In other words, software written specifically for the i486 microprocessor will still run on today's 386 CPU-based machines, although it won't be able to run on 86 or 286 machines.

Intel has made a commitment. It has promised that each successive microprocessor generation through the year 2000 will be compatible with the Intel 386 architecture.

Motorola's 68000 Family Evolves

The Motorola 68000 family started in 1979 with the introduction of the 68000, which had an external 16-bit data bus. But internally it had a 32-bit arithmetic unit and used 32-bit buses to move data around the chip. As the family evolved through the 68010 and up to the 68020, the transistor count more than doubled from about 68,000 devices to about 195,000 for the full 32-bit CPU with virtual memory support. Higher integration levels continued with the 68030, which added about 105,000 more transistors in the form of on-chip memory management and dual 256-byte caches to speed execution.

Not only has Motorola added more features to speed up execution, but they also fine-tuned the instructions so that many of them execute in fewer clock cycles than in previous generation versions. In addition, improved processing makes it possible to run the chips at higher clock frequencies. Nearly a 300 percent throughput improvement is possible on the 68030 alone, thanks to a boost of the clock rate from its original 16.67 MHz to the most recently released version, which can run as high as 50 MHz. Though not released until later, the 68040 processor combines all of the 68030 features with another 900,000 transistors that are used for greatly enlarged caches (4 KB each), an on-chip floating-point coprocessor, and bus-monitoring circuitry for multiprocessor systems.

Intel and Motorola 32-bit CISC Microprocessors

The Intel and Motorola 32-bit microprocessors have similar architectures consisting of internal fetch, decode, execution, and bus interface units. These architectures support pipelined operation, multitasking, privilege and protection schemes, coprocessor operation, and advanced memory management.

The ability of the fetch, decode, and execution units to operate semi-autonomously permits the prefetching of instructions in parallel. Thus as a previously fetched instruction is being executed, another instruction can be fetched, decoded, and loaded into an instruction queue. This overlapping of fetch, decode, and execution operations is referred to as pipelining.

These CPUs support multitasking. Multitasking is the ability to run several tasks concurrently and is achieved by interleaving the execution of different tasks. By supplying dedicated registers, such as task and process control block pointers, support for rapidly switching tasks into and out of the CPU is provided.

The register sets of each 32-bit CPU are conveniently separated into a program and system model set. Program model registers consist, at a minimum, of a program counter register, a stack pointer register, a flags register, and multiple general-purpose data and address registers. These registers are typically used by applications programs for data calculation and manipulation. The system model register set provides task, status, and interrupt registers required in the support of operating system functions.

An essential feature of the 32-bit CPUs is their ability to address a large address space (2^{32} KB). To effectively manage this space requires advanced memory management techniques. The central feature of all memory management is the ability to translate the program addresses manipulated by the CPU into physical memory addresses. Additional capabilities include subdividing programs into smaller sections called segments and then allocating each program segment to physical memory locations. In paged memory management, program segments are further subdivided into fixed divisions called pages. This facilitates moving small sections of code into and out of memory as needed.

Intel's 80386

The program model register set for the Intel 80386 microprocessor consists of eight 32-bit general-purpose registers, six 16-bit segment registers, one 16-bit flag register, and one 32-bit instruction pointer. The 80386 CPU constructs off-chip addresses as a combination of a segment and offset component. The six segment registers individually specify the starting address of a code, stack, and up to four external data segments. The offset address specifies the position of the desired location relative to the segment's starting address.

The value stored in a segment register is formally referred to as a selector. In protected addressing mode the selector is an index into a table of segment starting addresses. In real addressing mode the selector is the segment's starting address.

Offset addresses are also referred to as effective addresses. By selectively including or excluding the various components of an effective address, the conventional direct, indirect, base relative, indexed, base-indexed, and base-index-relative addressing modes are supported by the CPU.

The 80386 system model register set consists of system flags, four memory management registers, four control registers, eight debug registers, and two test registers. These registers are accessed by privileged operating system instructions.

The 80386 CPU supports signed and unsigned byte, word (16-bit), and double-word (32-bit) memory fetches. In addition, the CPU recognizes bit-field, packed and unpacked BCD digits, and pointer addresses. Within memory, the CPU assumes that lower-order bytes are stored at lower-order memory addresses.

An individual 80386 CPU instruction can explicitly reference, at most, two operands. Operands can be stored either directly within the general-purpose data registers, as immediate data encoded within an instruction, or external to the CPU in memory and I/O ports.

All 80386 CPU instructions are encoded using a maximum of 16 bytes. Typically a machine instruction consists of fewer than 12 bytes. Each instruction may contain one or more opcode bytes, one or more addressing mode bytes, a displacement address, and a 32-bit immediate datum.

The ModR/M byte is the primary addressing mode byte in an 80386 machine language instruction. This byte encodes addressing mode information for both single- and double-operand instructions. An additional SIB (scale, index and base) byte is used for addressing modes requiring a scale factor, index value, or base register.

The instruction set of the 80386 CPU consists of data transfer, arithmetic, logical, shift, rotate, string, program control, and high-level language and systems level instructions. Two-operand instructions within each group require that the destination operand be listed before the source operand.

The wide variety of instructions and addressing modes make possible efficient data manipulation routines. Subroutines that total and average, move and verify, and sort numerical tables require no more than ten to fifteen assembly language instructions.

Motorola's 68020 and 68030

The program model register set of the Motorola 68020 Microprocessor consists of eight general-purpose data registers, seven address registers, a user stack pointer, a program counter, and a condition code register. Except for the 8-bit condition code register, each of these registers is 32 bits wide. The 68020 system model register set consists of a status register, an interrupt stack pointer, a master stack pointer, a vector base register, two alternate code function registers, a cache control register, and a cache address register.

The 68020 CPU supports byte (8-bit), word (16-bit), long word (32-bit), packed and unpacked BCD, individual bit, and bit field data. Floating-point data is supported by a separate Motorola MC68881 floating-point coprocessor. The CPU instruction set provides that all data can be interpreted as either signed or unsigned quantities.

The 68020 CPU assumes data stored within memory is stored with its lower-order bytes at higher-order addresses. Although byte, word, and long word data can be stored starting at any byte address, more efficient data transfers occur when words are stored at even byte addresses and long words at addresses evenly divisible by four.

The addressing modes supported by the 68020 CPU include absolute, immediate, register direct, register indirect, indexed register indirect, memory indirect, and program counter relative addressing.

The instruction set of the 68020 CPU consists of data transfer, arithmetic, logical, bit management, program and system control, coprocessor, and multiprocessor instructions. An individual 68020 assembly language instruction consists of a label field, a mnemonic, and up to two operands. In two-operand instructions the source operand is always listed before the destination operand.

Machine language instructions are encoded using at least one 16-bit word and at most 11 words. The majority of instructions in which operands or their addresses are stored in registers are encoded using a 16-bit single effective address format. Up to five additional extension words may be required for each operand requiring immediate data.

The 68030 maintains the 32-bit registers available with the entire 68000 family as well as the 32-bit address and data paths. The coprocessor interface that was first defined on the 68020 is brought forward. Important additions to the 68020 architecture are an on-chip instruction cache, an improved bus controller, multiple internal instruction buses, and an on-chip paged memory management structure.

Clock Frequencies Rise

Various versions of the 80286 were pushed up in clock frequency specifications from 8 MHz to 20 MHz. However, the desire for more speed was such that some work was done at higher frequencies despite the fact that such operation was out of spec. One of the first designs along this line was done in Taiwan by Peter Chang. In December of 1987 he announced an AT main board based on an AMD 80286 CPU (specified for a 12-MHz clock) running at a clock of 27.5 MHz. For a board shown at the June 1988 Computex Show in Taipei, he pushed the clock speed up to 40 MHz. Later he changed chip sets and dropped back to 30.8 MHz. He promised new designs at 40 MHz for the 80286 and even a 100-MHz board using Intel's 80386SX. However, it is not known if these were realized.

Chang claimed the secret of running at such high clock frequencies above values specified by the chip manufacturers was the layout of the board. However, a Chicago-based company, Systems Integration Associates (SIA), which has developed boards with 25-MHz Intel 80386 CPUs running at 33 MHz, claims that the greatest difficulty in running these chips beyond their listed ratings is heating of the chip. To overcome this problem they have mounted a cross-flow blower just above the processor and the coprocessor sockets to keep the chip case very close to ambient temperature. The CPU is mounted in a slightly elevated position to allow air to flow underneath it.

SIA has also found that careful board layout is vital due to stray capacitance and transmission line effects. Also, a good design requires a memory subsystem that will not bottleneck the processor.

With 80486 and 68040 types specified at higher clock frequencies, it should not be necessary to build boards that exceed the chip specification limits. However, the temptation to do so will undoubtedly still be there to provide better performance ratings. The gain in speed that can be realized by the use of higher clock frequencies is illustrated by Intel's comparison of the 80486 at 25 MHz and 33 MHz. At 25 MHz the CPU can execute 15

VAX Mips, 37,000 dhrystones/s, and 6.1 double-precision megawhetstones/s. At 33 MHz it can execute 20 VAX Mips, 49,000 dhrystones/s, and 8.2 double-precision megawhetstones/s.

What Intel Learned from RISC

To design a 32-bit CISC microprocessor that would compete more fully with RISC microprocessors than would the 80386, Intel engineers studied RISC techniques to see if they could apply any of these to a CISC design. While the simple load/store architecture, the easily decoded fixed-length instructions, and the absence of microcode in RISC designs made it easier to build a one-instruction-per-clock execution engine, they concluded that these were not absolute requirements. They noted that most of the fastest RISC engines used an instruction pipeline that integrates a cache, but that this technique was not pioneered by RISC research and that it was first used in mainframe computers many years ago.

For example, the IBM 360 Model 85, introduced in the late 1960s, exhibited many characteristics of current RISC designs. It used a cache and a pipelined instruction execution engine, and most instructions were executed in one clock cycle. However, the Model 85 of the System 360 had a microprogrammed execution unit, which included many "non-RISC" instructions. Intel's CISC microprocessor technology had not yet reached the one-instruction-per-clock execution level, but it had made progress. Average clocks per instruction (CPI) had gone from 13 for the 8088 μP down to 4 for the 80386. Figure 3-1 shows the progress in reducing clocks/instructions from

FIGURE 3-1
The decrease in clocks/instruction over a decade. Reprinted courtesy of Intel Corporation.

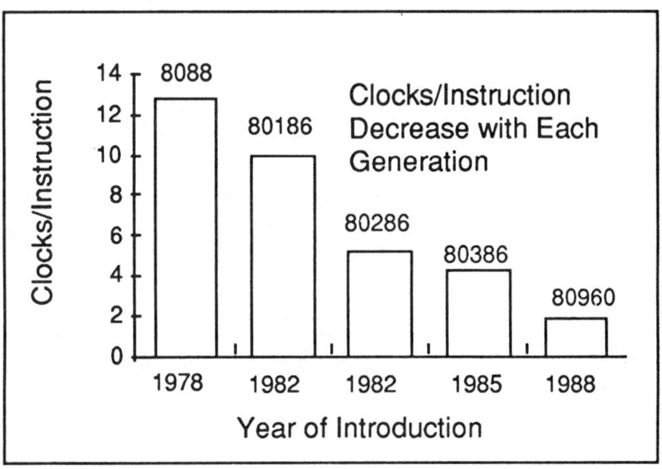

the 8088 to the 80386 designs. The key question was whether CISC could achieve the number of clocks/instruction equivalent of RISC types.

This comparison had two aspects. First, "equivalent instructions" is a difficult thing to measure. Second, the number of clocks/instruction for microprocessors has shown a steady decline. A further reduction in the number of clocks/instruction would obviously improve performance. When the number of clocks/instruction reaches one, will more work be done by one more powerful instruction per clock than by one RISC instruction/clock? The argument centers on whether the simpler instructions can be executed at a faster enough rate to compensate for the increased number of instructions that are needed to perform complex functions.

Another RISC technique is the availability of a large number of registers, on the order of 32 to 128. This large register set is closely associated with an optimizing compiler. An ambitious global register allocation scheme in the compiler is required to make use of a large number of registers. The same argument can be applied to normal computers as well. For example, the techniques needed to optimize for small numbers of registers is also well understood and negates the advantages of larger register sets. Some of the best compilers for the 386 processor use a register coloring technique for register allocation. Furthermore, many other optimization techniques used in RISC compilers apply to any computer and are being incorporated into CISC compilers.

The Trade-offs

With the above criteria as a background, Intel's engineers concluded that RISC designs give up three things, relative to comparable CISC designs, that are important. They are system support, binary program compatibility, and code size and instruction fetch bandwidth. Their engineers looked at these factors in more detail and reached conclusions described below.

System Support

There is more work to do than execute small benchmarks. For example, UNIX and other complex software systems require other system services to efficiently support the end user's needs. The operating system must be efficient. Task switching should not take lots of time. The operating system needs protection from programs that have bugs in them. Certain operations that are needed by important applications, for example, floating-point arithmetic or graphics display operations, must be supported. Where possible, the processor should support these needs. RISC machines, by their very definition, meet only a small subset of these needs. It is clear that RISC machines are evolving to become more like normal machines.

Mainstream machines are also evolving in obvious ways, in most cases incorporating the very good design techniques popularized by some RISC machines. RISC machines introduced during the last 18 months feature floating-point math and extensive cache and memory management hardware. CISC machines introduced recently feature floating-point math, memory management, and cache and other complex functions.

Binary Program Compatibility

Two important standards areas that have received recent attention are the Application Binary Interface (ABI) and the System V Interface Definition (SVID). These are both important standards levels. The SVID is a source-code-level standard, and the ABIs are a collection of object or binary code standards. Everyone is interested in code compatibility. If the same object code will run across different manufacturers' equipment, that is best, but at least the same source code should be able to be used, with recompilation. Source code compatibility depends on a common compiler language and on a common system call interface, for example, file system operations. Companies like Sun support common languages and system surface across their differing CPU architectures. Systems based on the Intel 386 architecture will all execute the same binary programs when running either DOS or UNIX System V Release 3. Figure 3-2 illustrates the two classes of standards and some examples.

FIGURE 3-2
Binary compatible standards versus UNIX system call standards. Reprinted courtesy of Intel Corporation.

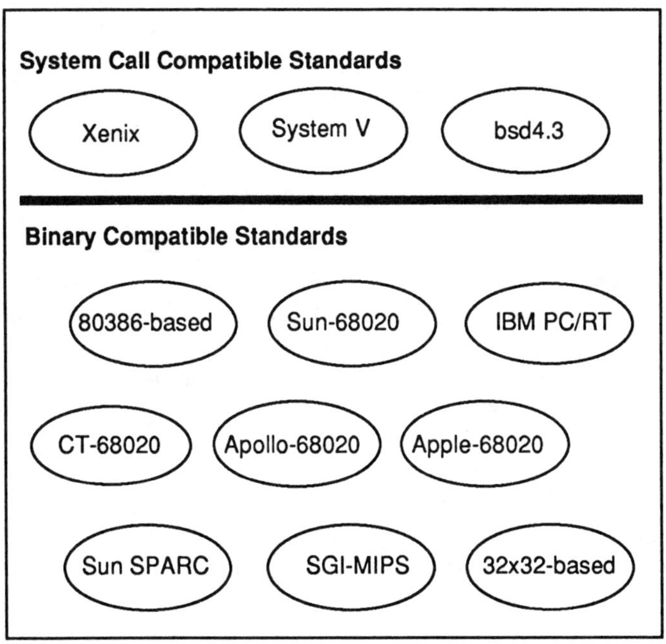

A current problem is that there are several standards in popular use for the system call interface, and a surprisingly large number of ABIs, as shown in Figure 3-2. The application software industry has been successful in the DOS area because of the large number of hardware platforms that all run the same binary programs—shrink-wrapped software—compatible with a single DOS ABI. The UNIX community would also like to share in this abundance of software. To date, however, the systems built on some CPUs, for example, the 68020/30, have all had different ABIs. That means that the application vendor has to do a port and sell a different version for each 68020-based system from different manufacturers.

The 386-based systems are unusual, in that they all have a common UNIX System V ABI. This is because of the hardware floating-point and memory management commonality provided by the Intel 386 architecture and by adherence by the system suppliers to hardware conventions promoted by Intel and AT&T. The result is a very large installed base of binary-compatible systems. The next result will be an increasing supply of shrink-wrapped UNIX applications software. This is an everybody wins situation. The application suppliers see a larger installed base of compatible systems. The system suppliers see a larger supply of applications software (which sells their systems), and the system suppliers do not give up useful differentiation in their products.

Why is this a RISC issue? Because the RISC machines are all different. They all have different instruction sets, hence, different ABIs. In some cases, as multiple-system suppliers using the same RISC integer unit implement the memory management unit (MMU) in different ways, they may lose binary compatibility. Their motivations should be in favor of compatibility, but they might make a mistake and discover it too late. This is less likely if the CPU contains the MMU and other required system functions, that is, if it is less RISC-like.

Code Size and Instruction Fetch Bandwidth

The code size for computers designed with RISC techniques is about 30 to 40 percent greater to implement the same functions. This comes from two sources. First, the individual instructions are longer, on the average. Second, it takes more of the longer instructions to implement complex functions. There are two implications. The memories for instruction storage must be larger by the corresponding amount. This is true for the instruction cache as well as for main memory. Disk space is not as critical since the main effect is on program loading time. The second implication is the instruction bandwidth to the CPU needs to be 30 to 40 percent greater for a RISC in order to support the same throughput rate. Instruction bandwidth costs a significant amount of money to support in higher-speed (greater than 10 Mips) computer designs.

Modern Computer Design Techniques

Intel engineers viewed the controversy about RISC versus CISC as misplaced. The name CISC was invented to describe all machines, except those that featured a small set of RISC design principles. Experience has shown that this set of design techniques is a good set. Not a complete set, but a good set. If we apply what we now know to be a good set of techniques to the design of any new computer, then there are two basic paths. Case 1 is to ignore backward binary compatibility and simply design the fastest possible computer using all available knowledge, including RISC techniques. Case 2 is to retain backward binary compatibility with an existing (CISC) architecture and design the fastest possible computer using the same available knowledge. See Figure 3-3.

Intel engineers believed that these case comparisons raised the following two important questions:

> Do new RISC machines have enough of a performance advantage over new CISC machines to compensate for the lack of backward binary program compatibility? Will RISC machines have an inherent performance advantage over CISC machines in the future?

FIGURE 3-3
Approaches to the use of RISC techniques. Reprinted courtesy of Intel Corporation.

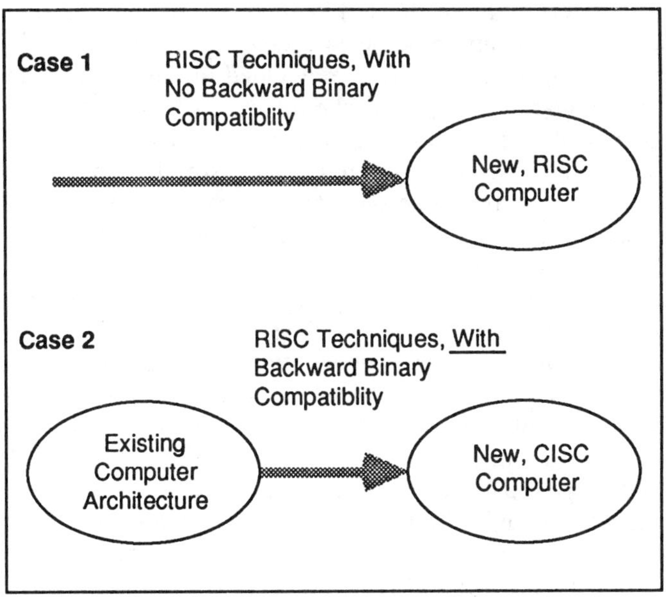

They concluded that many of the computer engineering techniques known collectively as applicable to RISC designs were also applicable to CISC designs. They projected that more of these techniques will be applied to microprocessors in the future, bringing high levels of performance while retaining binary compatibility. They contended that if the bottom-line performance of RISC techniques is one clock per instruction, a CISC design that executes more powerful instructions in one clock will execute programs faster than a RISC design that requires more instructions/function. Whether that is so or not, they concluded that it is certain that the next generation of microprocessors will be much faster and that some part of that credit goes to the researchers who have developed or rediscovered, and then proven, the advantages of a set of computer design techniques known as RISC.

Intel's 80486

The 486 microprocessor is a 32-bit architecture with on-chip memory management and floating-point and cache memory units. The 486 microprocessor contains all the features of the 386 microprocessor with enhancements to increase performance. The instruction set includes the complete 386 microprocessor instruction set along with extensions to serve new applications. The on-chip MMU is completely compatible with the 386 microprocessor MMU. The 486 microprocessor brings the 387™ math coprocessor on-chip. All software written for the 386 microprocessor, 387 math coprocessor, and previous members of the 86/87 architectural family will run on the 486 microprocessor without any modifications.

Several enhancements have been added to the 486 microprocessor to increase performance. On-chip cache memory allows frequently used data and code to be stored on-chip, reducing accesses to the external bus. RISC design techniques have been used to reduce instruction cycle times. A burst bus feature enables fast cache fills. All of these features combined lead to performance greater than twice that of a 386 microprocessor.

MMU consists of a segmentation unit and a paging unit. Segmentation allows management of the logical address space by providing easy data and code relocatibility and efficient sharing of global resources. The paging mechanism operates beneath segmentation and is transparent to the segmentation process. Paging is optional and can be disabled by system software. Each segment can be divided into one or more 4-KB segments. To implement a virtual memory system, the 486 microprocessor supports full restartability for all page and segment faults.

Memory is organized into one or more variable-length segments, each up to four gigabytes (2^{32} bytes) in size. A segment can have attributes associated with it that include its location, size, type (i.e., stack, code, or data), and protection characteristics. Each task on a 486 microprocessor can have a maximum of 16,381 segments, each up to four gigabytes in size. Thus each task has a maximum of 64 terabytes (trillion bytes) of virtual memory.

The segmentation unit provides four levels of protection for isolating and protecting applications and the operating system from each other. The hardware-enforced protection allows the design of systems with a high degree of integrity.

The 486 microprocessor has two modes of operation: Real Address Mode (Real Mode) and Protected Mode Virtual Address Mode (Protected Mode). In Real Mode the 486 microprocessor operates as a very fast 8086. Real Mode is required primarily to set up the processor for Protected Mode operation. Protected Mode provides access to the sophisticated memory management paging and privilege capabilities of the processor.

Within Protected Mode, software can perform a task switch to enter into tasks designated as Virtual 8086 Mode tasks. Each virtual 8086 task behaves with 8086 semantics, allowing 8086 software (an application program or an entire operating system) to execute.

The on-chip floating-point unit operates in parallel with the arithmetic and logic unit and provides arithmetic instructions for a variety of numeric data types. It executes numerous built-in transcendental functions (e.g., tangent, sine, cosine, and log functions). The floating-point unit fully conforms to the ANSI/IEEE Standard 754-1985 for floating-point arithmetic.

The on-chip cache is 8 KB in size. It is four-way set associative and follows a write-through policy. The on-chip cache includes features to provide flexibility in external memory system design. Individual pages can be designated as cacheable or noncacheable by software or hardware. The cache can also be enabled and disabled by software or hardware.

Finally, the 486 microprocessor has features to facilitate high-performance hardware designs. The 1X clock eases high-frequency board-level designs. The burst bus feature enables fast cache fills.

Register Set

The 486 microprocessor register set includes all the registers contained in the 386 microprocessor and the 387 math coprocessor. The register set can be split into the following categories:

Base Architecture Registers

General-purpose Registers
Instruction Pointer
Flags Register
Segment Registers

Systems-level Registers

Control Registers
System Address Registers

Floating-point Registers

Data Registers

Tag Word

Status Word

Instruction and Data Pointers

Control Word

Debug and Test Registers

The base architecture and floating-point registers are accessible by the applications program. The system-level registers are only accessible at privilege level 0 and are used by the systems-level program. The debug and test registers are also only accessible at privilege level 0.

Motorola's 32-bit CISC Microcontroller

In 1987, the majority of microcontrollers shipped were 4-bit MCUs, comprising 53 percent of the total available market. Eight-bit MCUs came in second with 47 percent of the market. Less than 1 percent of the market was 16-bit. The 4- and 8-bit MCUs generally ran at bus speeds of 1 to 4 MHz. By 1991, projected sales of 16-bit and 32-bit MCUs will have grown from less than one percent now to 20 percent of the market. These 16/32 bit MCUs will likely be operating in the 16 to 24 MHz range, similar to current MPU products, or higher.

Motorola has been very successful with its 68000 family in microcontroller applications, so it has been logical for the company to build on the core of its 68020, a 32-bit microprocessor, to develop a microcontroller that could meet the developing demand described above. Convenient progressions could follow by utilizing a MC68030, '040, and the like core with on-chip memory management units or floating-point processors. Having to sacrifice CPU power for the benefits of an MCU has become a thing of the past. Future CPU enhancements continue to be made possible by shrinking densities and increased integration.

Smart Peripheral Subsystems

Motorola studied the problem of peripherals slowness and resolved that the only permanent solution for real-time control was to increase the intelligence of the CPU's peripheral subsystems. The solution is smart peripherals—subsystems that contain all the logic necessary to perform their normal functions without requiring CPU intervention or interrupts. This solves both the context switching dilemma and the loading problem on the CPU.

Smart peripherals take us to a new level of sophistication. In effect, the MCU now contains a CPU surrounded by dedicated MCUs and memory. The individual smart peripherals may include a local microengine, hardware scheduler, and execution unit. Microcoding may be used for application flexibility.

The first implementation of this concept was introduced by Motorola on a 32-bit MCU in early 1989. An example taken from this MCU is shown in the timer block diagram in Figure 3-4. The timer has 16 timer channels for I/O. These I/O have individual interrupt vectors and priorities all managed by the timer. A microengine runs the timer system independent of the CPU. It contains its own execution unit, memory (for microcode), and scheduler for interrupt handling. It interfaces with the CPU via dual access and memory-mapped status and control registers. CPU intervention is limited to register initialization or modification and to user requests for status or information on performance.

Smart Serial Communications Interface

The user of smart peripherals is inherently a better solution to the interrupt response problem than either interrupt vectors or bank switching, as both methods must still execute an interrupt service routine at some point. The example shown in Figure 3-4 for a timer is logical and elegant. However, some peripherals may be better served by a combination approach that would take less silicon to implement.

A good example might be found in the serial communications interface (SCI) on an MCU. Heretofore, most MCUs employed a byte-oriented SCI. That is, if there was a message being sent to the MCU via the SCI or UART (universal asynchronous receiver transmitter), you would get an interrupt upon arrival. The CPU would stop, stack, and jump to the interrupt service routine. This routine would normally transfer a byte from the SCI received data register to a RAM buffer, check to see if it was the end of the message, and then return to the main program until the next byte arrived.

Taking this common example of transferring a byte of data and then checking for a limit or boundary condition, we may find a way to optimize the MCU performance for this task. Some assumptions can be made. First let's assume that most messages are more than one byte long, and could be quite large, but the length is known. Second, let's say that no action is taken by the CPU until the entire message has been received.

Now some logical enhancements turn the SCI into a smart peripheral. The SCI is given a dual-access (or dual-ported) RAM of its own. This way the SCI can transfer data to the RAM without needing to request the internal data and address bus from the CPU, that is, without interrupting the CPU. Additionally, this allows the CPU to access the RAM independent of or asynchronous to the operation of the SCI.

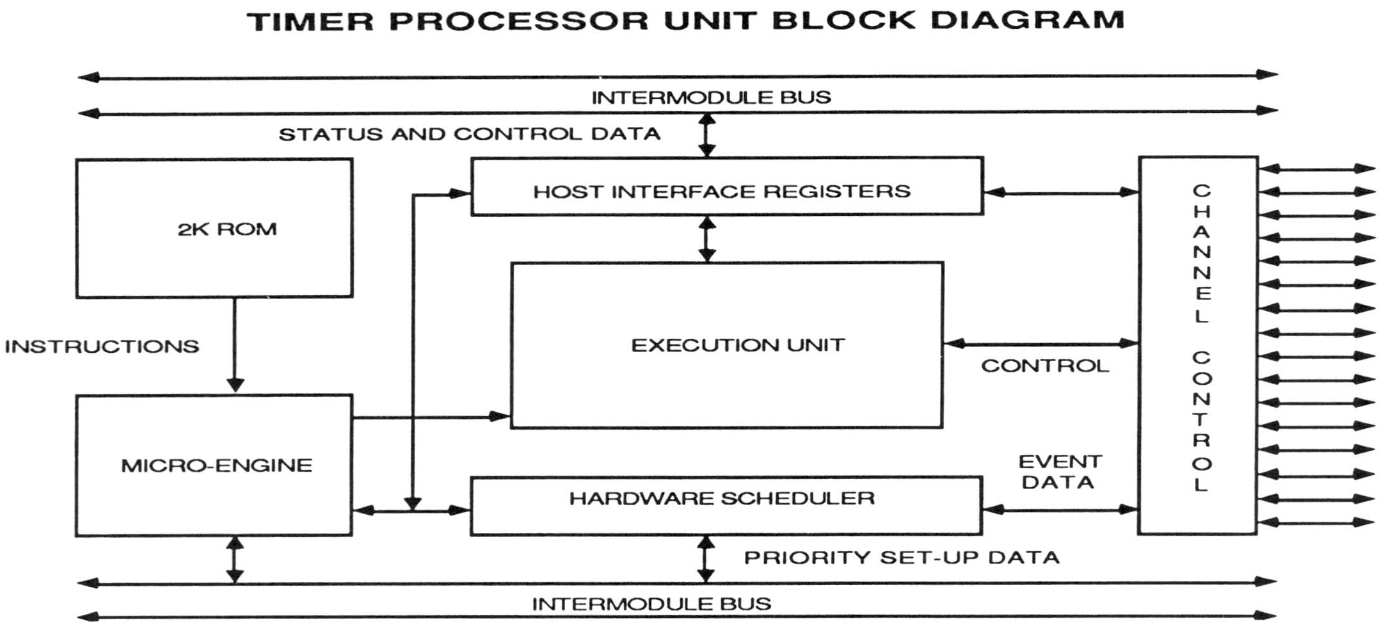

FIGURE 3-4 Timer processor unit block for Motorola CISC microcontroller. Reprinted courtesy of Motorola Semiconductor, Advanced MCU Operations.

This approach also requires some slight modifications to the SCI module. It must now be able to place incoming data into its RAM as directed by a pointer, increment the pointer, and check for a boundary condition, such as the end of its RAM or the end of the message. If neither condition occurs, then the SCI continues to receive new data. If there is a match with the boundary register, the SCI finally interrupts the CPU with a service request. Although in real life this example (Figure 3-5) will be complicated by other details, such as checking for parity or transmission error flags, the concept may be easily applied to both transmitting as well as receiving messages.

Internal Data Transfer Function

The previous SCI example exemplifies Motorola's philosophy regarding solving the real-time interrupt control problem. Another less elegant approach to the problem exists, but it has some different advantages. This approach uses the CPU to run a type of internal DMA in response to an interrupt or request for service.

This approach may be easily applied to the SCI example. Instead of the SCI having its own RAM, it could utilize the main on-chip RAM. This would keep the basic SCI design from being modified, but requires the user to reserve some RAM for a buffer. The registers that would point to the location in the RAM for storing the byte of data now belong to the CPU's register space, as well as another register to check for the end-of-RAM or

FIGURE 3-5
Serial communications interface (SCI) block diagram. Reprinted courtesy of Motorola Semiconductor, Advanced MCU Operations.

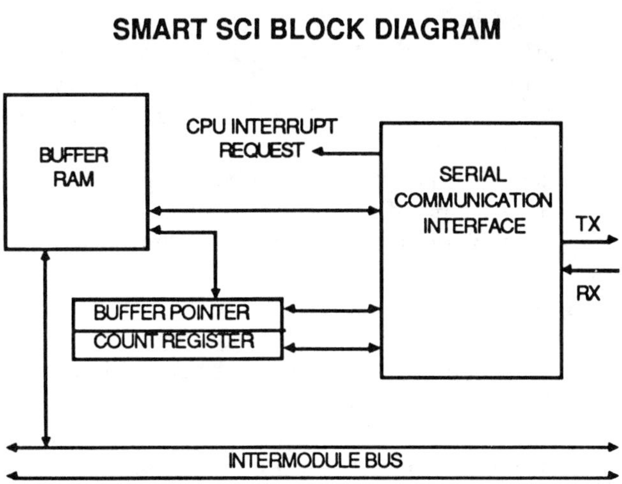

the end-of-message condition. This could be called a data transfer count (down) register.

When the SCI receives a byte, it no longer interrupts the CPU for a stack and vector operation. Instead it requests this internal DMA transfer function. The CPU stops normal processing in order to process the data transfer, decrement the data transfer counter, and check for the boundary condition (equal to zero) and then returns to normal processing. As no other registers are utilized, there is no loss of data or need to stack or swap registers. If the data transfer function is created to run asynchronously to the CPU, then the CPU is only interrupted on the byte that is the end-of-message byte.

This approach to interrupt handling is well suited to serial communications. However, it can be utilized by any possible interrupt source, since the DMA logic is part of the CPU, not part of any particular peripheral module. Also, it may be smaller in size to implement than providing additional power and RAM to the SCI module.

To make this data transfer control more generic (that is, applicable to other on-chip subsystems other than just the SCI), some additional logic must be implemented. First, the addition of a source address register will be needed to define where the data will originate. Second, some structure must be employed by the interrupt control logic to ascertain whether an interrupt source will stack or switch and vector, or will utilize the data transfer feature. Ultimately, the registers associated with the data transfer function could be overlayed on RAM and pointed to by a sort of interrupt vector, thus giving each interrupt source availability of the data transfer function independent of other sources.

Increased Performance

Freeing up the CPU from being burdened by interrupt handling should make a significant increase in system throughput. Other common methods of increasing performance include hiking the bus operating speed and increasing the width of the data bus. Historical trends show a migration over time to faster and more powerful MCUs.

Once thought of as the ultimate approach to providing custom MCUs for every application, application-specific integrated circuits (ASICs) were hailed as the wave of the future. Many semiconductor manufacturers formed ASIC groups along with new companies, to grab hold of the growing business. What they found was that real ASIC business existed only in a limited range of potential quantities. Hemmed in above by the cost effectiveness of custom MCUs in the higher quantities, and below by a minimal level of profitable business, ASIC groups ended up struggling to eke out an existence in between.

CSIC, or customer-specific integrated circuits, thus became all the more popular. Companies that didn't have the required volume to go ASIC started looking for a variation of a standard part that more specifically met the requirements of their type of application. The need for better and quicker methods of producing CSICs became critical. Modularity and core standardization were identified as the best ways to achieve these goals.

Modularity

Modularity is a concept that has been around for awhile, but it has not been seriously implemented until just recently. A modular design is one that may truly be divided into submodules, each of which may be designed separately (see Figure 3-6). All submodules, including the CPU and memory, would interface to each other via a standardly defined intermodule bus.

The benefits of a modular design go much further than decreasing the actual design cycle time. Debugging and testing are also improved, further shortening the design-to-production phase. Modular designs allow custom or derivative MCUs to be designed much more quickly and dramatically reduce the number of design errors. Motorola is committed to 100 percent modular designs and currently has a 16- and a 32-bit MCU designed in this manner.

Core Standardization

Core standardization is a technique that has worked well in the past. Custom or derivative MCUs are created using the core (a minimum of the CPU), and custom circuitry is added or exchanged for existing circuitry.

FIGURE 3-6
Modules of Motorola's 68332 CISC microcontroller. Reprinted courtesy of Motorola Semiconductor, Advanced MCU Operations.

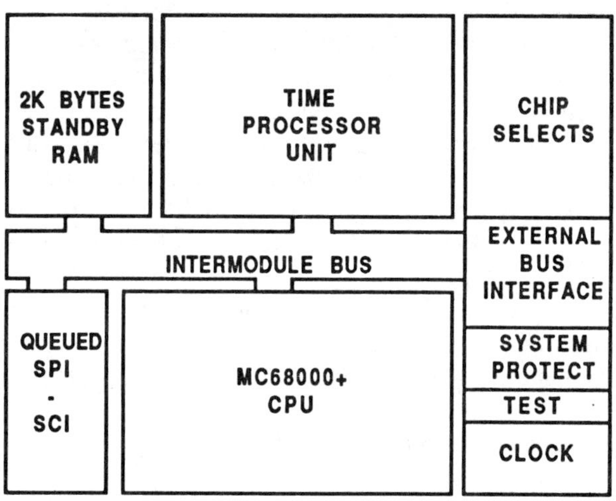

Prior to the advent of modular designs, MCU testing required that the CPU function (to some extent) in order to test the on-chip subsystems. Therefore, the less silicon was modified, the better the chances were of producing a working MCU.

Using modular design techniques alleviates the necessity of making the CPU untouchable. As long as there is an access to the intermodule bus externally, all peripheral subsystems, memory, and even the CPU may be debugged in parallel. However, using an existing CPU has the advantage of already being debugged. Software and hardware development (emulation) tools already exist, and in many cases they will be code-compatible with the user's existing software. In fact, the software design effort has now become the limiting factor in the total design time, making core standardization around an existing CPU the best way to save time and money.

Tying together modularly designed peripheral subsystems with memory and a standard CPU core via a standard intermodule bus allows derivative MCUs or CSICs to be more quickly designed and brought to production. This meets the goals of the SIC designs for both the manufacturer and the customer. This type of solution effectively lowers the cost of entry to the CSIC marketplace, squeezing the ASIC market window even more. All future MCU designs at Motorola will utilize these concepts.

Motorola's 68332 Microcontroller

The 68332 is a 32-bit integrated microcontroller, combining high-performance data manipulation capabilities with powerful peripheral subsystems. The 68332 is the first member of the 38300 family of modular embedded controllers, featuring fully static, high-speed CMOS technology. Based on the 68020, the CPU32 instruction processing module provides enhanced system performance and utilizes the extensive software base for the Motorola 68000 family. The major components of the MC68332 are similar to those of Figure 3-6, with CPU32 in the 68000 + CPU block.

The 68332 contains intelligent peripheral modules such as the Time Processor Unit (TPU), which provides 16 microcoded channels for performing time-related activities from simple input capture or output compare to complicated motor control or pulse-width modulation. High-speed serial communications are provided by the Queued Serial Module (QSM) with synchronous and asynchronous protocols available. Two kilobytes of fully static standby RAM allow fast two-cycle access for system and data stacks and variable storage with provision for battery backup. Twelve chip selects enhance system integration for fast external memory or peripheral access. These modules are connected on-chip via the intermodule bus (IBM).

The major features of the 68332 are as follows:

Low-power operation: 600 mW maximum; 500 μW in Standby Mode

Extended temperature operation: −40 to 125°C

Frequency: 16.78 MHz at 5-V supply

Technology: 1-micron HCMOS (high density complementary metal-oxide semiconductor), static design

Transistor count: 422,000

Package: 132-pin FP-PLCC (PQFP)

Modular architecture in a single chip

CPU: 32-bit 68000 family (upward object-code compatible)

—New instructions for controller applications intelligent 16-bit timer:

—16 independent, programmable channels and pins

—Any channel can perform any time function (input capture, output compare, PWM, etc.)

—Two timer count registers with 2-bit programmable prescalers

—Selectable channel priority levels

Two serial I/O subsystems:

—Enhanced 68HC11-type SCI (UART) with parity

—Enhanced 68HC11-type QSPI with I/O queue

On-chip memory: 2K standby RAM

On-chip programmable chip select logic

—Up to 12 signals for memory and peripheral I/O select

System failure protection:

—68HC11-type computer operating properly (watchdog) timer

—68HC11-type periodic interrupt timer

—68000 family spurious interrupt, halt, and bus time-out monitors

Central Processor Unit

The CPU32 is upward-compatible with the 68000 family, which excels at processing calculation-intensive algorithms and supporting high-level languages. All of the 68010 and most of the 68020 enhancements such as virtual memory support, loop mode operation, instruction pipeline, and 32-bit mathematical operations are supported. Powerful addressing modes provide compatibility with existing software and increase the efficiency of high-level language compilers. New instructions such as table lookup and interpolate and low-power stop support the specific requirements of controller applications.

Intelligent Peripherals

To improve total system throughput, the MC68332 features intelligent, stand-alone subsystems. These include the microcoded Time Processor Unit (TPU), the Queued Serial Module (QSM), the System Integration Module

(SIM), and the 2K Standby RAM Module. These modules work together with the CPU32 to reduce part count, size, and cost of system implementation.

Time Processor Unit (TPU)

The TPU provides optimum performance in controlling time-related activity. The TPU drastically reduces the need for CPU intervention with its dedicated execution unit, trilevel prioritized scheduler, data storage RAM, dual time bases, and microcode ROM. The TPU controls 16 independent, orthogonal channels, each with its associated I/O pin and capable of performing any time function. Each channel also contains a dedicated event register, allowing both match and input capture functions. The timing algorithms presently available in microcoded ROM include the following:

- Discrete input/output
- Pulse-width modulation
- Input capture/input transition counter
- Period measurement with additional transition detection
- Period measurement with missing transition detection
- Position-synchronized pulse generator
- Stepper motor control
- Output match
- Period/pulse-width accumulator

Each channel can be synchronized to either of two 16-bit free-running counters with a prescaler. One counter is based on the system clock and provides resolution to 500 ns. The second counter, based on an external reference, provides resolution to 250 ns. Channels may also be linked together, allowing the user to reference operations on one channel to the occurrence of a specified action on another channel to provide intertask control.

Queued Serial Module (QSM)

The Queued Serial Module (QSM) contains two serial ports. The Queued Serial Peripheral Interface (QSPI) provides easy peripheral expansion or interprocessor communications via a full-duplex, synchronous, three-line bus: data-in, data-out, and a serial clock. Four programmable peripheral select pins provide addressability for up to 16 peripheral devices. The QSPI has been enhanced with the addition of a queue contained in a small RAM. This allows the QSPI to handle up to 16 serial transfers of 8 to 16 bits each, or to transmit a stream of data up to 256 bits long without CPU intervention.

A special wraparound mode allows the user to do continuous sampling of a serial peripheral, automatically updating the QSPI RAM for efficient interfacing to serial analog-to-digital converters.

The Serial Communications Interface (SCI) provides a standard NRZ (non-return to zero) (mark/space) format. Advanced error detection circuitry catches noise glitches to $\frac{1}{16}$ of a bit time in duration. Word length is software selectable between 8 or 9 bits, and its modulus-type, baud rate generator provides baud rates from 64K to 524K baud based on a 16.78-MHz system clock. The SCI features full- or half-duplex operation, with separate transmitter and receiver enable bits and double buffering of data. Optional parity generation and detection provide either even or odd parity check capability. Wake-up functions allow the CPU to run uninterrupted until either a true idle line is detected or a new address byte is received.

System Integration Module

The System Integration Module (SIM) includes an external interface and various functions that reduce the need for external glue logic. The SIM contains the external bus interface (EBI), 12 chip selects, system protection, test, and clock submodules.

External Bus Interface. Based on the MC68020 bus, the external bus provides 24 address lines and a 16-bit data bus. The data bus allows dynamic sizing between 8- and 16-bit data accesses. Read-modify-write cycles are provided for via the RMC signal. External bus arbitration is accomplished by a three-line handshaking interface.

Chip Selects. Twelve independent programmable chip selects provide fast, two-cycle external memory or peripheral access. Block size is programmable from a minimum of 2KB to 1MB in length. Accesses can be preselected for either 8- or 16-bit transfers. Up to 13 wait states can be programmed for insertion during the access. All bus interface signals are automatically handled by the chip select logic.

System Protection Submodule. System protection is provided on the 68332 by various monitors and timers, including the bus monitor, halt monitor, spurious interrupt monitor, software watchdog timer, and the periodic interrupt timer. These system functions are integrated on the microcontroller to reduce board size and the cost required by external components.

Test Submodule. The test submodule consolidates the microcontroller test logic into a single block to facilitate production testing, user self-test, and system diagnostics. Scan paths throughout the MC68332 provide signature analysis checks on internal logic. User self-test is initiated by asserting the test pin to enter Test Mode. This test can provide a good/not-good response to various externally supplied test vectors.

System Clock. The system clock is generated by an on-chip phase-locked loop circuit to run the device up to 16.78 MHz from either high performance or low power consumption under software control. With its fully static CMOS design, it is possible to completely stop the system clock using a low-power stop instruction, while still retaining the contents of the registers and on-board RAM.

Standby RAM Module

The standby RAM module contains 2K of the fast static RAM powered by Vdd (supply voltage) in normal operation. During power-down, the RAM is especially useful for the CPU as system stack space or variable storage, due to its fast two-cycle access time. Alternately, it can be used by the TPU as emulation RAM for new or additional timer algorithms. Data can be read or written by byte, word, or long-word length. The RAM can be mapped to any 2-KB boundary in the address map.

Chapter 4

A Comprehensive Development System for 32-Bit CISC Microprocessors

Prior to the expanded activity in RISC microprocessors, most development systems supporting 32-bit CISC microprocessors utilized combinations of emulators and logic analyzers. A typical system of this type was the WE-321DS used by AT&T to support developers of systems based on their 32100 and 32200 CISC processors. These processors were used by AT&T primarily in their 3B computers. This computer is the heart of their No. 5 ESS digital telephone switching system. Tens of thousands of these microprocessors have been produced for this application, despite the fact that they have not competed with Intel's 80386 and Motorola's 68000 family in the merchant market. AT&T has also sold a large quantity of the 3B computers to the U.S. Air Force.

There are a variety of microprocessor development systems available for use with CISC microprocessors using clock frequencies up to 25 MHz. However, most of them do not provide for the options available from the system described here. In this chapter a detailed description of this system is given as an example of a comprehensive development system having all of the options that a developer might desire.

The WE-321DS provided an effective environment for both hardware and software developers. Designed to support developers from the earliest stages of system design (e.g., where hardware is not yet sane) through the final stages of system test (e.g., premanufacture certification of the hardware and software together), the HSDS (as shown in Figure 4-1) was embodied in a modular, user configurable set of three building blocks and a terminal, namely:

1. An analysis pod to provide in-circuit emulation capabilities for both the CPU and MMU
2. A UNIX system to host the processor (e.g., the AT&T 3B-2 computer)
3. A commercially available logic analyzer (e.g., the HP64000 Logic Development System

The analysis pod (AP) was the central or key component of this architecture. It provided accurate and nonintrusive (for example, no extra circuitry and no extra wait states) in-circuit emulation of the CPU and MMU bus protocols up to the rated speed of these devices.

To maximize the utility and ease of use of this fundamental capability, the pod also contained the following:

1. An interface to the logic analyzer so that state sequences of the signals in the target could be traced. Further, this interface was used to set up, tear down, and symbolically control hardware breakpoints.
2. An interface to the UNIX host to provide access to a C-language symbolic debugger. Using this tool, software designers could symbolically reference procedures, variables, and the like in their program. Further, this

FIGURE 4-1
The major components of AT&T's Microprocessor Development Laboratory. Reproduced with permission © 1986 AT&T.

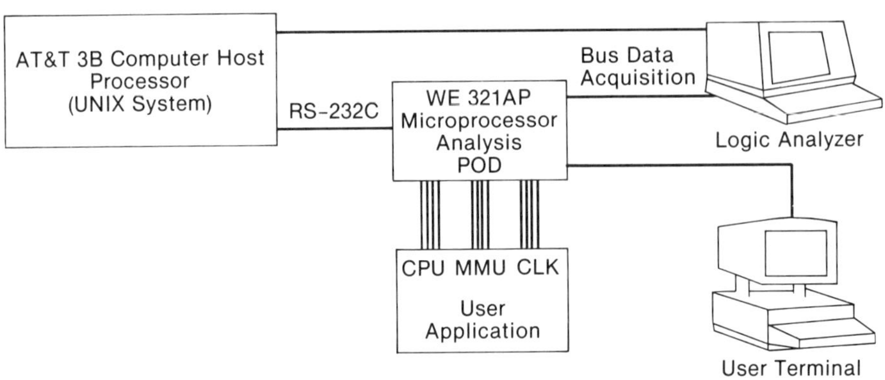

debugger was process-oriented and had the ability to execute command scripts interactively. Finally, this interface also provided a link back to the UNIX environment with its software development and project management tools.

3. An IEEE 488 bus interface for fast access to disk and printer resources for memory trace dumps. This channel could also be used for up/down loading large programs to/from the target.

To complete the picture, the pod contained resident hardware and firmware to access and control the in-circuit emulation capability and to coordinate communications among its interfaces. For example:

1. There was hardware that, under firmware control, allowed the user to enable and disable selected signals in the CPU/MMU complex. This was particularly useful for doing software development early in the system design cycle when the hardware was not completely debugged.
2. There was a self-contained resident low-level debugger to provide upload/download, display/modify memory and software breakpoint capabilities for simple programs (e.g., hardware debug test). The pod also had a built-in self-test program.

The pod, logic analyzer, and UNIX host could be used in a number of configurations, depending on the needs of the development organization. For example, the pod could be used as a stand-alone for simple system development. When the logic analyzer was added, the pair became an effective team for debugging hardware. By adding the UNIX host, one had a complete software development system with access and control of the target hardware. Depending on the size of the system software development team, the host could vary from a relatively small, few-user machine (e.g., an AT&T 3B-2) to a relatively large, many-user computer (e.g., an AT&T 3B-20).

Emulation and Debugging

The WE321DS Microprocessor Development System was an integrated collection of hardware and software components used to develop, test, and debug WE32100 microprocessor (CPU), WE32101 memory management unit (MMU), and WE32106 math acceleration unit (MAU) based applications. Components of the system included the WE321AP microprocessor analysis pod, the WE321SD development software programs, a UNIX system host computer, a high-level symbolic debugger, bus state analysis software, and a logic analyzer. In certain configurations of the system the host computer and the logic analyzer were not required. Using the modular development architecture, the developer could configure the components into any of four workstations to emphasize hardware testing, software testing, or hardware/software integration and testing.

The main function of the WE321DS Microprocessor Development System was to provide in-circuit emulation of the WE32100 microprocessor and the WE32101 MMU, the devices around which the target system (the product or system under development) was being developed. The in-circuit emulation allowed the user to evaluate the microprocessor-based target system in a highly controlled environment. The WE321AP microprocessor analysis pod was the means through which the development system provided in-circuit emulation of the CPU and MMU devices. Its operating frequency was normally 14 MHz. However, the WE32100 was available in 10-, 14-, and 18-MHz frequency versions. The pod could be adapted to these frequencies by changing components within its assembly.

The analysis pod contained footprint connectors that replace the actual CPU and MMU devices in the target system. These connectors plugged directly into the CPU and MMU sockets in the target system. Through the footprint connectors, the analysis pod accurately emulated (with zero wait states) the CPU and MMU bus protocols. The analysis also contained a third footprint connector for the WE32102 clock to provide a clock source for the target system. By using the analysis pod in place of the actual CPU and MMU devices, the developer could easily exercise control over the emulation in the target system. Control gained by using the analysis provided the necessary features to aid the three design phases of the target system: hardware development, software development, and system integration. The analysis pod also provided software support for the WE32106 MAU. Floating-point data could be stored and displayed through software emulation. If the target system contained an MAU, the MAU registers could be displayed or modified.

The analysis pod contained an interactive monitor program (IMP), an assembly-level debugger resident in ROM that provided the necessary features to aid both hardware and software debugging. These features included the following:

- Object file downloading from the UNIX system host computer to either the target system memory or the RAM available to the developer on the analysis pod.
- Ability to display and modify memory and CPU/MMU/MAU registers.
- A program trace mechanism that allowed the user to execute one or more assembly instructions and then return to IMP to examine the machine state.
- Generation of breakpoints allowing control to be transferred from a target process to IMP, upon the occurrence of a predefined event, during the execution of the process.

Disassembly of instructions stored in memory.

An option of debugger controlling responses to all CPU and MMU exceptional conditions.

A transparent mode that allowed the user to access the UNIX system host computer via communication ports on the analysis pod. This feature allowed the developer to access the UNIX system resident software generation programs for program development.

All IMP capabilities were available to the developer processes executing in either the physical mode (MMU inactive) or the virtual mode (MMU active).

An optional high-level symbolic debugger, called Ferret, was also available. Ferret resided partially in the analysis pod firmware and partially on the UNIX system host computer. It permitted the developer to examine and modify the execution of the application program interactively at the C language level. The main features provided by Ferret included the following:

Symbolic referencing of procedures and variables

Ability to interpret commands noninteractively from a script file

Support of process-specific software breakpoints with action lists

Memory and register examination and modification

Symbolic stack back-traces

Block structured command language with C-like constructs

Controlled response to unexpected external events

Like the IMP debugger, Ferret could control virtual and physical mode processes and provide transparent access to the UNIX system host.

An I/O library, consisting of routines that software developers could use when writing application code, was included.

The analysis pod provided access to all CPU and MMU signals via access connectors. This access was provided so that logic analyzers could easily observe the signal activity at the CPU and MMU footprints and optionally in the target system hardware. The analysis pod also provided two stimulus inputs that could be used by a logic analyzer to generate hardware breakpoints in response to system events.

A bus state analysis option was also available. This option provided software utilities for symbolic tracing and disassembly with a Hewlett-Packard Model 64000 Logic Development System. Development was under way on an events monitor, which would allow the developer to use other logic analyzers.

The above utilities included two configuration files, upload and download utilities, a symbol table utility, and a program trace disassembler. The configuration files assigned descriptive names to the signals available at the access connectors, which provided easy signal recognition when performing signal observation. Of the two configuration files available, one supported 10-MHz analysis pod operation and the other supported 14-MHz operation. The upload utility provided test result uploading from the HP64000 system to the UNIX system host, and the download utility provided symbolic definition downloading from the UNIX system host to the HP64000 system for symbolic debugging. The symbol table utility translated object file symbolic information for use by the HP64000 for symbolic debugging. The program trace assembler provided disassembly of hardware bus traces allowing HP64000 system logic traces to be displayed in WE32100 microprocessor assembly language format.

Analysis Pod Hardware

The analysis pod consisted of a CPU and MMU board and an I/O and memory board and contained the following components:

A WE32100 microprocessor, WE32101 MMU, and a WE32102 clock, which formed a board-level computer.

User controls for abort, reset, memory-mapping option, and an external clock option.

A configuration register which disabled faulty target-generated signals to the CPU.

Interactive monitor program (IMP) firmware.

Symbolic debugger (Ferret) firmware.

System self-test (SST) firmware.

64K of static RAM, of which 55K were available to the developer when the active debugger was in the IMP mode.

EEPROM (electrically erasable programmable read-only memory) memory that stored the status of registers and could be set and displayed by the developer. These registers contained the baud rate of the user terminal, the baud rate of the UNIX system link, the default debugger (IMP or Ferret), the addresses of external MMUs, and the state (on or off) of the configuration bits.

An LED display used for monitor and diagnostics messages.

Access connectors that allowed a logic analyzer to monitor up to 120 signals of the microprocessor, MMU, and clock.

An external clock input.

External hardware interrupt inputs compatible with the HP64000 stimulus signals.

Two RS-232C interface connectors for communication with a UNIX host and a user terminal.

Three ribbon-cable umbilicals that terminated with the footprint connector of the emulated device (CPU, MMU, and clock).

A block diagram of the analysis pod CPU and MMU board is shown in Figure 4-2. The analysis pod I/O and memory board block diagram are shown in Figure 4-3.

Development System Configurations

As mentioned previously, the WE321DS Microprocessor Development System could be arranged in four separate configurations. Each configuration supported a particular phase of product development (e.g., software development, hardware development, and hardware/software integration). The four configurations were stand-alone operation of the analysis pod, a software development workstation (analysis pod with UNIX system host), hardware development workstation (analysis pod with a logic analyzer), and a hybrid workstation (analysis pod, UNIX system host, and a logic analyzer). All four configurations required a user terminal for interaction with IMP and Ferret.

The Stand-Alone Configuration

The analysis pod and a user terminal comprised the stand-alone configuration, as shown in Figure 4-4. This configuration provided the software and hardware developer with the features of the IMP assembly-level debugger. IMP allowed the developer to examine and modify memory, set and clear breakpoints, control program execution, and perform software debugging at assembly level.

Software Development Workstation

The software development workstation consisted of the analysis pod, a link to a UNIX system (host computer), and a user terminal. This configuration supported all of the operations provided by the stand-alone configuration. In addition, the software development workstation allowed the user to develop software on the UNIX system for downloading into either the analysis pod (if the active debugger is IMP) or into the target system memory. Symbolic debugging was supported only by the software development workstation configuration and the hybrid workstation described below. The arrangement of the software development workstation configuration is shown in Figure 4-5. The system architecture was slightly different for the IMP and Ferret modes of software development. When using IMP,

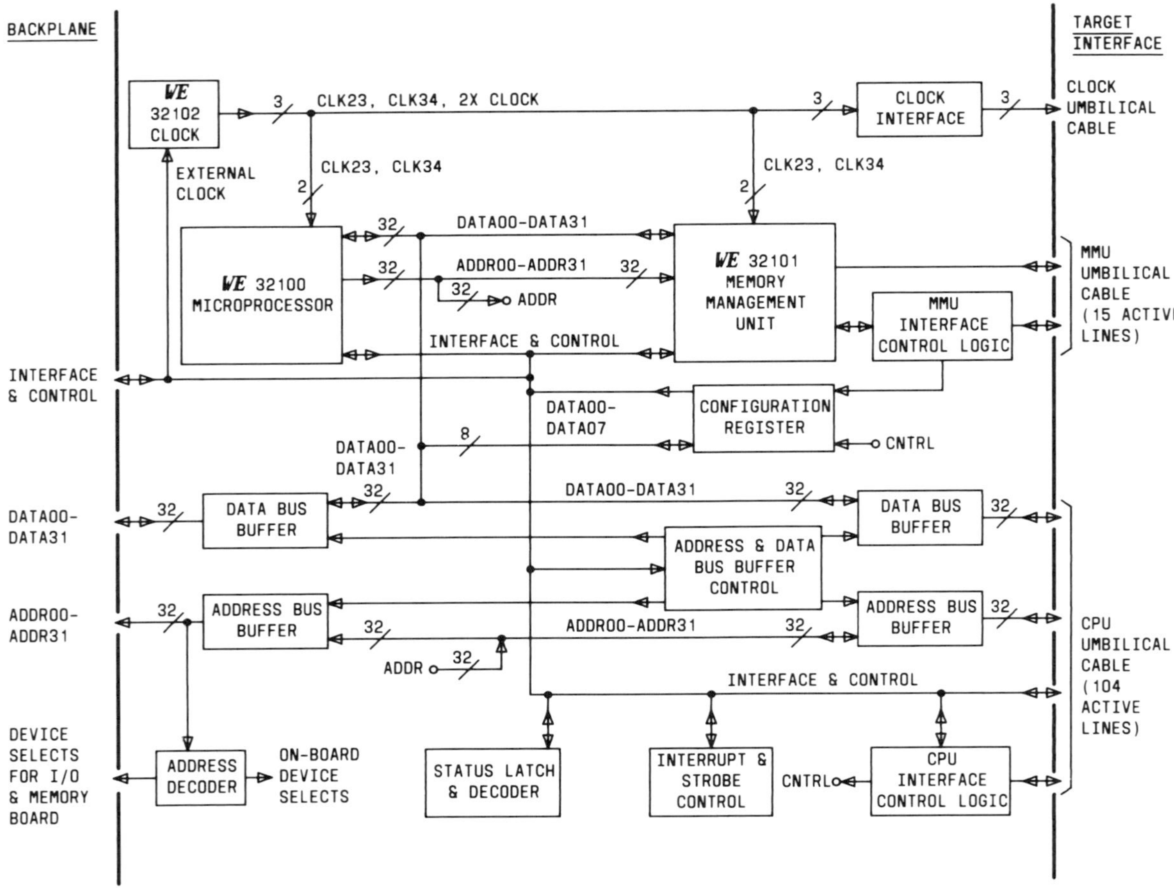

FIGURE 4-2 Analysis pod CPU and MMU board block diagram. Reproduced with permission © 1986 AT&T.

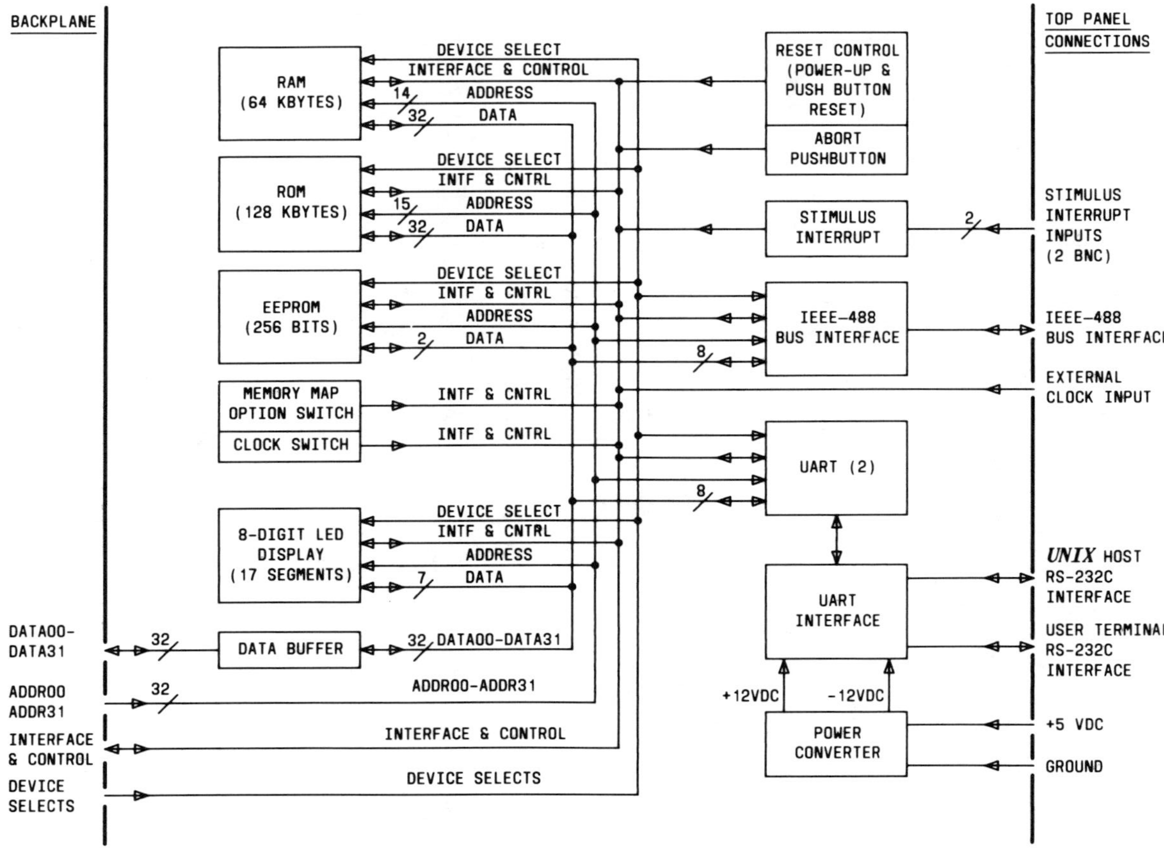

FIGURE 4-3 Analysis pod I/O and memory board block diagram. Reproduced with permission © 1986 AT&T.

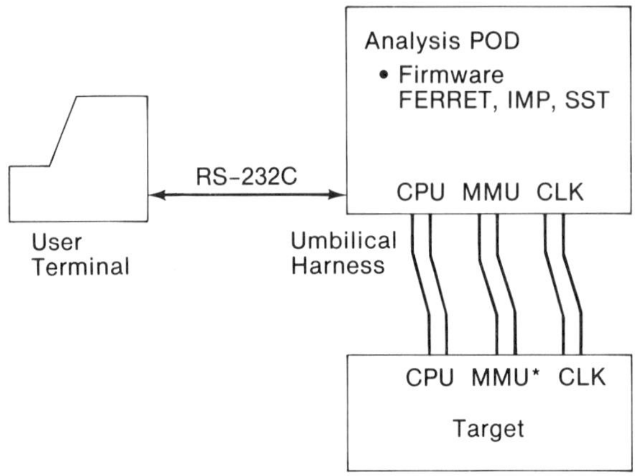

*Connection not required for CPU-only emulation.

FIGURE 4-4
Stand-alone configuration. Reproduced with permission © 1986 AT&T.

FIGURE 4-5
Software development workstation—IMP or Ferret mode. Reproduced with permission © 1986 AT&T.

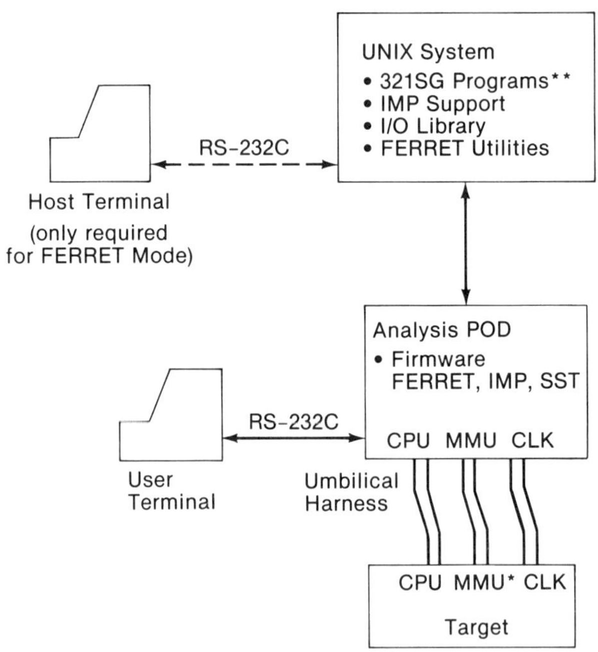

*Connection not required for CPU-only emulation.

**WE 321SG Software Generation Programs.

the user terminal was connected to the analysis pod and use of the UNIX system host was optional. When using Ferret, the user terminal and the analysis pod were connected to the UNIX system host.

Hardware Development Workstation

The hardware development workstation (see Figure 4-6) consisted of the analysis pod, a logic analyzer, and a user terminal. This configuration supported all of the operations provided by the stand-alone configuration. The addition of a logic analyzer enhanced hardware development support. Target hardware could be monitored by a logic analyzer interfaced to the target system through the analysis pod. The logic analyzer could monitor up to 120 CPU, MMU, and clock signals.

A bus state analysis option that provided software utilities for use with the HP64000 system was also available. The utilities included in this option were two configuration files, upload and download utilities, and a program trace disassembler. The configuration files were provided to support the HP64000 system in defining signal names and polarities of the 120 CPU, MMU, and clock signals available at the analysis pod access connectors.

FIGURE 4-6
Hardware development workstation. Reproduced with permission © 1986 AT&T.

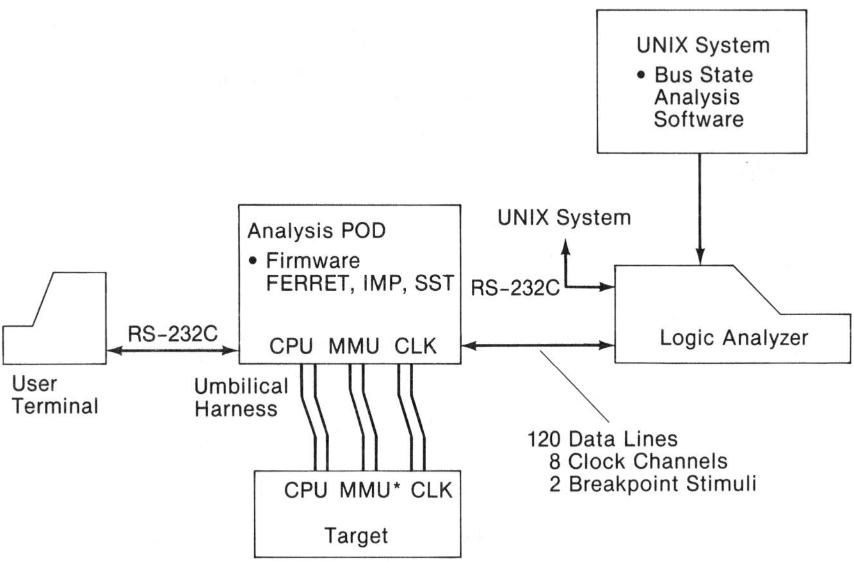

*Connection not required for CPU-only emulation.

**WE 321SG Software Generation Programs.

Two configuration files were available—one that supported 10-MHz analysis pod operation and another that supported 14-MHz operation. The upload utility provided test result uploading from the HP64000 system to the UNIX system host, and the download utility provided symbolic definition downloading from the UNIX system host to the HP64000 system for symbolic debugging. The program trace disassembler provided disassembly of hardware bus traces, allowing the 64000 system logic traces to be displayed in the WE32100 microprocessor instruction format.

Hybrid Workstation—IMP or Ferret Mode

The hybrid workstation configuration provided all of the development system's features and was most useful for the hardware/software integration phase. The link of the UNIX system host to the analysis pod provided the system with all of the capabilities of the software development workstation, while the logic analyzer provided hardware development needs. Figure 4-7 shows the configuration of the hybrid workstation, which could be operated in either IMP or Ferret mode.

FIGURE 4-7
Hybrid workstation—IMP or Ferret mode. Reproduced with permission © 1986 AT&T.

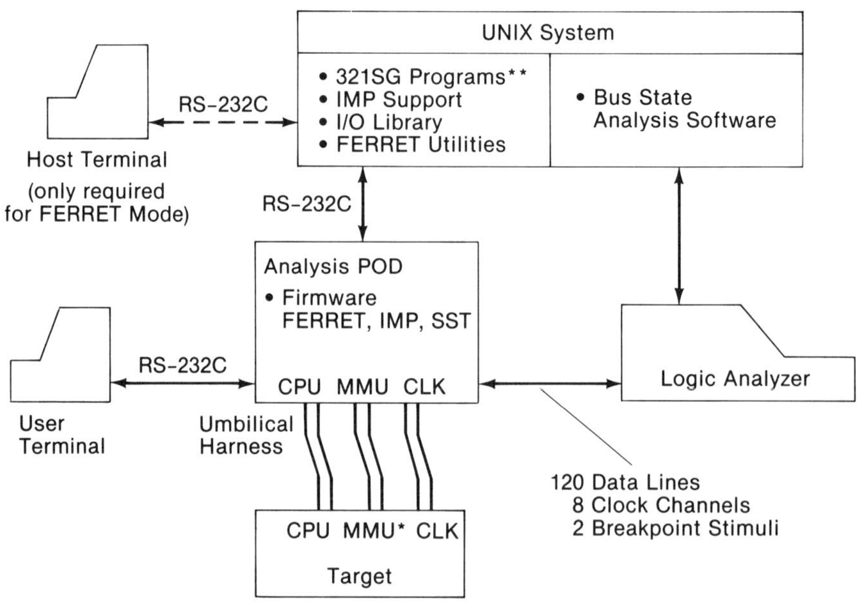

*Connection not required for CPU-only emulation.

**WE 321SG Software Generation Programs.

Development System Software and Firmware

Software support for the WE321DS Microprocessor Development Station was provided by the WE321SG Software Generation Program (SGP) and the WE321SD Development Software Programs. The SGP allowed the user to create and debug application programs before they downloaded to the analysis pod or target system. The WE321SD Development Software Programs supplied with the analysis pod consisted of IMP, an assembly-level debugger, a download program for IMP, and an analysis pod I/O library. Options of the Development Software Programs included Ferret, a high-level symbolic debugger, and bus state analysis (BSA) software. The BSA option included symbolic tracing and disassembly routines, a symbol table utility, download and upload utilities, two HP64000 Logic Development System configuration files, and a terminal setup file. Figure 4-8 shows how the available software was associated with the various components of the development system.

The IMP and Ferret debuggers were used to control and modify the execution of the application program by single-stepping execution, setting and displaying breakpoints, and setting and displaying registers and memory. IMP operated at the assembly level and Ferret operated at the C language level. Other features of Ferret included the capability of multilevel access (i.e., stack back-trace and memory dump, referencing variables by their symbolic names and absolute addresses, and software breakpointing with action lists).

A system self-test (SST) program resident in firmware on the analysis pod determined the hardware integrity of the analysis pod when the pod was powered up. This self-test could also be invoked through IMP.

The symbol table utility provided information for symbolic debugging. The configuration command files provided symbolic definition of CPU and MMU signal names, user program locations, and measurement experiment specifications for the HP64000 system.

The WE321SD Development Software Programs provided for the analysis pod were contained on magnetic tape, floppy disks, or disk cartridges, depending on the system host used. Instructions were provided for loading these programs onto a UNIX system host.

Interactive Monitor Program (IMP) Firmware

IMP was an assembly-level, hex-oriented debugger contained in EEPROM on the analysis pod. IMP offered an optional data link to a support processor (i.e., a UNIX system host) that could be used for reading object and ASCII files from an established file system before downloading them into the analysis pod or the target system memory. The support processor was required only if the programs were to be downloaded. IMP could be used during the development of an operating system kernel and could also

SOFTWARE DESCRIPTION

Software Description

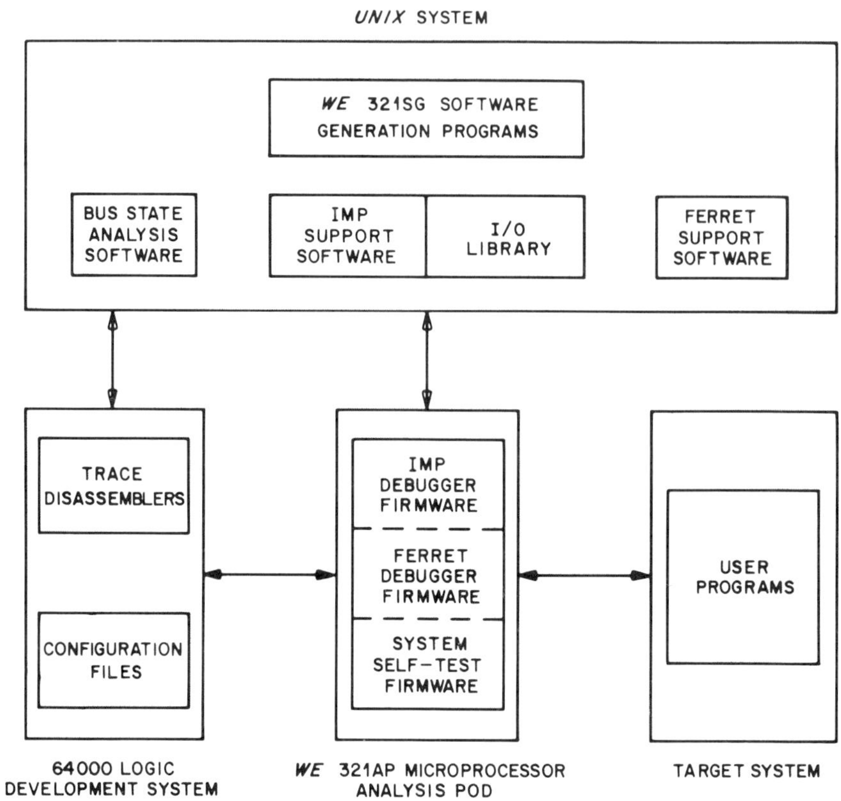

FIGURE 4-8
Available software/firmware for the development system.
Reproduced with permission © 1986 AT&T.

be used for verifying the operation of hardware and the debugging of firmware.

IMP provided the user with all of the commands needed to control a debugging session. IMP controls used hardware and software running in virtual and physical modes through interrupts and exception handlers. It interacted with the user software through a set of software control mechanisms shared in common with the Ferret debugger. IMP provided commands for the following:

Displaying and setting the current state of WE32100 microprocessor registers, pod or target system memory, or memory-mapped I/O

Displaying and setting current data-link information and controlling the transparent mode to a host system for program development

Displaying and setting breakpoints

Tracing the execution of user application programs

Downloading application programs from the UNIX system host

Addressing physical and virtual memory

Execution control (initiating program execution, calling a procedure, and performing a process switch)

MMU control (when an MMU is included in the target system)

MAU control (when an MAU is included in the target system)

IMP used approximately 8K of analysis pod RAM for its workspace.

Ferret Symbolic Debugger (Optional)

Ferret was a symbolic debugger for C language programs which had been written for the target systems of the WE321DS Microprocessor Development System. Ferret permitted the developer to interactively examine and modify the execution of the application program at the C language level. Ferret also provided a noninteractive mode of operation for debugging programs from a test script file (the file providing the information for the debugging session). Ferret provided commands for downloading object files, single-stepping program execution, setting breakpoints with action lists, and exercising the program for the discovery of errors. Ferret was specifically designed to symbolically test C programs in the target environment in which the programs would actually run. The programs had to be generated using the WE321SG Software Generation Program.

Ferret had code running on three systems: a host processor running under the UNIX operating system, the WE321AP microprocessor analysis pod, and the target system under test. The host processor was responsible for reading and parsing input, looking up symbols, and printing output. Commands were executed on the target system. The analysis pod acted as an interface between the host processor and target system to control the target system and protect Ferret's sanity. Communication between the host processor and the analysis pod was over an RS-232C data link. Message packets were sent between both systems. The data link also provided the host with control over the analysis pod processes under test executing in the target system maintaining the target system environment. Emulation cables (CPU, MMU) and a clock cable, connecting to the target system, provided control over the target.

Some of the features of Ferret were as follows:

A block-structured command language similar to C language.

Automatic symbolic lookup of procedures, variables, arguments, C source statements, and stack back-traces.

Process-oriented debugging in both physical and virtual mode.

Software breakpoints with action lists to stop the program and execute a list of debugger commands.

Ferret statements that could be interpreted interactively as well as from a file (test script).

A syntax-checking mode that allowed a developer to check a test script file before entering the lab.

Ability of the developer to call any function in the test process, passing to the function any set of expressions as arguments, and having the function return to a value within any Ferret expression. Ferret guaranteed that the test process function call did not change the state of any of the process-oriented registers for this process.

Control of the target system through analysis pod emulation of the CPU and MMU with access to internal registers.

Controlled response to unexpected events (e.g., aborts, stimulus inputs, interrupts, and exceptions).

The major debugging feature of Ferret provided breakpoints with action lists. These action lists consisted of user-defined code that performed intelligent actions at the breakpoint location in the process. Such actions included variable checking, message printing, and code patching. Other debugging features of Ferret included symbolic stack tracing of the current process, and symbolic examination and modification of code. Ferret had both interactive and noninteractive modes. Normally, all invocations of Ferret were in the interactive mode from the host terminal that was connected to the UNIX system. However, it may have been desirable to execute in a noninteractive mode if the test scripts were extensive. This feature was useful for soaking test equipment and regression testing. Log files of the test session were produced by Ferret. These session log files could be compared to "golden" log files (a file that contains the expected results) to produce regression testing reports. Before a debugging session with Ferret was started, requirements for compilation, link-editing, listing, and script checking had to be taken into account. Guidelines for these requirements were given in the user manual for the development system.

Firmware Control Mechanisms

The analysis pod firmware provided handlers for all types and modes of user interrupts, gates, exceptions, and microprocessor resets. The interrupt and exception handler detected whether an interrupt or exception was generated within the analysis pod or by the target system. The analysis pod classified all target interrupts and exceptions as "stray" and printed a message to that effect on the user terminal. The analysis pod accessed its vector table to fetch the vector to the appropriate handler for a given interrupt or exception. Once the interrupt had been handled, control was returned to the active debugger, IMP or Ferret. The user could then perform debugger actions and resume the suspended user process.

Interrupt and Exception Handling Environments

The user had the option of having interrupts and exceptions handled by either the analysis pod or by the target system. The choice of handler was provided for target systems operating in either physical or virtual mode. In addition, target systems executing in virtual mode were not restricted to the location of the target vector table (that is, the vector table could be located at any available address outside the 0 to FFF address range).

The analysis pod provided four interrupt and exception handling environments that were designed to aid in the debugging of the target system's hardware and software. Each environment was defined by several variables that included the following:

The handler of exceptions and user interrupts, analysis pod or target

The mode application software executes, physical or virtual

The location of the target system vector table, physical address 0 through FFF or elsewhere in the target system memory

Environment One: Analysis Pod Responses to Target-Generated Interrupts and Exceptions

In the first environment, the analysis pod interrupt and exception handlers responded to all interrupts and exceptions generated by either the analysis pod or the target system. Environment One was intended primarily for use during the early stages of hardware debugging where the emphasis was on writing software to test individual hardware components. The analysis pod handlers were used to provide sane responses to target-system-generated interrupts and exceptions, thus allowing the user to concentrate solely on hardware testing and not to worry about the sanity of the target handlers.

To operate in Environment One the user had to place the memory option switch in the pod position. The user should not have placed executable code in the target system memory at addresses 0 through FFF, since this region was not accessible in the target system if the memory map option switch was set in the POD position.

Figure 4-9 illustrates the actions that took place when a process executing in Environment One was interrupted. As shown in the figure, all interrupts were serviced by the analysis pod handlers, since the memory

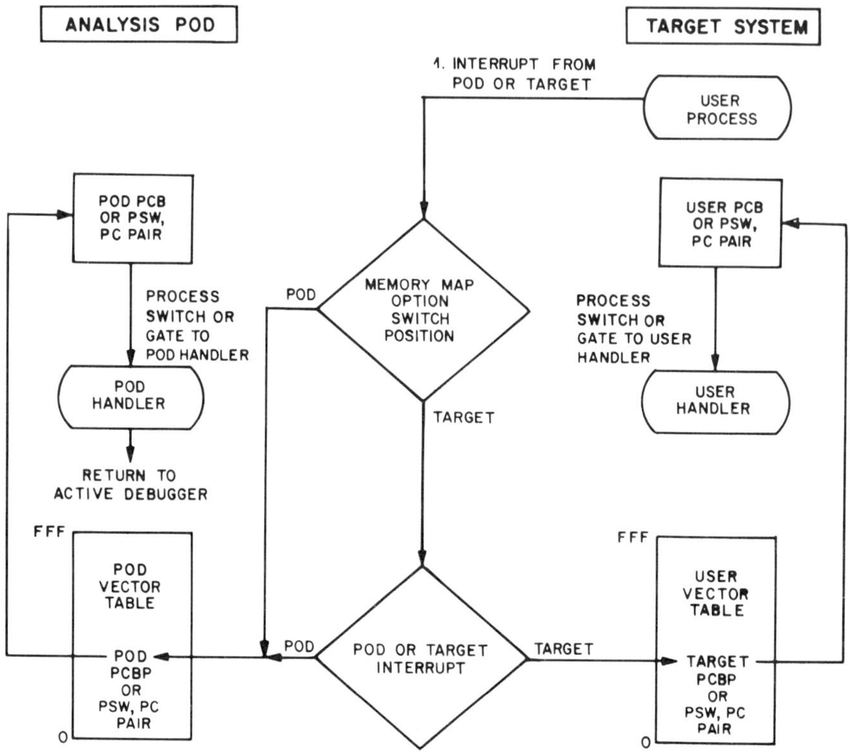

FIGURE 4-9
Environments One and Two interrupt handling mechanism. Reproduced with permission © 1986 AT&T.

map option switch was in the POD position. Once the interrupt was handled by the analysis pod, control was returned to the active debugger (IMP or Ferret).

Environment One supported both physical and virtual execution modes. The target system application software had to reserve the physical and virtual (if the target system is operating in the virtual mode) addresses between C0F00000 and C0FFFFFF to the analysis pod. The target system memory and I/O devices residing in the analysis pod address range were not accessible.

The developer had to also supply memory management descriptors covering the analysis pod's 1-MB address space and the address space containing the interrupt and exception handlers vector table (address range 0 to FFF) if the target system was operating in the virtual mode. These descriptors were to translate the virtual address range C0F00000 to C0FFFFFF to physical address range C0F00000 through C0FFFFFF and the virtual address range 0 through FFF to physical address range 0 through FFF (i.e., null translation). In addition the developer had to program the MMU's section RAM A (SRAMA) section RAM B (SRAMB) registers to reflect the locations and lengths of the analysis pod and target system descriptor tables. This could be accomplished by using the IMP commands srama and sramb or the Ferret commands $srama and $sramb.

Environment Two: Target System Responses to Target-Generated Interrupts and Exceptions

Environment Two provided hardware separation (in real time) or analysis pod and target system interrupts, allowing target system response to target-generated interrupts and exceptions. This environment supported both hardware and software testing. Hardware testing could consist of system-level tests where the user assumed that the individual components were working properly. The emphasis was then on generating software to test communication between the system components.

Software testing could consist of debugging the target system interrupt and exception handlers. The developer had to supply a target system vector table beginning at physical address 0 in the target memory. The target vector table could contain vectors to user interrupt and exception handlers, a copy of the analysis pod vectors, or a combination of target system and analysis pod vectors.

A combination of target system and analysis pod vectors might have been desirable to users who were debugging interrupt and exception handlers. The user could place the vector(s) to the target system handler(s) to be debugged in the appropriate location(s) in the target system vector table and copy the analysis pod vectors for the handlers that were yet to be debugged into the remaining locations. Thus the user could concentrate on debugging a subset of handlers while the analysis pod handlers were invoked for interrupts and exceptions whose handlers were yet to be written or debugged.

Environment Two, like Environment One, supported both physical and virtual execution modes. To operate in Environment Two the developer had to set the memory map option switch in the TARGET position and to supply a target system vector table beginning at physical address 0 in target memory.

Figure 4-9 illustrates the actions that take place when a process executing in Environment Two is interrupted. As shown in the figure, the analysis pod first determined the source of the interrupt (analysis pod or target system). If the source was the analysis pod, the vector was fetched from the analysis pod vector table and a process switch (for full interrupts) or gate (for quick interrupts) was performed on the corresponding analysis pod handler. Control was returned to the active debugger (IMP or Ferret) once the interrupt was serviced. If the target system was the source of the interrupt, the vector to the handler was fetched from the target system vector table and a process switch or gate was performed to the corresponding target system handler.

The target system application software had to reserve the physical and virtual (if the target is operating in the virtual mode) addresses between C0F00000 and C0FFFFFF to the analysis pod. Target system memory and I/O devices residing in the analysis pod address range were not accessible.

Additionally, the user had to supply memory management descriptors covering the analysis pod's 1-MB address space and the address space containing the target system vector table (address range 0 through FFF) if the target system was operating in the virtual mode. These descriptors were to translate the virtual address range C0F00000 through C0FFFFFF to physical address range C0F00000 through C0FFFFFF and the virtual address range 0 through FFF to physical address range 0 through FFF (i.e., null translation). In addition the user had to program the MMU's section RAM A (SRAMA) and section RAM B (SRAMB) registers to reflect the locations and lengths of the analysis pod and the target system descriptor tables.

The analysis pod debuggers (IMP and Ferret) used breakpoint and trace trap exceptions to control user software execution. IMP and Ferret wrote the user's breakpoint and trace trap vectors automatically. Only Environment Two users that had ROM vector tables and desired analysis pod responses to breakpoint and trace trap conditions had to place copies of the analysis pod second-level normal exception handler and gate handler entries in the appropriate locations of the target system vector table. These entries transferred control of the CPU from the target system to the analysis pod resident monitor, where the facilities of IMP and Ferret were available. If these entries were not supplied, the monitor facilities could not be used after such a trap.

Environment Three: Virtual Vector Table for Target System Responses to Target-Generated Interrupts and Exceptions

Environment Three, like Environment Two, allowed for target system responses to target-generated interrupts and exceptions. The differences between these two environments was that separation of analysis pod and target system interrupts was done with software in Environment Three, instead of with hardware as in Environment Two. Hardware separation was done in real time while software separation incurred some overhead. Software separation of analysis pod and target system interrupts provided the additional flexibility of having the target system vector table in RAM instead of booting ROM at physical addresses 0 through FFF. Environment Three was intended for debugging of the target system kernel.

To operate in Environment Three, the developer had to provide an interrupt and exception handler vector table beginning at virtual address 0 in the target system memory. Virtual address 0 had to be translated to a physical address outside the 0 through FFF address range in order to operate in this environment. This vector table, called the primary vector table, had to contain a copy of the analysis pod interrupt handler vectors. This vector table must also contain vectors to the exception handlers. These vectors could be to user exception handlers or to the analysis pod exception handlers. The user also had to supply a secondary vector table that was located outside the 0 through FFF address range in the target memory and a pointer to the beginning of the secondary vector table. This vector table contained vectors to the actual interrupt handlers. The pointer to the secondary vector table (the audit pointer) can be set by using either the IMP command audit addr or the Ferret command audit (addr-exp).

The primary vector table allowed transfer of control to be passed from a target process to an analysis pod audit routine when an interrupt occurred. All interrupts caused transfer of control to the analysis pod, since the primary vector table contained a copy of the analysis pod interrupt vectors. The audit routine then determined the source of the interrupt (analysis pod or target system) in order to invoke the appropriate handling routine. If the source was the analysis pod, the interrupt handler vector was fetched from the analysis pod vector table. If the source of the interrupt was the target system and the user had provided an audit pointer to a secondary vector table, then the audit pointer was used to access the secondary vector table for the appropriate interrupt handler vector. However, if the user had not provided an audit pointer (audit pointer value = 0), the interrupt vector was fetched from the analysis pod vector table.

The secondary vector table could contain vectors to user handlers, copies of analysis pod vectors, or a combination of both. This allowed Environment Three to provide the same interrupt and exception handler debug feature as Environment Two. This enabled the developer to debug a subset of handlers while the analysis pod handlers were invoked for those handlers that were yet to be debugged. Environment Three allowed the developer to have more than one secondary vector table in the target memory. The user could then select different handlers by simply changing the audit pointer value. This feature allowed the user to choose different interrupt and exception handlers without having to rewrite the secondary vector table. All the user had to do was to change the audit pointer to the address of another secondary table. This table contained vectors to the appropriate handlers for the application that was to be debugged.

Figure 4-10 illustrates the actions that took place when a process executing in Environment Three was interrupted. The following actions are numbered 1 to 6 on Figure 4-10.

1. The user process was interrupted by either the analysis pod or the target system.

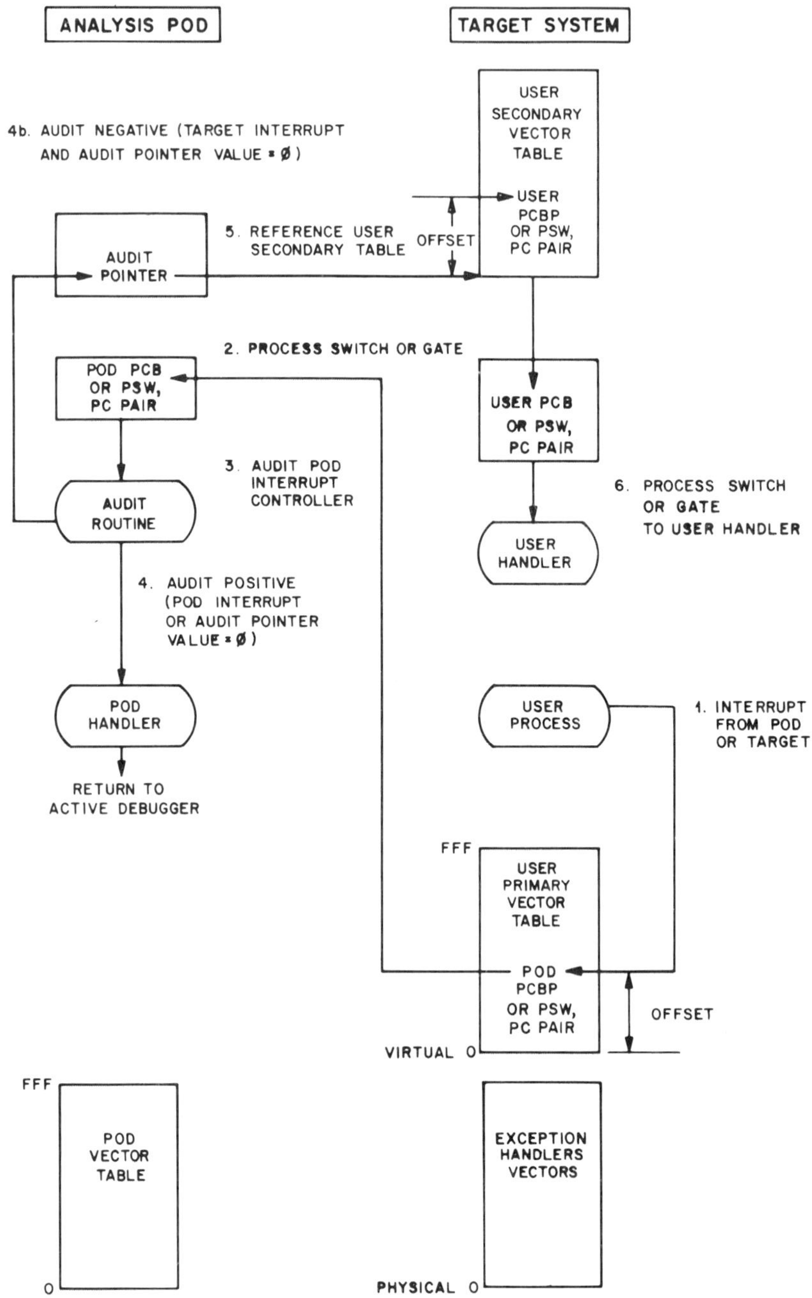

Note: Exception handler vectors must be provided at their appropriate physical addresses in order to support reset exceptions.

FIGURE 4-10
Environment Three interrupt handling mechanism. Reproduced with permission © 1986 AT&T.

2. Control was passed from the user process to the analysis pod. (Control was always passed to the analysis pod because the primary vector table contained a copy of the analysis pod interrupt vectors.) In addition, the offset from the bottom of the primary vector table to the vector for the generated interrupt was saved for use in steps 4a and 5.
3. The analysis pod invoked an audit routine that determined the source of the interrupt (pod or target). If the source was the target, the analysis pod checked the audit pointer value.

4a. If the audit was positive (i.e., the analysis pod generated the interrupt or the audit pointer value is equal to 0), the appropriate analysis pod handler was invoked. The appropriate handler was determined from the offset saved in step 2.

4b. If the audit was negative (i.e., the target generated the interrupt and the audit pointer value was not equal to 0), the audit pointer value was used to access the target system secondary vector table and steps 5 and 6 were performed.

5. The audit pointer was used to reference the secondary vector table, and the offset saved in step 2 was used to find the appropriate vector for the interrupt.
6. A process switch (for full interrupts) or gate (for quick interrupts) was invoked to the user handler that corresponded to the interrupt generated.

If the interrupt was handled by the analysis pod, control was returned to the active debugger (IMP or Ferret) after the interrupt had been serviced. The interrupted process could then be resumed at the point at which it was interrupted by entering the IMP command go or the Ferret command goto( ).

The user also had to provide exception handler vectors at their appropriate physical addresses (for example, a reset process control block pointer at physical address 80 hexadecimal). This was necessary to support reset exceptions, since the CPU switched to physical mode operation upon reset and executed in the physical mode during the reset sequence. If an exception occurred during this time, the CPU would then be able to fetch the appropriate exception handler vector from physical memory.

Environment Three supported applications executing in virtual mode only. The target system software had to reserve the physical and virtual addresses between C0F00000 and C0FFFFFF for the analysis pod. Target system memory and I/O devices residing in the analysis pod address range were not accessible.

The user had to supply memory management descriptors covering the analysis pod's 1-MB address space and the address space containing the target system primary vector table (address range 0 through FFF). The analysis pod address descriptors translated the virtual address range C0F00000 through C0FFFFFF to physical address range C0F00000 through

USER EXECUTION DEFINING PARAMETERS

Memory map option switch set to pod (P) or target (T)
User software execution in physical (P) or virtual (V) mode
Target interrupts and exceptions handled by pod (P) or target (T)
Vector table at physical address 0 (P) or virtual address 0 (V)

ACTIONS REQUIRED BY THE USER

Supply descriptors for the analysis pod address space
Supply descriptors for the target vector table
Place copy of pod vector table at target virtual address 0
Provide breakpoint and trace trap exception vectors (1)
Supply secondary vector table
Provide pointer to secondary vector table

USER SOFTWARE EXECUTION ENVIRONMENT

Memory map option switch	User software execution	Target interrupts and exceptions handled by	Vector table at	Supply descriptors for the analysis pod address space	Supply descriptors for the target vector table	Place copy of pod vector table at target virtual address 0	Provide breakpoint and trace trap exception vectors	Supply secondary vector table	Provide pointer to secondary vector table	User software execution environment
P	P	P	3	N	N	N	N	N	N	Environment One
P	V	P	3	Y	Y	N	N	N	N	
T	P	T	P	N	N	N	Y	N	N	Environment Two
T	V	T	P	Y	Y	N	Y	N	N	
4	V	2	V	Y	Y	Y	N	Y	Y	Environment Three
4	V	T	V	N	Y	N	N	N	N	Environment Four

Notes:

- IMP and Ferret write the user's breakpoint and trace trap vectors automatically. Only users having ROM vector tables must copy the analysis pod second level normal exception handler and gate handler entries into the target system vector table.
- In environment three, all interrupts first pass through an analysis pod interrupt handler which audits the analysis pod's internal interrupt sources. If the audit is negative, control is passed to the target system interrupt handler via the audit pointer and the secondary vector table, both provided by the user.
- A target system vector table is not required in environment one since all interrupts and exceptions are handled by the analysis pod.
- Memory map switch has no effect in determining which handler responds to interrupts and exceptions in this environment.

FIGURE 4-11
Simplified summary of interrupt and exception handling environment. Reproduced with permission © 1986 AT&T.

C0FFFFFF (i.e., null translation). In addition, the user had to program the MMU's section RAM A (SRAMA) and section RAM B (SRAMB) registers to reflect the locations and lengths of the analysis pod and target system descriptor tables.

Environment Four: Target System Responses to All Interrupts and Exceptions

Environment Four essentially disassociated the target system from the analysis pod. This environment supported only virtual mode execution. It also offered the advantage that the developer needed not to comply with any of the requirements regarding interrupt vectors, exception vectors, and arbitration of analysis pod interrupts that were placed on Environments One, Two, and Three. The target system had to respond to all interrupts and exceptions (that is, supply all interrupt and exception handle vectors at the target system virtual address 0). The only constraint is that the analysis pod address space could not be used. Thus Environment Four makes the execution environment as close as possible to that of the chip set itself.

Because Environment Four did not adhere to the interrupt and exception vector requirements, all analysis pod interrupt-based features (such as Ferret, the stimulus inputs, and abort button) were inoperable. In addition, breakpoints and tracing by the analysis pod debuggers were not supported in Environment Four.

Figure 4-11 provides a simplified summary of the four interrupt and exception handling environments described above. It also summarizes what the developer must do to operate in each environment and summarizes the constraints of each interrupt and exception handling environment.

Chapter 5

Logic Analyzers and Monitors

When logic analyzers were brought into the microprocessor scene in the mid-1970s, they were generally viewed as troubleshooting tools. Today many are used in the development and testing of microprocessor-based products and systems. Some companies, specializing in logic analyzers, have combined them with in-circuit emulators, debuggers, and personal computers to provide development systems. Although Hewlett-Packard appears to have been the first to include logic analysis capability in a development system, Intel has been credited with the introduction of the concept of combining a logic analyzer with an in-circuit emulator in 1983. This combination was referred to as I^2ICE (Integrated Instrumentation and In-Circuit Emulation). While they referred to the flexibility of high-speed triggering and glitch detection to be realized by combining a logic timing analyzer with an in-circuit emulation system (ICE), they offered it only as optional equipment. It remained for specialists in the logic analyzer field to make it a centerpiece in their development systems. These specialists are now adding onto logic analyzers additional elements to produce software analyzers.

Logic Analyzer Operation

Logic analyzers are an extension of the oscilloscope family. They display timing data with time along a horizontal axis and amplitude along a vertical axis. In another mode they display numbers and letters much the same as the display screen of a personal computer. They are multichannel, handling hundreds of input signals in some designs. They are also storage devices capable of holding trace information. A simplified block diagram of a single-channel analyzer is shown in Figure 5-1. It is seen to have provision for collecting data from multiple channels, a memory to store trace signals, memory control triggering, and display generating and formatting sections. In addition it has an internal clock and provisions for an external clock. It is this availability of both internal and external clock signals that enables a logic analyzer to function in two separate modes. In one of these modes it is a logic timing analyzer and in the other mode it is a logic state analyzer. In the timing mode the input waveforms are sampled to determine whether

FIGURE 5-1
Block diagram of single-channel logic analyzer

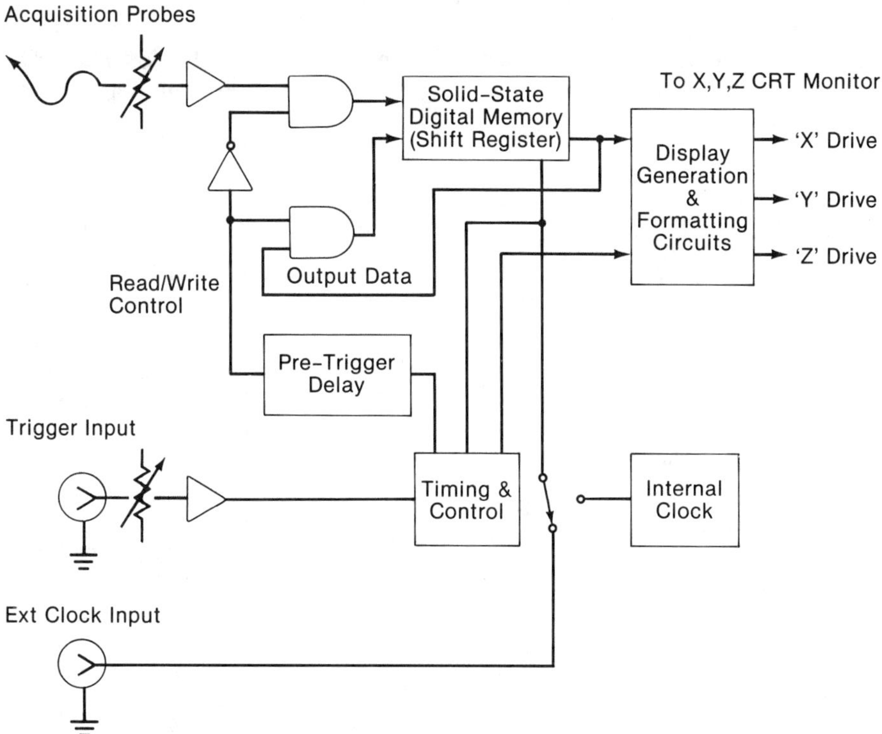

they are high or low. Acting like a digitizing oscilloscope with only one bit of vertical resolution, the logic analyzer can display only two states—high and low.

Signals above threshold are displayed as 1 or high, while signals below threshold are displayed as 0 or low. The upper display in Figure 5-2 illustrates timing output of a logic analyzer. Since digital information takes only these values, the logic analyzer provides the output required for an examination of channels in the digital domain. Figure 5-2 shows the two types of output that can be obtained from this instrument.

The essential digital information is displayed, but the actual voltage values and their waveshapes are not shown, leading some to call them "false waveforms." When one needs to view the actual amplitudes and waveforms of the signals, the logic analyzer is not the proper instrument to use. Such data can be seen and analyzed through the use of an oscilloscope. Figure 5-3 shows a block diagram of the various steps in the development of microprocessor-based systems. The dotted lines surround those sections where the oscilloscope and the logic analyzer have their application domains. They are seen to overlap in hardware development. In the hardware domain of the logic analyzer the timing display mode of the logic analyzer has its principal application, and in the software domain the state display mode has its usefulness. Figure 5-4 shows various parameters of pulse analysis, which can be provided by oscilloscopes or by an analyzer like Hewlett-Packard's Model 16500. This logic analyzer provides for analog and digital analysis as well as timing and state analysis.

FIGURE 5-2
Block diagram of logic analyzer sampling. Reprinted with permission of AEU, Dempa Publications, Inc.

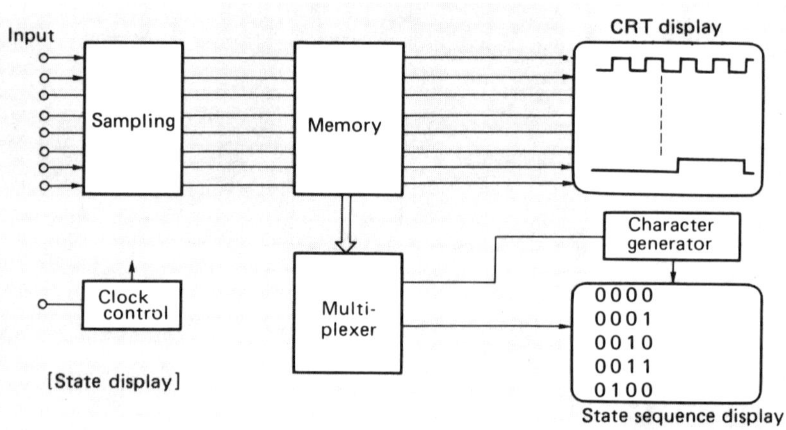

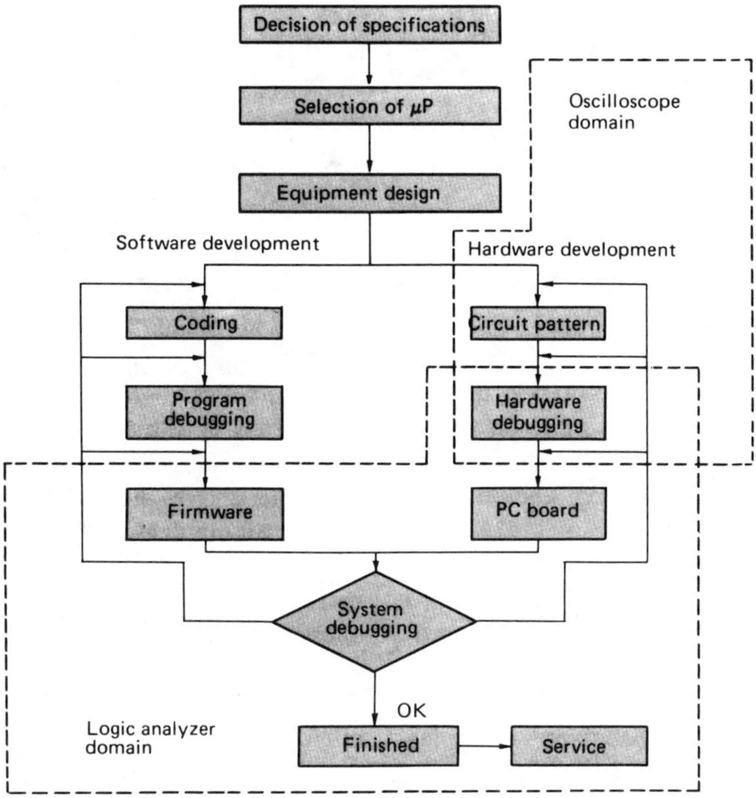

FIGURE 5-3
The structure of microprocessor-based development showing domains of the logic analyzer and the oscilloscope. Reprinted with permission of AEU, Dempa Publications, Inc.

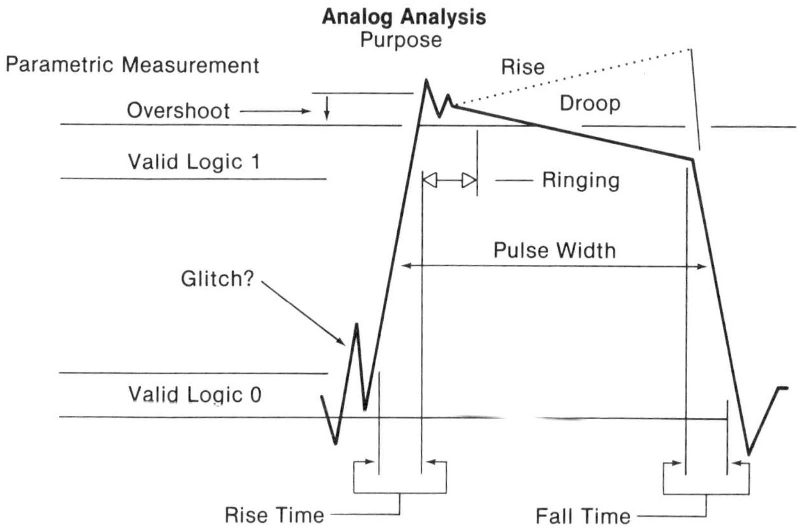

FIGURE 5-4
Parameters of pulse analysis provided by oscilloscopes

Timing Analyzers

The logic timing analyzer was the answer to the limitations of a multichannel storage oscilloscope in finding glitches (short noise pulses that upset digital operation) and in checking signal timing. The scope was unable to display signals with extremely low recurrence rates, as is the case with glitches, and it could not provide a backward look in time. Logic timing analyzers overcome both of these limitations. It samples the incoming data and can keep track of any transition that crosses logic threshold more than once between samples. By keeping track of all multiple transitions, it can display glitches. In addition, due to its storage capacity, it can not only trigger on a glitch but also display data that occurred before it. This information can make it much easier to determine the cause of the glitch.

Timing logic analyzers provide the only effective and efficient means of capturing timing information in a digital system. With such an analyzer the signal activity taking place at a given time at a number of different points within a system may be viewed and the time between logic events may be measured.

The usefulness of a timing analyzer in hardware design can be appreciated by considering the areas where it can be applied in a system bus shown in Figure 5-5. The control bus is composed of lines that most frequently operate asynchronously, making it desirable to view them for a continuous period of time. The logic timing analyzer with its own internal

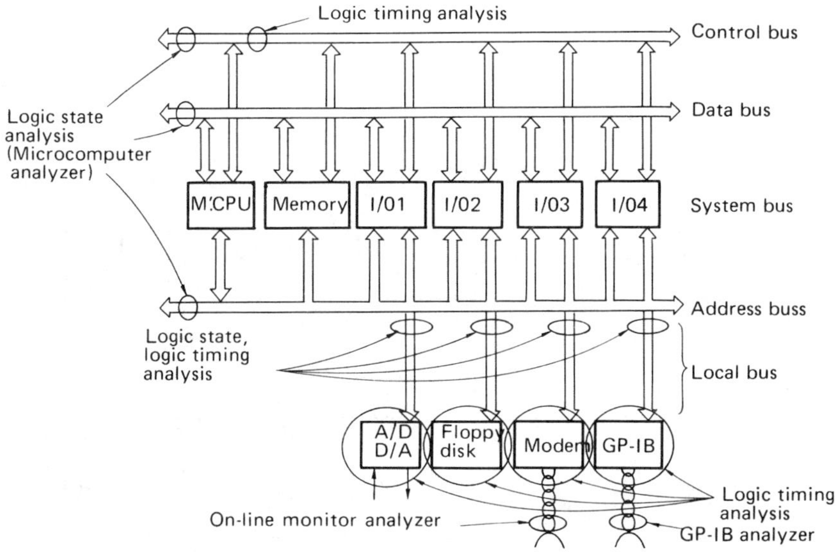

FIGURE 5-5
Points of state and timing analysis in a microprocessor-based system. Reprinted with permission of AEU, Dempa Publications, Inc.

clock is well suited for this asynchronous operation. Analyzing such operating requires the ability to trigger on unique events. Control line signals may be transitional in nature for only a few nanoseconds, like narrow intermittent spikes caused by crosstalk or noise. To capture such phenomena, the hardware engineer needs a logic timing analyzer that not only recognizes these events but that triggers reliably when they appear.

When the control bus is operating properly, the hardware engineer can perform further tests on the hardware relative to its response to the processor's modes of addressing and data fetching and its execution of instructions. To do so, she or he can generate some simple code and put it in a programmable read-only memory that plugs into the system. At this point the interest is in the actual timing of signals as in the flow of data in the system, like the transfer of information versus strobe signals such as memory enables or read/write pulses. However, at this point the engineer has entered the logic state domain, where it is better to read data in machine code. Once the hardware engineer is satisfied that the system is operating properly, the software designer can enter the picture and the integration phase can begin.

Figure 5-6 shows microprocessor control line relationships in a timing analysis example provided by Hewlett-Packard's Model 16500.

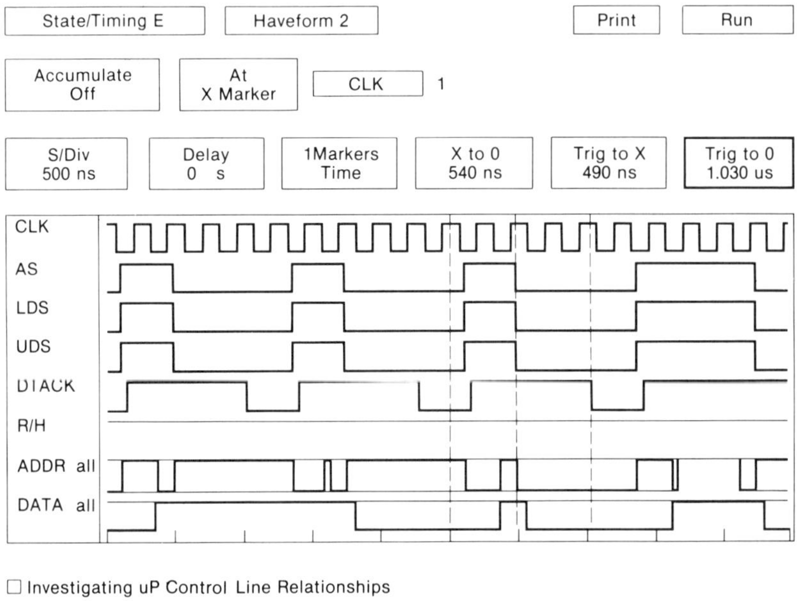

FIGURE 5-6
An application example of timing analysis

State Analyzers

State analyzers are used to trace the execution of instructions through a processor system. Data, address, and status codes are captured and displayed as they occur on the corresponding buses. A state for a logic circuit is a sample of a bus or line when it is valid. Data is displayed in list format because the software designer is not particularly interested in waveforms. No internal clock is used, and the operation is asynchronous depending upon the clocks in the system being tested. Thus logic state analysis is more applicable to the address and data buses, as shown in Figure 5-7.

As in the case of timing analyzers, logic analyzers have special areas where they fit in testing a system. Some of these are selective sampling and storage, and disassembly. One can trigger on a specific address in the RAM on a chip or board and display the data stored there. Selective storage permits the storing of only specific routines of an entire program. Segments where problems exist may be stored and examined in detail. Most logic analyzers provide a software package that acts as an inverse assembler to translate hexadecimal instruction codes into mnemonic or assembly code. With mnemonic code the instructions can usually be interpreted more readily than when it is presented in hex code.

Expanded state-analysis capabilities, which have been provided in

State/Timing E | Listing 1 | Print | Run

Markers Off

Label>	ADDR	DATA	STAT	Time
Base>	Hex	Hex	Hex	Relative
0	008930	B03C	30	1.28 us
1	008932	00FF	30	1.24 us
2	008934	67F8	30	1.24 us
3	008936	B03C	30	1.24 us
4	00892E	61FA	30	1.76 us
5	008930	B03C	30	1.24 us
6	0004F4	0000	29	2.00 us
7	0004F6	8930	29	1.48 us
8	00892A	4EFA	30	1.28 us
9	00892C	FF9A	30	1.24 us
10	0088C6	61E6	30	1.72 us
11	0088C8	B03C	30	1.28 us
12	0004F0	0000	29	2.00 us
13	0004F2	88C8	29	1.48 us
14	0088AE	103C	30	1.24 us
15	008880	00FF	30	1.24 us

☐ Monitoring Microprocessor Code Execution

FIGURE 5-7
An application example of state analysis

logic analyzers of enhanced design, have given both software writers and hardware designers significant new features. One of these relates to program-trace. Commands can be actuated to allow one to tell the analyzer which parts of a software routine to follow and which portions to capture. One can instruct the analyzer to monitor parts of a program's flow without capturing data, to follow a program's loop for a given number of iterations, to capture multiple windows of data, and to perform other conditional operations. Earlier analyzers had only straight-line sequential triggering.

Many advanced analyzers allow one to assign symbolic labels in the trace-sequence setup menus. In this way the same terms can be used for programs and logic analyzer setups without reverting to raw hex numbers or binary bit patterns, which can be hard to remember. Some of the newer analyzers also provide some form of off-line memory onto which you can save and restore elaborate setups. In Figure 5-7 microprocessor code execution is monitored in the state analysis example provided by Hewlett-Packard's Model 16500.

Hardware and Software Debugging

Gaining access to software information can be invaluable to hardware designers in the hardware/software integration phase. However, digital hardware still centers on signal timing, and one of the primary purposes of a logic analyzer is to resolve timing problems. Therefore, it is crucial that the instrument provide timing and state information simultaneously, but independently. To understand the need for such independence requires an understanding of the difference between hardware and software troubleshooting.

One such difference lies in the speeds at which one needs to collect state and timing information. State information deals with instructions that occur every few clock cycles, while timing information depends on events—such as glitches—that occur between clocks. If one collects state and timing information at the same speed to track the software slowly enough to follow a few jump commands, it would not be fast enough to resolve critical timing relationships. If one collects timing data at a higher speed to capture glitches and collects software data at the same speed, one would probably run out of analyzer memory before many software instructions had been traced through. Thus a logic analyzer suitable for hardware/software debugging must be able to trigger in two ways. It must recognize a high-speed timing event and capture lower-speed program activity in the vicinity of the high-speed event. It must also recognize a lower-speed program-related event and capture high-speed hardware activity in the vicinity of the lower-speed event.

Hardware/software integration is undoubtedly the most difficult phase of product development. There are approaches to assist in distinguishing between hardware and software problems. One approach is to use state analysis to examine the software first to see what the hardware should be doing. Another technique is to start the integration process with simple test software. This approach allows one to make sure that the hardware works correctly before the situation is complicated by the implication of the complete software package. The fewer routines one has to deal with, the easier it is to trigger an analyzer on a section of code that appears to be causing problems.

Although software debugging has gotten the reputation of being the area in which most of the work is required, logic analyzer vendors have found an increasing demand for timing analyzers, which can benefit the hardware designer by keeping pace with higher and higher logic speeds. Very fast complex circuits, based on 32-bit microprocessors and ECL devices, have emphasized the need for faster timing analyzers with more data channels and deeper memories. To a logic analyzer vendor, a glitch is any signal shorter than the analyzer's sample period. In these terms, a glitch is any even number of signal transitions between two of the logic analyzer's sample points. Short glitches can go undetected by analyzers that do not provide special glitch-capture circuitry.

Because A/D (analog/digital) converters in logic analyzers are not fast enough to capture short glitches, many logic analyzers employ latches on the inputs to detect and store any short glitches. One analyzer stores the glitch data in its memory by causing the logic state at the time of the next sample point to switch to the opposite level for one period and then switch back to its original state. However, with this technique the logic analyzer does not display any glitch that occurs just before a valid logic-level transition takes place. To overcome this limitation, some analyzers store the captured glitches in a second memory. After the data capture is completed, the analyzer merges the two sets of data and displays them. This two-memory method of storing glitch information is found in Hewlett-Packard's Model 1630/31 logic analyzers. Because the glitch information is stored in a second memory, it can also be highlighted on the display screen. In this model the second memory is provided without adding memory chips. Instead it is borrowed from adjacent channels. Unfortunately, with this method half of the machine's data-acquisition channels are given up, when the analyzer is in the glitch-capture mode. Using another approach, some Tektronix analyzers give up half of each channel's memory depth to store glitches.

To store glitches by either method one faces a trade-off. If glitch data is stored in the main memory, one has full use of all of the measurement channels, but the analyzer will miss glitches that occur near a valid signal transition and one must find the glitches in the display without any help from the analyzer. If the analyzer stores glitch data in a second memory, all of the captured glitches can be separately spotted, including those next to a signal transition. However, only half of the probes catch glitches in some analyzers, so one has to be aware of this limitation and connect the correct probes to those parts of the circuit where glitches are expected to be found.

Glitch-capture circuitry is useful for detecting short glitches on a single data channel, but it cannot be used to resolve high-speed timing events that occur over two different channels. Such events can only be resolved by sampling at a faster rate. However, once again there is a limitation. The faster sampling rate limits the size of the window that can be stored. In some fast analyzers this window drops under 1 μs. More high-speed memory can be added to increase the length of the window, but the price of the analyzer rises dramatically.

Monitors, Debuggers, and Preprocessors

AT&T's WE312DM Device Monitor

The WE321DM CPU & MMU Device Monitor provided a low-cost solution to debugging the 32100 microprocessor and 32101 MMU based on applications operating at frequencies up to 18 MHz, with a logic analyzer that was limited to a maximum sample rate of 10 MHz. The device monitor, whose configuration is shown in Figure 5-8, supported zero wait-state block fetch accesses up to its maximum frequency. This monitor consisted of a

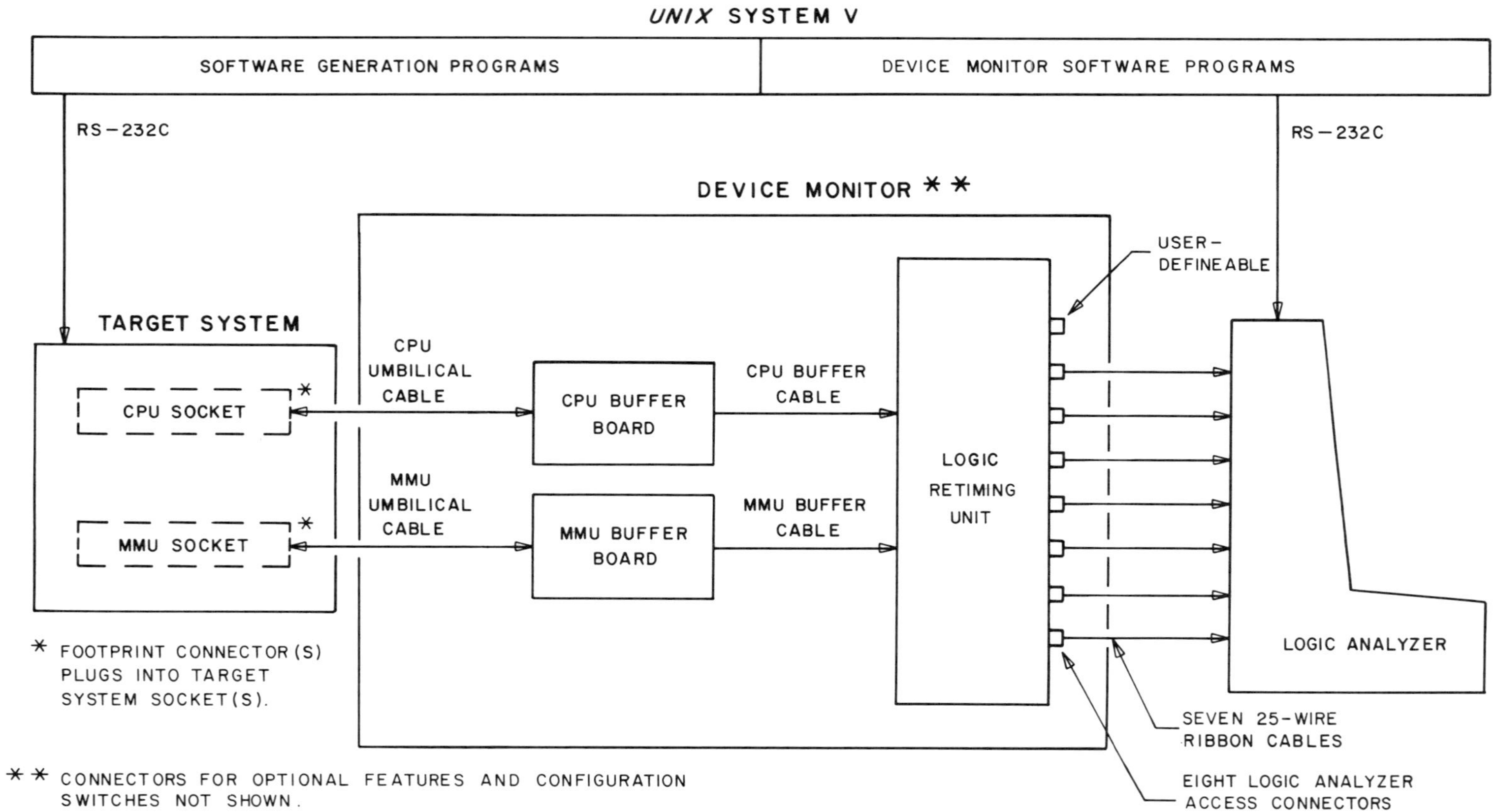

FIGURE 5-8 Configuration showing application of the device monitor. Reproduced with permission © 1986 AT&T.

logic retiming unit, a CPU cable system, an MMU cable system, and seven 25-wire ribbon cables that provided a standard interface to a logic analyzer. The CPU and the MMU cable systems contained footprint connectors that replaced the CPU and the MMU devices in the target system. However, this device monitor could be used regardless of whether the target system contains an MMU. The footprint connectors plugged directly into the CPU and MMU sockets in the target system. The footprint connectors were linked via 6-inch umbilical cables to two buffer boards (one for the CPU and one for the MMU). The device(s) replaced in the target system by the footprint connector(s) plugged directly into zero-insertion-force sockets on the buffer board(s). The buffer boards connected via 12.5-inch cables to a logic retiming unit. The overall length of each cable was 24.5 inches.

This device monitor allowed signal observation of high-speed target systems with a logic analyzer that was limited to 10-MHz sampling rates by the following actions:

Generating logic analyzer sample points based on bus cycle activity and guaranteeing a minimum of 100 ns between sample points at target system operating frequencies as high as 18 MHz

Sampling and holding signals as necessary to assure 30-ns signal setup and 0-ns signal hold times with respect to selected sampling points and a target system operating frequency up to 18 MHz

Providing the flexibility of observing a narrow window of bus activity with reduced detail but with a greater perspective of system bus activity

A Hewlett-Packard 64000 Logic Development System was recommended by AT&T for signal observation.

UNIX System V support could be obtained from at least five different computers: AT&T's 3B2/300, 3B2/310, and 3B2/400, the WE 321SB VMEbus Single-Board Computer, and the VAX 11/780 computer. WE321DM-SP Device Monitor software supplied with the device monitor provided symbolic tracing and disassembly with the 64000 system. This software consisted of three program trace disassemblers, three configuration files, 64000 system upload and download utilities, a symbol table utility, and a communication file for communicating with the host UNIX system from the 64000 system console. The program trace disassemblers allowed hardware bus traces obtained by using the 64000 system to be displayed in the 32100 microprocessor assembly language format. The configuration files assigned descriptive names to the signals available at the logic analyzer access connectors, providing easy signal recognition when performing signal observation. The upload utility provided the transfer of files from the 64000 system to the host UNIX system, and the download utility provided symbolic defined downloading from the host UNIX system to the 64000 system for symbolic debugging. The symbol table utility translated object file symbolic information for use by the 64000 system for symbolic debugging.

The logic analyzer of the Hewlett-Packard 64000 Logic Development System was recommended for signal observation. This system made full use of the development features provided by the device monitor. Other logic analyzers could be used by connecting user-supplied adapter cables between the state data and clock pods of the logic analyzer and the logic analyzer access connectors on the logic retiming unit of the device monitor.

Operational Modes

The device monitor generated various signal sampling points for debugging target systems operating at frequencies greater than 10 MHz with a logic analyzer limited to a 10-MHz sample rate. For example, the 64000 system required a minimum of 100 ns between sample points and a minimum of 30 ns signal setup time. The falling edge of the target system clock could be used as the sample clock for accurate data sampling of all CPU and MMU signals at target system operating frequencies of 10 MHz or less but was not used at operating frequencies above 10 MHz. For frequencies greater than 10 MHz, the sample points generated by the device monitor were used to ensure 100 ns between sample points.

Users initially debugged their target systems at 10 MHz using the transparent mode and the falling edge of the target system clock for data sampling. This facilitated debugging, since all signals could be viewed and data samples could be taken every clock cycle. Once the target system was functional at 10 MHz, the user could switch to the latched mode and use sample points generated by the device monitor for debugging at higher target system operating frequencies.

The two switch-selectable modes of operation, transparent and latched modes, are described below.

Transparent Mode

Although the transparent mode was for target systems operating at frequencies less than or equal to 10 MHz, greater operating frequencies could be used. In this mode the signals were not latched and some did not meet the 30-ns setup time required by a 10-MHz logic analyzer. The user had to determine which signals met this time constraint. In addition, the disassembler did not work above 10 MHz in the transparent mode. In this mode the falling edge of the target sample clock was used, and all CPU and MMU signals went through at least one buffer.

Latched Mode

The latched mode was used for target systems operating at frequencies from 10 MHz to 18 MHz. Latching of some signals was necessary because after they were asserted some signals might be negated before they could be sampled. The latch guaranteed that these signals would be held long enough to ensure accurate data sampling. The latched signals were cleared after sampling.

Software Description

The software support for the device monitor was provided by its own special programs and optional Software Generation Programs (SGP). SGP had to be installed on the host UNIX system prior to debugging with the device monitor, if it was to be used. The SGP provided utility programs that enabled programmers to write application software in C or assembly language and then translate it into machine code before the programs were downloaded to the target system. The software supplied with the device monitor consisted of three program trace assemblers, three 64000 system configuration command files, and a terminal setup file, all of which was downloaded to the 64000 system. In addition, a symbol table utility and download and upload utilities were provided on the host UNIX system.

The program trace disassemblers could display 64000 system logic traces in the microprocessor instruction format, allowing the user to employ the mnemonics for debugging application programs. The configuration files provided symbolic definition of CPU and MMU signal names, user program locations, and measurement experiment specifications for the 64000 system. The terminal setup file configured the 64000 system console for communication with the host UNIX system. The symbol table utility took the relocated object code load module (created with SGP) and generated a 64000 system command file. This file was then downloaded to the 64000 system to create a symbol map. The 64000 system used this symbol map to replace raw addresses in trace lists with symbolic names from the source code. Download and upload utilities were provided for program transfers between the host UNIX system and the 64000 system for downloading files to the 64000 system and for uploading measurement data from the 64000 system to the UNIX system.

The software for the device monitor was provided on magnetic tape or floppy disk depending upon the UNIX system host being used. A release description and installation guide for loading software onto a UNIX system host was supplied with the package.

Logic Analyzer Configuration

To best support the device monitor in providing a full set of hardware and software observation features, the Hewlett-Packard 64000 Logic Development System required the following configuration:

120 channels of state analysis with optional overview capability

16 channels of timing analysis

Dual flexible disk drive and system hard disk

120 channels of state analysis allowed the user to monitor all footprint signals of the microprocessor, the memory management unit, and the clock. An overview system, which provided for enhanced software performance measurement, monitored the 100 channels most useful for software analysis.

The 100 CPU and MMU signals most useful for software state analysis were grouped in the first five logic analysis access connectors for support of both the 100- and 120-channel configurations.

If the 64000 system timing subsystem was used in conjunction with the state analyzer, the user had to connect any of the 16 timing channel probes to the target system in the conventional manner.

As part of the 64000 system, the user received an operating system that included the Hewlett-Packard file system and editor. The operating system allowed the 64000 system to communicate with the host UNIX operating system. This feature was required for downloading configuration and symbolic debugging information and uploading measurement data.

The hard-disk subsystem was necessary to obtain reasonable command response time and to facilitate storage of trace and configuration data. Up to six 64000 Logic Development System stations could share the same hard disk.

Signal Observation

The 64000 system provided for the observation of bus activity by means of state and disassembly traces. The state trace provided a listing of the logic state of the captured bus activity. The disassembly trace provided a descriptive listing of the CPU's internal activity as well as the bus operations performed. The CPU's internal activity and bus activity were determined based on the logic state information contained in the captured bus activity. This internal activity was disassembled into the microprocessor's instruction format to indicate the instruction the CPU had been executing. Bus operations are described in words providing easy recognition of the bus operation being performed.

In order for the 64000 system to sample and store data, the user had to define trigger conditions (when data was to be saved).

Symbolic debugging was performed using the symbol table utility software. This utility resided on the host UNIX system and was used to generate a 64000 system command file, which, when downloaded to the 64000 system, could be used for symbolic debugging.

A comparison of the symbolic trace list output with a trace list captured using single sample mode is shown in Figures 5-9 and 5-10. A symbolic trace list including the STACK variable is shown in Figure 5-11.

64620 State Analyzer

	Trace List			State 1,120 channel, general probes				
Label:	ADDR		DATA	R1W0	VAD0	SRDY0	DRDY0	SAS
Base:	hex		hex	hex	hex	hex	hex	hex
Map:	ADDR_MAP							SAS_MAP
trigger	abs	00100000	10499C4F	1	1	0	0	IFPCD
+001	abs	00100004	0C000000	1	1	0	0	PREFTCH
+002	abs	0010F000	0000E2A5	0	1	0	0	WRITE
+003	abs	0010F008	0010F000	0	1	0	0	WRITE
+004	abs	00100008	4C840559	1	1	0	0	FETCH
+005	abs	0010000C	840964A0	1	1	0	0	PREFTCH
+006	abs	0010F024	00000005	0	1	0	0	WRITE
+007	abs	0010F028	00000009	0	1	0	0	WRITE
+008	abs	00100010	59A0642C	1	1	0	0	FETCH
+009	abs	00100014	CCF8AF0D	1	1	0	0	PREFTCH
+010	abs	0010F024	00000005	1	1	0	0	OPRFTCH
+011	abs	0010F030	00000005	0	1	0	0	WRITE
+012	abs	0010F028	00000009	1	1	0	0	OPRFTCH
+013	abs	0010F034	00000009	0	1	0	0	WRITE
+014	abs	0010F03C	0010F000	0	1	0	0	WRITE
+015	abs	00100018	00844068	1	1	0	0	FETCH
+016	abs	0010001C	18490870	1	1	0	0	PREFTCH
+017	abs	00100020	10499C4F	1	1	0	0	IFPCD
+018	abs	00100024	04000000	1	1	0	0	PREFTCH
+019	abs	0010F038	00100019	0	1	0	0	WRITE
+020	abs	0010F040	0010F024	0	1	0	0	WRITE

FIGURE 5-9
Trace list captured using single sample mode. Reproduced with permission © 1986 AT&T.

Before using the device monitor/logic analyzer combination in actual product development, the user could be familiarized with it and the capabilities of the device monitor software could be demonstrated by applying them to a microprocessor evaluation board and a VMEbus Single-Board Computer. With the evaluation board the capability of observing bus activity of the physical mode processes could be demonstrated, while with the SBC the capability for signal observation of virtual mode processes could be demonstrated.

In-Circuit Debugger for Intel's 80486

The ICD-486 in-circuit debugger provides real-time hardware and software debug capabilities for i486 microprocessor-based designs. Since the ICD-486 debug hardware resides on a printed circuit board that plugs directly

64620 State Analyzer

	Trace List		State	1,120 channel,	general probes		
Label:	ADDR	DATA	R1W0	VAD0	SRDY0	DRDY0	SAS
Base:	hex	hex	hex	hex	hex	hex	hex
Map:	ADDR_MAP						SAS_MAP
trigger	Tmain+00000000	10499C4F	1	1	0	0	IFPCD
+001	Tmain+00000004	0C000000	1	1	0	0	PREFTCH
+002	abs 0010F000	0000E2A5	0	1	0	0	WRITE
+003	abs 0010F008	0010F000	0	1	0	0	WRITE
+004	Tmain+00000008	4C840559	1	1	0	0	FETCH
+005	Tmain+0000000C	840964A0	1	1	0	0	PREFTCH
+006	abs 0010F024	00000005	0	1	0	0	WRITE
+007	abs 0010F028	00000009	0	1	0	0	WRITE
+008	Tmain+00000010	59A0642C	1	1	0	0	FETCH
+009	Tmain+00000014	CCF8AF0D	1	1	0	0	PREFTCH
+010	abs 0010F024	00000005	1	1	0	0	OPRFTCH
+011	abs 0010F030	00000005	0	1	0	0	WRITE
+012	abs 0010F028	00000009	1	1	0	0	OPRFTCH
+013	abs 0010F034	00000009	0	1	0	0	WRITE
+014	abs 0010F03C	0010F000	0	1	0	0	WRITE
+015	Tmain+00000018	00844068	1	1	0	0	FETCH
+016	Tmain+0000001C	18490870	1	1	0	0	PREFTCH
+017	Tadd+00000000	10499C4F	1	1	0	0	IFPCD
+018	Tadd+00000004	04000000	1	1	0	0	PREFTCH
+019	abs 0010F038	00100019	0	1	0	0	WRITE
+020	abs 0010F040	0010F024	0	1	0	0	WRITE

FIGURE 5-10
Symbolic trace list output. Reproduced with permission © 1986 AT&T.

into the target system, the user can run at the full speed of the component (see Figure 5-12).

The ICD-486 allows users to debug applications that utilize the on-chip caching capabilities of the i486 microprocessor.

ICD-486 can be used by both hardware and real-time software developers, at any stage of development. Early in the development process, ICD-486 allows prototype development and software debugging. Later in the design cycle, the ICD-486 can be used when integrating hardware and software modules.

With symbolic debugging, memory locations can be examined or modified using symbolic references to the original program, such as a procedure or variable name, line number, or program label. Microprocessor data structures, such as registers, descriptor tables, and page tables, can also be ex-

64620 State Analyzer

	Trace List	State 1,120 channel, general probes					
Label:	ADDR	DATA	R1W0	VAD0	SRDY0	DRDY0	SAS
Base:	hex	hex	hex	hex	hex	hex	hex
Map:	ADDR_MAP						SAS_MAP
trigger	Tmain+00000000	10499C4F	1	1	0	0	IFPCD
+001	Tmain+00000004	0C000000	1	1	0	0	PREFTCH
+002	STACK+00000000	0000E2A5	0	1	0	0	WRITE
+003	STACK+00000008	0010F000	0	1	0	0	WRITE
+004	Tmain+00000008	4C840559	1	1	0	0	FETCH
+005	Tmain+0000000C	840964A0	1	1	0	0	PREFTCH
+006	STACK+00000024	00000005	0	1	0	0	WRITE
+007	STACK+00000028	00000009	0	1	0	0	WRITE
+008	Tmain+00000010	59A0642C	1	1	0	0	FETCH
+009	Tmain+00000014	CCF8AF0D	1	1	0	0	PREFTCH
+010	STACK+00000024	00000005	1	1	0	0	OPRFTCH
+011	STACK+00000030	00000005	0	1	0	0	WRITE
+012	STACK+00000028	00000009	1	1	0	0	OPRFTCH
+013	STACK+00000034	00000009	0	1	0	0	WRITE
+014	STACK+0000003C	0010F000	0	1	0	0	WRITE
+015	Tmain+00000018	00844068	1	1	0	0	FETCH
+016	Tmain+0000001C	18490870	1	1	0	0	PREFTCH
+017	Tadd+00000000	10499C4F	1	1	0	0	IFPCD
+018	Tadd+00000004	04000000	1	1	0	0	PREFTCH
+019	STACK+00000038	00100019	0	1	0	0	WRITE
+020	STACK+00000040	0010F024	0	1	0	0	WRITE

FIGURE 5-11
Symbolic trace list including STACK variable. Reproduced with permission © 1986 AT&T.

amined and modified using symbolic names. Optimal symbolic debugging can be achieved when using the ICD-486 with Intel languages, which produce not only address and memory locations but also variable and procedure typing information.

For advanced hardware debugging, the ICD-486 has been designed to work with high-speed logic analyzers. The ICD-486 is shipped with a logic analyzer interface board that provides the control signals to trigger the logic analyzer. With the logic analyzer, the ICD-486 can be used to monitor or recognize bus activity and to gather execution traces. A block diagram of the ICD-486 is shown in Figure 5-13.

Unfortunately, the cable length and buffering can contribute timing delays that make it difficult to run the microprocessor at full speed. Intel designers overcame this problem by literally turning the ICD into the pod;

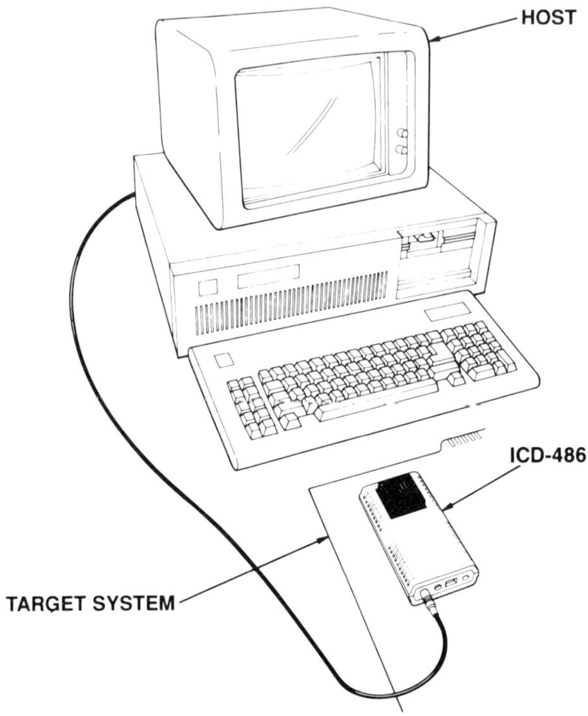

FIGURE 5-12
ICD-486 in-circuit debugger with host computer.
Reprinted courtesy of Intel Corporation.

the ICD plugs directly into the board. This was accomplished by keeping the ICD's size to 3.5 × 6.0 × 0.5 inches (0.85-inch deep with the plug for the 486).

The single, eight-layer ICD-486 board contains a 486 (to replace the target processor) and an 80C188 as an ICD controller. Additionally, it uses a dense field-programmable gate array to hold the majority of control logic. The test unit relies on special bond-outs to monitor the 486 internal states and pipelines; these signals feed the control logic for logic-analyzer triggers and timer controls.

Up to four hardware breakpoints can be set, using the 486's hardware-test debug registers. Additionally, designers can specify up to 16 software breakpoints on execution address or data-read or data-access addresses. Intel has taken some criticism for the limited number of hardware and software breakpoints; some advanced ICEs offer far greater numbers, with sophisticated, multilevel triggers.

With the ICD, internal registers can be displayed and modified from

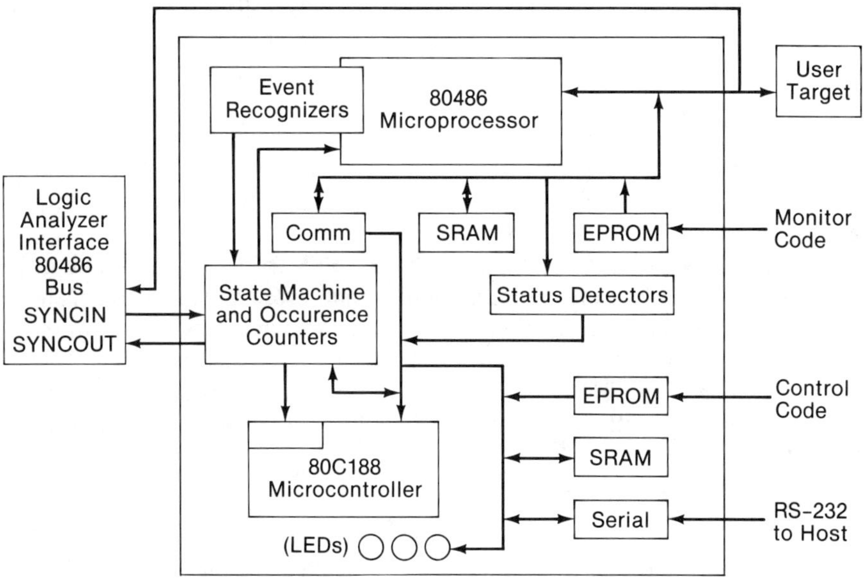

FIGURE 5-13
Block diagram of ICD-486 in-circuit debugger. Reprinted courtesty of *EE Times*.

the host processor controlling the ICD. Target-system status is also monitored and displayed on the host: flags are set for execution and for bus and clock inactivity.

Additionally, the ICD supports analysis of the 486's internal microprocessor data structures. Designers can examine and change descriptor tables and page tables using symbolic names. The ICD supports symbolic, source-level debugging with C, ASM-386/486, FORTRAN, and PL/M.

Intel supplies a set of software utilities (RLL-386/486) and a software debugger (DB-386/486) for the 486 target that runs on a 386 or 386-SX-based development host.

Engineers intending to use the ICD-486 should design their circuits to handle the ICD's additional loading, which plugs directly into the address and data buses. Doing so will add up to 30 picofarads (pF) to the clock signals and up to 20 pF to all other 486 chip signals.

The ICD-486 is powered by the target system through the chip power and ground pins and will draw up to 1.3 A more than the target 486.

A host development system—an IBM PC AT (or compatible) or PS/2—controls the ICD for test via an RS-232 serial line. The serial port won't slow down the test, since the bulk of the test data—bus traces—are held in the logic analyzer and don't pass down the serial path.

A memory board that piggybacks onto the ICE-like plugs to the target 486 is provided. The board can be used to provide emulation memory—changeable memory substituted for the target board's nonvolatile memory (EPROM, ROM).

Also, a stand-alone soft-test box comes with the ICD-486 to ensure that if problems develop with the ICD, they will not complicate the target test. Nothing is more self-defeating than a combination of faulty test equipment and unstable target hardware.

An isolation board isolates the ICD circuitry from dangerous, aberrant target-circuit behavior during initial debug and protects the ICD from shorts and grounds. The isolation board introduces a 5-ns delay (with a 50-pF load) on the address and data buses, as well as byte enables and DS#, W/R#, MI/O#, Blast#, and HLDA signals. Other signals are passed through the board but are not buffered and have an additional capacitive loading (up to 10 pF).

Preprocessors

Another approach to connect logic analyzers to the target system or to system buses is to use preprocessors. If the logic analyzer is to be linked to the target system, the preprocessor is plugged into the CPU socket, after its microprocessor is removed. In this case, the preprocessor is tailored to the microprocessor it replaces. Bus preprocessors, such as Hewlett-Packard's 10342B, provide analysis capabilities for three popular interface buses, HP-IB (Hewlett-Packard's implementation of the IEEE-488-1978), RS-232C/CCITT1 V.24, or RS-449. A package gives convenient and reliable access to the signals on the bus, provides the software necessary to set up the logic analyzer for the desired measurement, and gives a complete view of the asserted control lines, as well as conversion of the captured data into mnemonics, which one can easily understand.

A Source Code Debugger in C

The increasing role of the logic analyzer as part of the development environment is seen to considerable effect in, for example, the case where it is necessary to debug high-level source code such as C or Ada. A typical problem would be when an embedded processor is running synchronously but it is affecting an asynchronous activity such as the output of a serial port of a timed loop.

A solution to this problem from Tektronix is the TekDB, a C source code debugger that consists of a host-based, windowed (up to four windows for: Source Code, Stack Registers, Memory, Variables) source-level debugging environment. It is used in conjunction with about 4K of target system EPROM monitor that is personalized to the application. The host system and the TekDB debugger talk to the in-system TekDB monitor through an

RS-232 serial port. In operation the target system operates synchronously until the logic analyzer is triggered by seeing corrupted data or the failure of an event to occur in a given time. The trigger is applied to the NMI (nonmaskable interrupt) part of the processor; this stops the processor and the operating source code is displayed in one window on the host. Using all the window source code, registers and memory can be examined and interactively changed via the monitor to detect the true source of the error. This technique allows the dynamic information within an embedded system to be examined, whereas under more normal circumstances only the static code and variables relating to program execution are available.

Logic Analyzer Application to RISC

The Sun SPARC, the MIPS R3000, and the Motorola 88000 RISC microprocessors have not been supported by high-speed emulation that operates transparently. It has been very difficult to develop emulators that do not impose severe restrictions on the microprocessor's performance. A different approach to debugging tools for RISC machines has been necessary. Only high-performance logic analyzers with customized interface and disassemblers for a given processor have been able to examine the operation of high-speed machines at full speed.

One of the reasons in-circuit emulators have found it difficult to operate with RISC processors is due to their heavy use of pipelining. As an example, the MIPS R3000 processor has a five-stage pipeline and, in general, each stage is simultaneously executing a different instruction. The five pipeline stages are: Instruction Fetch (I), Register Fetch (R), ALU (A), Memory Access (M), and Write-back (W). These five successive instructions in the R3000 pipeline are shown in Figure 5-14. For a given instruction, the instruction itself is present on the data bus during phase 1 of the register phase

FIGURE 5-14
Successive instructions in the R3000 pipeline

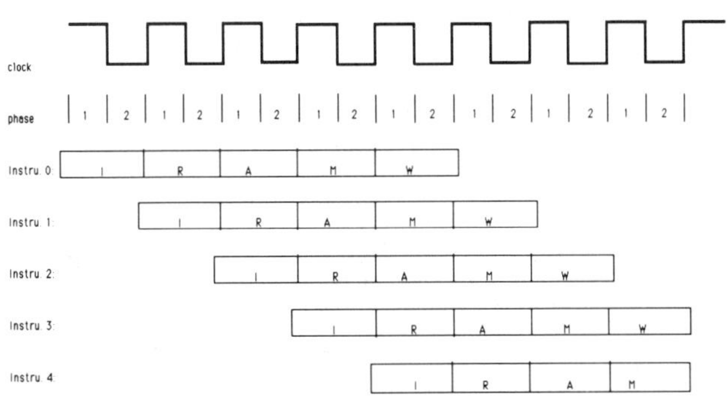

stage of the pipeline, and the data instruction, if any, occurs during phase 2 of the memory access pipeline stage. The data transaction could be Load or Store instructions. The bus transactions relative to the processor pipeline are shown in Figure 5-15. I-1 represents the instruction prior to the "0"th instruction shown as I0. The clock period is 40 ns.

The bus transactions when the pipeline is full appear as shown in Figure 5-16. It can be noted that at every phase (at 25 MHz, 20 ns) both instruction and data cache accesses take place. Figure 5-16 makes it clear why a traditional emulator will have difficulty analyzing the instructions. The 40-ns time slot available is not sufficient to accomplish the following:

a. Send signals through interface buffer delays, memory map labels, cables, etc.
b. Rearrange captured data to detect instruction boundaries.
c. Access emulator memory.
d. Store data in trace memory.

It should be noted that the Loads/Stores occur two cycles after the instruction is fetched. For example, D0 and I0 when the data and instructions of the instruction 0 are fetched on the data bus in Figure 5-16. These traces must be rearranged to obtain the correct information about program execution.

FIGURE 5-15
Bus transactions relative to the R3000 pipeline

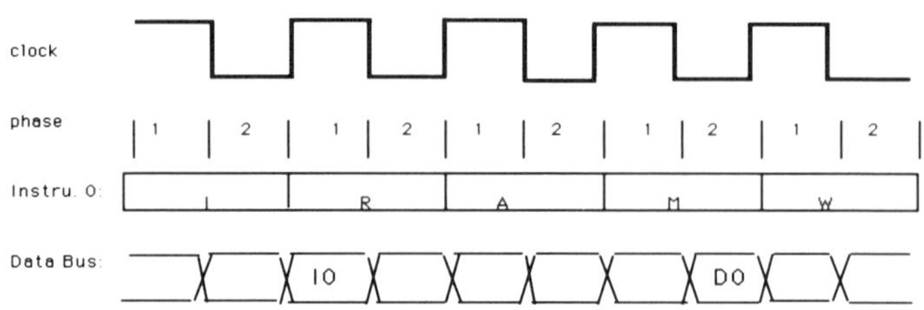

FIGURE 5-16
Bus transactions with a full R3000 pipeline

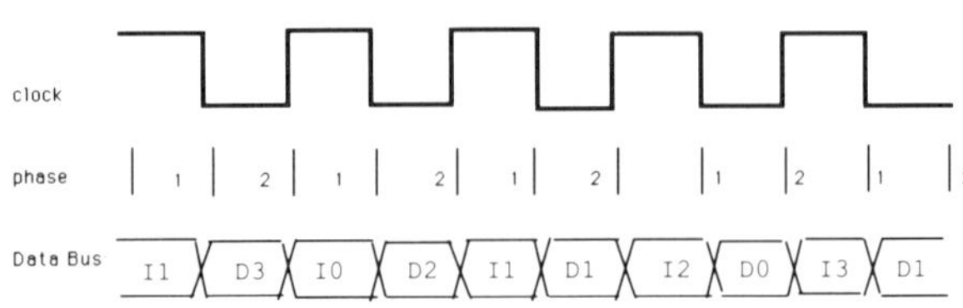

High-speed logic analyzers can be very effectively used to identify and solve problems if a customized probe for the processor and disassembler software is provided. As an example, the Biomation CLAS 4000 logic analyzer has been used to capture data on the MIPS R3000 even with 20-ns bus cycles, as shown in Figure 5-16. Here the probe was electronically passive and did not disrupt the operation of the microprocessor. It was also small in size, allowing connection to the R3000 microprocessor even in crowded designs. The greatest advantage of such tools is their flexibility in data capture and triggering. They can be used in conjunction with the monitor program's breakpoint capabilities to capture the execution of specific instruction sequences.

Logic Analyzer Application to Motorola's 88000 RISC Microprocessor

Another example of logic analyzer application to high-speed RISC microprocessor system design is found in the 88000 Microprocessor Analysis Package (MAP), which provides disassembly with symbolic address replacement in Motorola's standard mnemonics. MAP supports the fastest 88100 components, available to 50-MHz data rates, operating with the CLAS 4000 logic analyzer and its system.

MAP automatically configures the analyzer to match the 88100 architecture by collecting address and data bus plus control signals into channel groups with Motorola-defined names. The disassembled data display shows all bus cycles with type information and mnemonics with options to delete unexecuted instructions and data read/write cycles. Symbols from the source programming language may replace address information. The symbols may be entered directly into the logic analyzer or downloaded from a software development host.

Absolute addresses are calculated and displayed when relative address mode instructions are encountered. Branch instructions are evaluated to determine whether the branch was taken. Instructions that are fetched but not executed and illegal instructions or sequences are marked in the display. Disassembly may be started at the beginning of the captured data or aligned to the instruction flow on byte boundaries at any point within the recording.

The 88100 MAP provides data capture and μP-oriented triggering to a maximum data rate of 50 MHz. Trace control provides full-speed sequential triggering with 15 store and conditional branch decision steps. Instruction sequences that occur at each clock cycle of the 88100 can be identified and captured. Triggering may be performed on instruction, data address or address range, and data and control signal groups. Triggering can be further qualified by bus cycle type, time, and user-defined event and loop counts. A timer can be activated at any decision step to measure the time required to execute a software module.

Each decision step may be named. These names appear as trigger markers on disassembled data. Trace control can gate the data capture process on and off, allowing cause and effect relationships to be captured even if they are separated by a large number of instructions. Each captured instruction is time stamped, preserving the timing relationships.

Timing measurements, with up to 5-ns resolution, may be made directly through the MAP connection. All signals are brought directly into the analyzer without intervening electronics, which can distort timing relationships. In this mode, trace control triggering will recognize 5-ns events or 3-ns glitches at any of 15 decision steps. Timing may be displayed as both individual signals and bus structures.

Disassembly is displayed in a color windowing environment. Program disassembly, statc listings, timing diagrams, triggering conditions, and instrument status can all be viewed on the screen simultaneously. Coupled cursors and a time stamp correlate the data between windows. Data and instrument configuration may be recorded on the hard disk in the CLAS 4000 controller to be sent over a network connection or printed.

Two 96-channel CLAS 4000 Pyramid Measurement Modules are required to monitor all of the signals from the 88100 CPU. The analyzer-to-chip connection is made through an interface module, which accepts standard pyramid module high-impedance probes. The interface module is passive, placing a 1-megohm 10-pF load on the component. The low-profile logic analyzer-to-component connection (chip size) fits easily into dense board or card cage designs.

Beyond Disassembly

As the logic analyzer becomes a key tool in the development of high-speed microprocessor-based systems, microprocessor control and performance analysis capability become desirable features.

Microprocessor control allows the insertion of breakpoints in code execution. When control is asserted, register and memory contents may be displayed and/or altered. The user gains access to the fault trigger of the logic analyzer, as well as the emulatorlike trace control. An example of this feature is found in the Tektronix Prism 3000 analyzer. It comes with a Prototype Debug Tool (PDT), which provides emulatorlike control of the system under test without the intrusiveness of conventional emulators. Minimal logic-analyzer interference with normal microprocessor operation results. The PDT allows the target to run at full speed with no wait states. Registers and memory can be patched and examined, software breakpoints can be set, and the system can be stopped and restarted. The PDT's 8K samples of memory can be used as a trace buffer, due to the tight coupling with the state section. Complex hardware breakpoints can be set by using the full triggering capabilities of all modules installed in the Prism 3000.

The Prism 3000 provides four performance-analysis modes to give further verification, test, and debug capability. These modes include real-time timing overview, real-time count overview, state overview, and single-event measurement. The two real-time modes provide accurate, nonstatistical histograms of code execution times or event occurrences without missing any system activity. This feature is useful for real-time embedded control applications. The state overview and single-event measurement features are statistical modes, giving general performance overviews and spotting areas of code where optimization is needed.

Chapter 6

RISC and Simulation

Software simulators are being used as the primary development tools in system and product applications of SPARC, MIPS, and Motorola RISC microprocessors. They are used far less in the CISC processor world, because developing a simulator for a complex instruction is difficult. For the most part, software simulators are being used in lieu of in-circuit emulators. However, if tool vendors can keep pace with RISC processor speeds, these simulators will end up complementing in-circuit emulators. But if vendors cannot curb the price of in-circuit emulators, RISC in-circuit emulation may go by the wayside.

Simulators versus Emulators

Simulators offer several key advantages over in-circuit emulators. For instance, in addition to providing a test and debug environment for the application code, the simulator lets engineers perform architectural exploration and evaluate the code's performance and efficiency in the context of the target system. And software simulators are multiuser, which allows an entire development team to exploit the tool. This contrasts with some emulators, which serve single-user debug environments. Moreover, software engineers working with a software simulator can debug code without the target system hardware.

On the down side, software simulators can be slow at simulating code. Execution speed varies among different simulators, with rates depending on the complexity of the code being simulated, the target system's architecture, and the speed of the host computer. Also, software simulators do not offer the real-time debug capabilities of in-circuit emulation, a serious drawback for many RISC-based development projects.

In general the various software simulators available are quite similar in that they provide engineers with hardware-free and, hence, bug-free environments for developing and debugging code. Simulation is getting a lot of play for the upper-end systems, for which designers seek alternatives to full ICE-based design.

Nearly all RISC vendors, for example, have fielded instruction-set simulators, which have allowed software developers to integrate their software with the hardware before the hardware has been ready.

Many of the RISC designers haven't used ICEs, primarily because the emulators haven't been available with the early silicon. But a prevailing view for most RISC vendors is that development needs can be met with instruction-set simulation of key software combined with a sophisticated logic analyzer to examine signal and bus relationships.

Simulating SPARC

For developing SPARC-based code, Sun provides its SPARC Architectural Simulator (SPARCsim), which offers both instruction-set and architectural simulation capabilities. When debugging code on SPARCsim, users must work at the assembly level. Sun is now integrating SPARCsim with rdbxtool, the firm's source-code symbolic debugger.

The binary simulator and the STA comprise SPARCsim. Within the simulator are source-code modules that simulate the target system's integer unit, floating-point unit, memory management unit, cache, and main memory. Two versions of the simulator come with SPARCsim: an executable binary version simulates a Sun-4 system, and a relocatable binary version contains user-configurable source modules to simulate other systems.

Design Implementation

Valid Logic is one of the companies providing state-of-the-art computer-aided design tools for SPARC on Sun workstations. These tools allow a designer to develop schematic drawings and simulate the logic to verify the hardware before committing to printed circuit board layout. Full simulation of a SPARC-based design is supported by a SPARC H/W-S/W emulator, now part of Valid's Realproducts™ hardware simulation systems.

At this design level, changes impact only the schematic, and testing simply requires reexecuting the simulation. Thus the breadboard phase of a SPARC-based design can be eliminated. The designer can go directly to layout for a prototype, especially when using high-frequency chip sets performing at 20 VAX 11/780 Mips at 33 MHz.

Prototype

Once the prototype is verified by simulation, it undergoes layout and fabrication. The prototype is then used in the initial tests and validation of hardware. This phase is necessary to eliminate errors associated with board layout and fabrication. While microprocessor hardware emulators may speed up prototype debugging, their availability generally lags chip availability, especially in the case of very high frequencies of operation. To be competitive, systems designers must provide products long before emulators are available. System debugging is therefore partitioned into two phases. First, hardware is debugged to a minimal level of functionality and second, software monitors are used to bring the hardware to full functionality. Software monitors are inexpensive tools that can be used very effectively in the debug stage of new microprocessor hardware. Bringing the hardware up to the level of functionality required to use a software monitor is a simple task.

A number of different software monitors are available for use. The software monitors available with a number of SPARC evaluation boards perform many of the functions of a processor emulator. They allow uploading and downloading of code, setting of breakpoints, manipulation of data in registers and memory, single-stepping, and assembly/disassembly of user code. A logic analyzer provides a real-time trace capability allowing the designer to monitor the processor signals during software execution and compare them, if required, to the results of the same program running on either a CAD/CAE emulator or SPARCsim.

The software monitor-logic analyzer approach provides a low-cost debugging solution. It also has the ability to run the hardware at full processor speed, since none of the processor signals needs to go through any emulator hardware.

Software Development

Software development usually begins on a UNIX-based development system. Using the software development and management tools on UNIX, the engineering team can begin to design and implement the modules of the application. Developers are free to use the usual UNIX tools such as text editors, compilers, assemblers, make, RCS, SCCS, and so on. The application itself can make use of many of the libraries available from UNIX.

For SPARC-targeted development, engineers can choose from two platforms. Although the native environment is based on the Sun-4 workstation, code can also be developed on a Sun-3 in a cross-development environment based on SunOS, a converged version of UNIX System V and Berkeley 4.3. SunOS delivers state-of-the-art virtual memory management, shared libraries, and dynamic linking, as well as the distributed services, administration, and security services. A traditional software development cycle is shown in Figure 6-1(a). The unique software tools that assist in this cycle are discussed below.

Using the text editors available on the Sun platform, source modules can be created and compiled. Available compilers for SPARC-based development are: C, FORTRAN, Pascal, LISP, and Ada. C, FORTRAN, and Pascal share the underlying language base tools. The set of language-based tools includes a SPARC code generator, a global optimizer, a peephole optimizer, an assembler, a linker, and libraries.

High-Performance Optimization

The first phase of the compilation process involves scanning and parsing the source language. From this, the compiler generates an intermediate language called SunIR. SunIR was designed to remedy the shortcomings of the older PCC (processor condition code) intermediate representation, which

FIGURE 6-1(a)
Traditional system development cycle. Reprinted courtesy of David L. Weaver, Sun Microsystems, Inc.

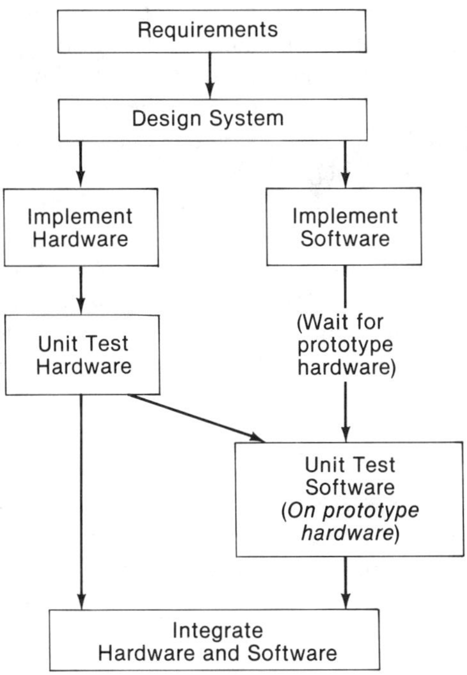

did not provide enough information to support the extensive control and data flow analysis required to perform global optimization.

The global optimizer, IROPT, reads the SunIR code produced by the compiler front end. A series of data-flow analyses and transformations are applied to the module, such as local and global common subexpression elimination, local and global copy propagation, loop-invariant code motion, tail-recursion elimination, dead code elimination, and global register allocation. IROPT fully handles memory aliasing, a difficulty in globally optimizing code from languages (such as C and Pascal) that have pointer variables. The use of the global optimizer has resulted in measured performance improvements of 10 to 80 percent, compared to unoptimized code.

Output of IROPT, in SunIR form, is ready by the code generator, which produces SPARC assembly language. This assembly language is then processed by the peephole optimizer. Here instruction scheduling, register coalescing, brand chaining, creation of "leaf" routines, utilization of machine idioms, and several other straight-line code optimizations take place. The resulting code is then assembled into relocatable object code and linked with run-time libraries, system libraries, or other compiler libraries, as needed.

Software Simulation

At this point, programmers can use SPARCsim to test and debug the object modules prior to the availability of actual target hardware. Not only does this save time, but the programmer does not have to contend with hardware implementation bugs. The simulator provides two levels of simulation: user-level and supervisor-level. User-level simulation is used to test programs such as compilers and linkers, where privileged operations are not required. Supervisor-level simulation is used to develop operating system software. SPARCsim is rich with features to help the programmer. It contains debugging tools similar to the adb debugger available under the UNIX operating system. The debugger is an assembly-level symbolic debugger that supports breakpoints, single-stepping, and set/display of memory and registers, dynamic execution tracing, disassembly, and the like. The simulator provides complete execution traces that can feed into other analysis tools, such as the SPARC Trace Analyzer (STA) tool. STA reports instruction counts, machine cycle counts, and the percentages of each. This feature helps the programmer pinpoint areas where execution time can be improved.

The SPARC Toolkit Environment

The environment provided by the SPARC Toolkit addresses both the system development cycle and the integration debugging environment. The two components of the Toolkit involved with debugging are the SPARCsim and the Remote Debugger.

The SPARC Architectural Simulator

With the SPARC Toolkit, the development cycle can become that of Figure 6-1(b). Elapsed development time can be reduced because new software can be debugged independently from new hardware, before it is complicated by interactions with partially tested hardware. This is accomplished by using SPARCsim, which executes binary SPARC software on (simulated) bug-free hardware. The simulator not only models any SPARC processor implementation, but it can model proprietary hardware attached to it—typically MMU, cache, memory, and memory-mapped devices.

The SPARC Architectural Simulator includes a symbolic debugger that can set breakpoints, single-step instructions, examine/modify memory, registers, and devices, simulate asynchronous events, symbolically disassemble code, and maintain a "real-time" trace buffer of the last N instructions executed. It also performs many functions not related to debugging (for example, producing traces for performance analysis) and is often used by system designers for hardware design iteration and debugging.

As an example of the utility of the simulator, the entire SunOS (UNIX) operating system for the Sun-4/2 × 0 was debugged and booted up to multiuser state before the first SPARC processor existed in hardware, using a simulator that was configured to model a Sun-4/2 × 0 system (MMU), cache,

FIGURE 6-1(b)
Improved system development cycle. Reprinted courtesy of David L. Weaver, Sun Microsystems, Inc.

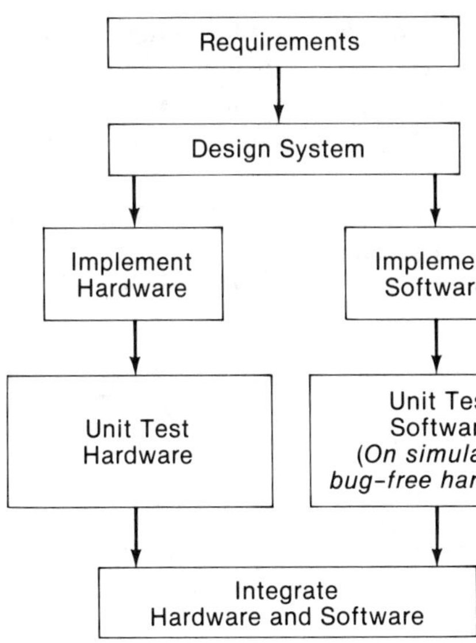

memory system, and so on. This work was a significant factor in Sun's ability to rapidly bring up SunOS on real hardware, shortly after the first CPU boards became available.

After the software has been tested and debugged on the simulator, it can enter software/hardware integration testing. The portion of the SPARC Toolkit which addresses this development phase is the Remote Debugger.

Remote Debugger: Host Component

The Remote Debugger requires two systems, host and target. The target system is the one to be debugged. The host system is one rich in computational, storage, and display resources and one on which the user runs the host component of the Remote Debugger. The host component, rdbxtool, communicates with a small piece of software on the target end (the debugger target component), which is capable of accessing the target system's registers, memory, and other devices. See Figure 6-2 for a diagram.

The host component is based on Sun's native dbx and dbxtool debugging software. The versions modified to operate with remote, non-UNIX target systems are known as rdbx and rdbxtool. The user sees rdbxtool, a bit-map window- and mouse-based, fully symbolic, source-code-level debugger.

Rdbxtool (see Figure 6-3) displays the current point of execution in its upper subwindow, with a surrounding context of source code. A small "arrow" icon to the left of a source code line indicates the place at which execution is stopped. A "stop-sign" icon to the left of a source code line annotates a breakpoint on the line.

Lower in the rdbxtool window is a "command" subwindow in which textual commands can be typed and in which most output is displayed. This subwindow is scrollable and retains a history of the entire debugging session. One can scroll through the session history at any time, using the mouse device.

The user can create command "aliases" (macros) for textual commands, so that commonly used textual commands can be executed with few keystrokes. One use for this is displaying many pieces of data; after executing

```
alias ds "print i, array [i]; call display_flags (ptr→flags)"
```

a simple "ds" command will execute all of the stored display commands. Single-character names are commonly aliased to the standard command names; for example,

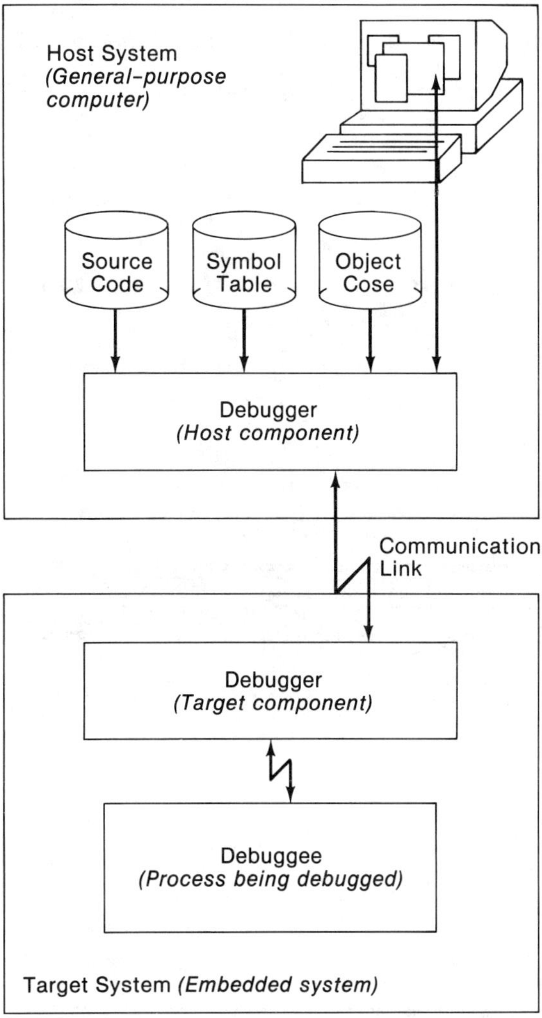

FIGURE 6-2
Typical remote debugger. Reprinted courtesy of David L. Weaver, Sun Microsystems, Inc.

alias p print

allows a "print i" command to be abbreviated as "p i".

Far more useful are "buttons," which can be selected by the mouse and appear in a narrow subwindow between the source code subwindow and the command subwindow. Simple mouse-clicks can select all common

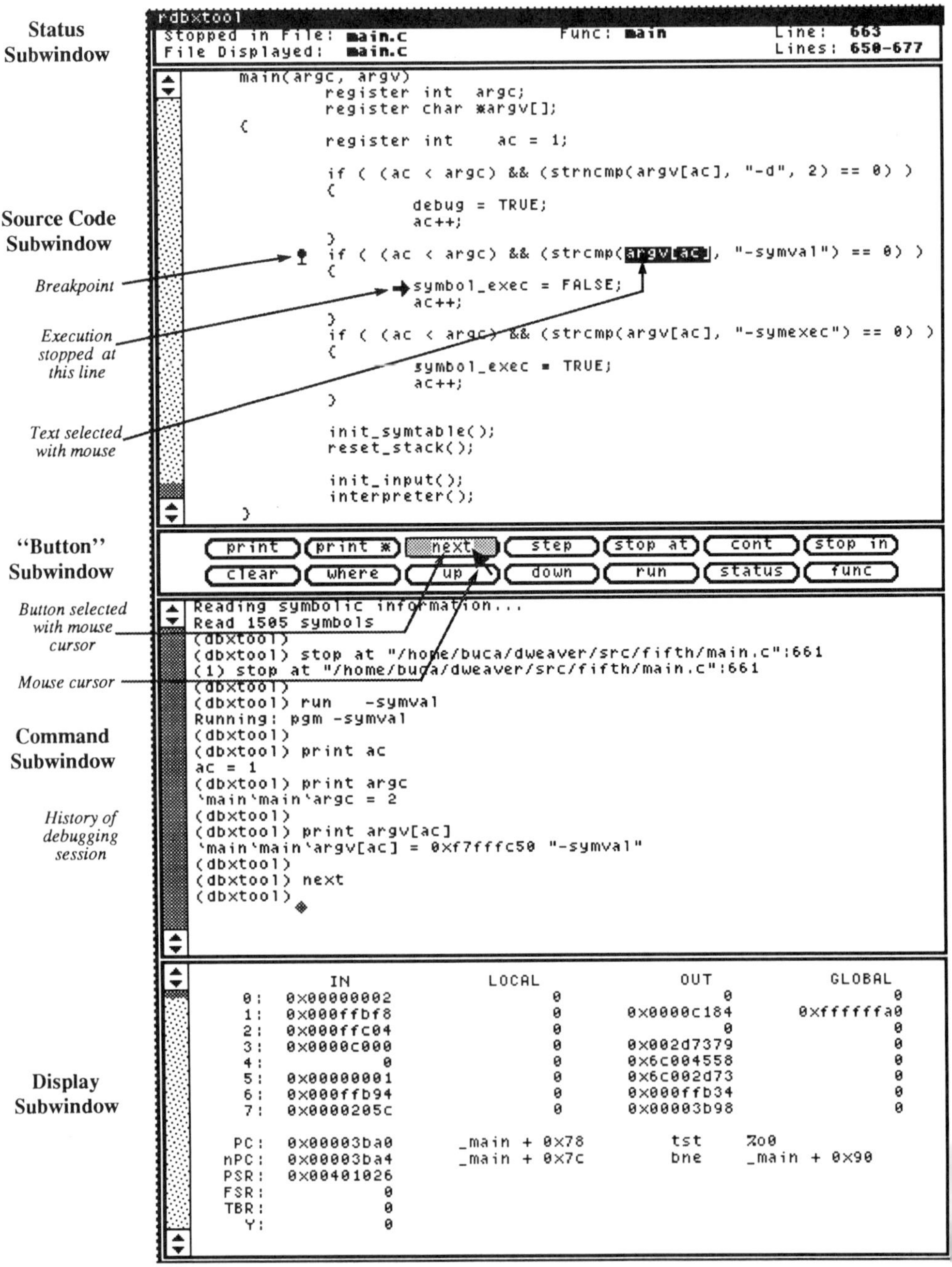

FIGURE 6-3
Rdbxtool user interface. Reprinted courtesy of David L. Weaver, Sun Microsystems, Inc.

debugging operations in a quick and intuitive manner. For example, the value of a variable can be printed by pointing at its name anywhere in the source code subwindow with the mouse, then clicking on the print button. This is how the values of ac, argc, and argv[ac] were displayed in Figure 6-3.

A breakpoint can be planted similarly, by first selecting (pointing at and clicking the mouse on) the source code line at which to stop, then clicking on the stop at button. Other common operations include starting execution (run), continuing from a breakpoint (cont), stopping upon entry to a function (stop in), single-stepping over source code lines (step and next), and displaying a function-call ("stack") back-trace (where). The user can associate an arbitrary sequence of debugging commands with a new button, at any time.

A breakpoint can be set conditionally using an if clause, so that it only occurs when the value of an arbitrary C expression is true, for example as follows:

```
stop at 73 if (i = 5) && (ptr→next = foo[n])
```

A breakpoint can also be set to occur only when the value of a specified variable changes. This is particularly useful for pinpointing when the value of a variable is getting unexpectedly "trashed" during execution. This simulates one of the more useful types of hardware breakpoints, without requiring special-purpose hardware.

Rdbxtool knows the types of all variables, so it can read, modify, and display them appropriately without the user needing to keep track. For example, if foo is a structure containing an int, a char, a character pointer, pointers to two functions, and another structure that contains an enumerated type and two floats,

```
print foo
```

would display something like the following:

```
foo – {
        size = 4
        type = 'o'
        title = 0x7098 "widget"
        create = & obj_create( ) at 0x5658
        delete = & obj_delete ( ) at 0x6834
        object = {
                color – GREEN
                x_coord = 3.0
                y_coord = 64.75
        }
}
```

The value of a variable can be displayed as if it has a different type by using the normal C "cast" operation. For example, to print the value of character variable ch as an integer,

```
print (int)ch
```

The debugger supports the entire edit-compile ("make")-link-debug cycle in a single integrated context. When it is time to correct a bug in the source code, it can be edited directly in the source-code subwindow. The user then types "make" in the command subwindow, which causes the changed module to be recompiled, and the resulting object code to be relinked into the program. Typing "run" (or clicking on the run button with the mouse) then reloads the program and restarts execution.

The debugger works with programs written in C, FORTRAN, Pascal, and SPARC assembly language, or any combination of these languages. A full C expression evaluator was built into the debugger, so that the user can select arbitrary expressions from the source code, for example,

```
(*fcn_ptr_array[index]) (value+1, "string")
```

and display them with the print button. Functions in user code can be implicitly called from the debugger by referencing them in any debugger command, for example,

```
print factorial (7)
```

The call command can be used to explicitly call a function in user code. The following are examples:

```
call reset_state( )
call prettyprint_data(my_struct)
```

Remote Debugger: Target Component

The target component of the Remote Debugger, rdbxmon, runs on the SPARC-based system that is being debugged. It implements the process-control services that the host component requires to be provided on the target end, such as reading and modifying memory and registers, and starting, stopping, and continuing execution of the program under test.

The Remote Debugger target component is written in C (and a modicum of SPARC assembly language) and can be reconfigured to run on any SPARC-based target system. It includes basic trap handlers (for example, for register window overflow and underflow), setup code for trap vectors, serial device driver(s), and code that implements the debugger's process-control functions.

It is also possible to access the target component from a TTY-style front end, much like a traditional PROM monitor. A table-driven version of such a front end is being developed. Since it is table-driven, the user can easily add his or her own monitor commands to this interface. Or, it can be configured out entirely, so that the target component is only accessed by the Remote Debugger and is as small as possible.

The communication protocol used between the host and target components is key to their operation. The ability to communicate over either high-speed or slow "lowest common denominator" media was desired—VMEbus, Ethernet, and RS-232 are all supportable. The first version of the Remote Debugger supports RS-232.

Since the communication protocol must support debugging of multiple processes on the target system, the communication protocol supports packet multiplexing.

Last, debugging a SPARC-based target system should not require a SPARC-based host system. Therefore, the protocol supports the exchange of binary data in Sun's External Data Representation (XDR) format. This enables any type of host to be used, even one with a different native byte ordering than SPARC (such as one based on a VAX or 80386 processor architecture).

The host component of the Remote Debugger currently runs on Sun-3 and Sun-4 workstations and could be ported to other platforms. All of the above functionality is implemented using the host machine's resources; it does not intrude upon the target system.

Remote Debugger: Client/Server Model

The Remote Debugger is designed using a client/server model; the user interacts with the target-independent client side of the Remote Debugger, while the service side implements the target-dependent debugging functions. Currently, the client side code is implemented in the host component and the server code is divided between the host and target components (see Figure 6-4). This choice was motivated by the desire to keep the code size of the target component small.

In alternative implementations, the client or server functionality could swing in other directions, with potential advantages to either the implementor or the user. For example, all target-dependent server-side code could migrate into the target component, as in Figure 6-5. This might allow the host component code to become target independent, able to control debugging of not only SPARC-based targets but also, for example, MC680x0 targets.

Given an appropriate communication protocol, the functionality of the client and server processes could migrate between the host and target systems as best suited for any particular debugging application.

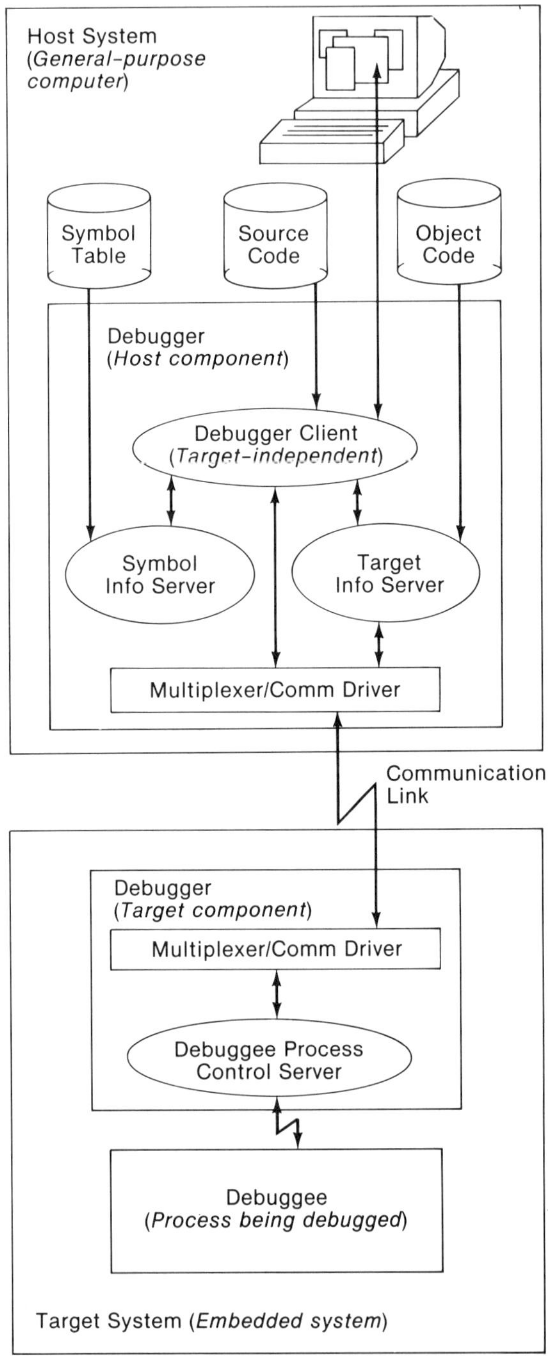

FIGURE 6-4

Remote debugger: client server versus host/target. Reprinted courtesy of David L. Weaver, Sun Microsystems, Inc.

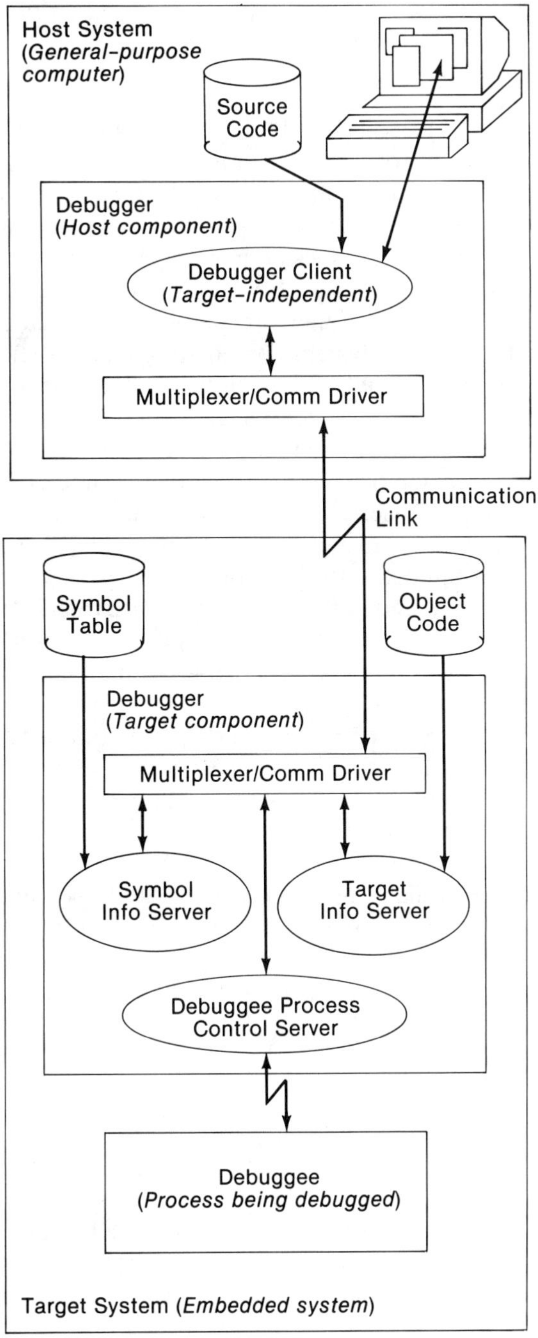

FIGURE 6-5

Alternate configuration of remote debugger. Reprinted courtesy of David L. Weaver, Sun Microsystems, Inc.

Simulating MIPS

System Programmer's Package (SPP) is a complete integrated tool kit for developing system software and integrating hardware/software on designs using the MIPS R3000 architecture. It is used by designers to write stand-alone software systems, create new operating systems, modify existing kernels, and write machine diagnostics all in a development environment prior to the existence of hardware. It is also used by hardware and software designers to download and debug the software on a bare machine in order to bring up a total functioning system immediately after the first hardware is built. And it is often used to help debug difficult system-dependent problems that appear only under heavy system loading conditions.

SPP is a source-level program that executes on MIPS M series development systems. A major feature is the software simulation environment that allows debugging of users' programs in a native execution mode on a known functioning system independent of the target hardware. Another feature of SPP is its architectural simulator, which allows users to quantify cache and main memory trade-offs efficiently. These trade-offs may be made early in the design using realistic large applications. SPP also contains UNIX-like utilities to facilitate the building of stand-alone software systems. And it contains monitors, program libraries, I/O drivers, diagnostics, and code segments to incorporate into the user's product or use as examples for code to be developed. These packages have been integrated into SPP with facilities designed to communicate with the target hardware starting at a primitive state and remotely building the system by adding capabilities and debugging one step at a time.

Users find SPP a reliable and powerful tool kit for keeping development projects on schedule. It was developed by MIPS computer designers who themselves are SPP users. It facilitates and simplifies the overall software development, and it is an indispensable aid in the critical integration and debug phases.

Design Considerations

While various details of competing processors receive the lion's share of attention in any architectural decision, most experienced designers realize that a total solution must be in place if they are to derive any practical benefit from the new processor. That total solution must include a path for them to get their product to market. If the development support is nonexistent, weak, or late, then they fail and the new processor technology fails, regardless of its underlying merits.

Once a project is committed, the critical need is to keep on schedule. Several important factors typically come into play in developing high-performance RISC systems to assure a project is completed on time.

First, the hardware and software developments must proceed independently with the expectation that each will be done on time and that they will work together when complete. As computer systems have become more sophisticated over time, hardware design has yielded up its title as the development bottleneck to software design. Hardware design has not become easier, but the industry has been able to develop a good complement of CAE and CAD tools that have kept up with the increasing requirements. CASE (computer-assisted software engineering) has been slower to deliver on its promises. The net result is that software is most often in the critical path of a development. This can be very important for designs where the software is a significant part of the added value, such as with embedded control applications or designs involving operating system kernel work. The new RISC processors from some companies are particularly subject to this need because they lack the basic software to make a working system. Therefore, any new processor technology that does not adequately address software development puts design schedules in jeopardy.

A second factor is the timely completion of the hardware/software debugging. This is a test point for all designers and a crucial element in the schedule. Often, the need is to bring up a sophisticated system where the software has been debugged only on a simulator and the hardware is functioning but not fully debugged. The task requires a well-developed methodology and a set of tools to facilitate the iterative process so that the job can be done efficiently. No toolkit can compensate for inadequate initial specifications or poor design practices, nor can one substitute for the understanding of the designer when it comes to chasing down intricate bugs, but a good tool kit can help the designer bring up the system in a logical and controlled progression, testing at each step in the process. Again, a technology that does not address these aspects leaves the designer in a position where there is a significant risk to the schedule.

The last point is that the designer should spend his or her time designing the product, not the tools to build the product. If the tools available from the vendor are incomplete or if the designer must integrate tools from multiple sources, the project schedule pushes out and the amount of uncertainty increases drastically. The brochures may indicate "compatible" development tools, but that cannot be relied upon when they are from different companies. A fact of any project involving new technology is that the number of interfaces the designer must work with greatly impacts how long it takes to solve each problem.

MIPS' answer to these needs is a product called System Programmer's Package. SPP directly addresses software development and hardware/software integration and it also provides routines to incorporate in a user's product. Figure 6-6 illustrates the uses of SPP. It is designed and supplied by the same people who developed the processor technology, so the fit

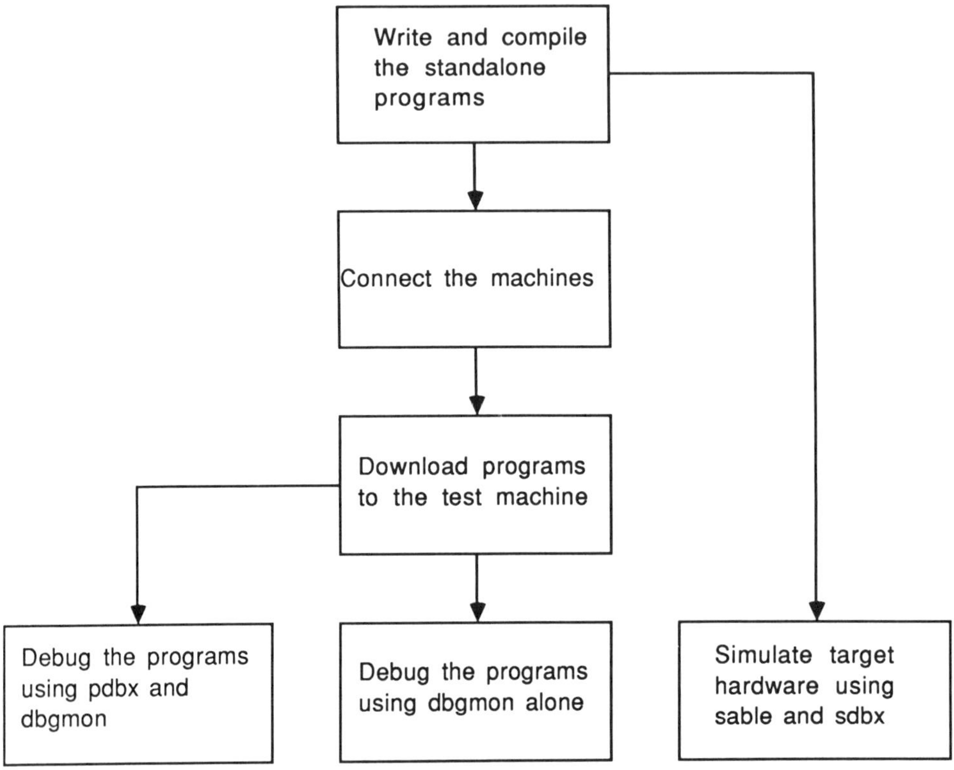

FIGURE 6-6
Typical software development cycle. Reprinted with permission of MIPS Computer Systems, Inc.

between the tool and the technology is as close as can be. MIPS also uses SPP to develop its own commercial computer systems, which means that SPP has stood the test of usability.

Development Environment

The starting point for any development is an appropriate platform. The M-series computer systems from MIPS are powerful UNIX multiuser systems with performance in the range of 10 to 20 times a VAX 11/780. MIPS has licensed versions of UNIX that are either System V or BSD compatible. These platforms are general-purpose servers designed to operate in a heterogeneous environment (with SUNs, VAXs, and the like) and are ideal for both software development and the hardware/software integration tasks at hand.

M-series computer systems may also be viewed as ideal "evaluation vehicles" for MIPS RISC processors, since they are actual commercial designs by experts on the MIPS architecture. Another important benefit of using these systems for development of target system software is that the code will be debugged on the native processor. This is preferable to development in a cross-system which leaves open the possibility of errors in the interpretation of how the processor behaves.

The SPP product is supplied in source form on the M-series platforms. There is documentation on the product as well as extensive commentary in the code. For code to be incorporated in the final product, many users take the SPP routines and customize the source to their application requirements. That may mean changing a few variables, rewriting sections, or simply using the SPP routines as examples for totally rewritten routines. In all cases, routines are customarily recompiled before they become a part of the final product, to assure that they are what the designer intended. Source format also allows the designer to extend the capabilities of SPP if so desired.

Figure 6-7 shows the SPP development environment on an M-series development system. The following sections give an overview of the modules composing SPP.

Software Simulator

The MIPS stand-alone software simulator, Sable, is comprised of the R3000 CPU, the R3010 floating-point accelerator, the cache and main mem-

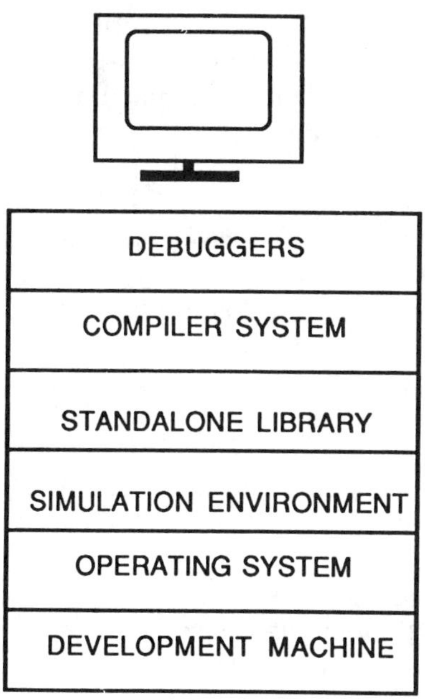

FIGURE 6-7 System Programmer's Package (SPP) options. Reprinted with permission of MIPS Computer Systems, Inc.

ory structures, memory management and the TLB, a simple disk, and a simple console. The function of Sable is to allow the execution and debugging of users' software before the target hardware is debugged.

Sable is an instruction simulator that understands some detail about the physical environment in which the program executes. It models the state of the hardware at the end of each instruction cycle. The level of detail in the hardware model is only that necessary to correctly predict program behavior. For example, the model of the disk is a simple file that can source or sink data files but that does not mimic latency or track seeking or error behavior.

On the other hand, Sable models the memory hierarchy so precisely that every element that is in the cache and the contents of the TLB are known after each cycle because they are intimately tied to the program. Sable is capable of debugging kernels, diagnostics, monitors, and any other software that requires knowledge of the R3000 privileged state. It is also capable of debugging applications programs residing on top of an operating system.

The value of a software simulator with this degree of sophistication is that very sophisticated programs can be constructed and debugged with confidence before the target hardware is finished. MIPS uses Sable to debug its operating systems for new computers and typically has them up and running months in advance of the hardware.

Utilities

A comprehensive set of utilities for building and verifying complete software systems is part of SPP. The stand-alone I/O library, saio, contains run-time routines often installed into the final product. Initialization modules that set the pointers and clear the cache and valid bits are available. Fault-handling modules that report the nature of a fault and transfer control are also included.

There is also an archive module containing disk and tape routines as well as stdio and libc modules familiar to UNIX programmers. String copy, printf, get, get string, and put character to console are examples of commands. These functions are similar to those from UNIX but do not require UNIX and therefore are particularly useful for embedded control applications. Having saio available means the designer can spend time doing design rather than recreating common functions.

The stand-alone compiler system consists of: a C compiler, sacc; an assembler, sass; and a loader and linker, said. These modules are similar to MIPS conventional compiler modules, but they are configured to develop stand-alone applications and they provide automatic linking to the stand-alone library. If a user wishes to use another language, for example Pascal, the conventional compiler is used and the linking in of saio is done explicitly.

The simulation debugger, sdbx, is a source-level debugging tool based on MIPS' enhanced version of dbx, a utility familiar to BSD programmers. It is a symbolic debugger as well, which means the user will troubleshoot the design in her or his terms. As a source-level tool it traces every action back to the original source statement. The user can set breakpoints, observe the disassembled code before and after the break, single-step, examine or alter the machine state at any point, and so on. An important point is that all the MIPS software traces the action back to the original source statement.

The development environment consists of the software simulator and the utilities described above. Once these elements have been configured on top of an M-series computer system, the designer has all the elements needed to create and verify a complete stand-alone software system. That verification can then be done in the same native environment as the target hardware but without the luxury of the hardware existing or being fully debugged.

There are two additional utilities associated with the development environment (or host system) that will be described before addressing the hardware/software integration facilities. These are programs residing on the host that are useful, when the hardware starts to come up, to do remote debugging from the host.

The PROM debugger, pdbx, is another version of dbx configured to operate remotely and debug software on the target system. It is also a source-level, symbolic debugger.

The bootfile server driver, bfsd, is a convenient utility to download bootable images across Ethernet for execution on the target hardware.

Hardware/Software Integration

The first builds of the target hardware should be available when the software system has been debugged. The hardware can be connected to a host, which is usually the same machine as that used for the development environment, and the process of hardware/software integration and debugging can begin. The idea is to use a known functioning system (the development environment) from which to control the integration and debugging process on the less certain target machine.

The objective during this stage is to bring up the target system so that it is capable of booting and running stand-alone programs. The debugging will be controlled from the host. Sometimes there is a need to return to the host-controlled system to help isolate particularly troublesome bugs that can only be observed under heavy system resource (I/O, foreground, background tasks) utilization.

Given that the software has already been debugged on the software simulator, the remaining task is to check for time-dependent bugs and verify the model.

Basic Hardware

The hardware design almost always proceeds in parallel with the system software development. Hardware design is done using development tools common in the industry, such as CAE workstations, schematic capture packages, logic and circuit simulators, and PCB layout software. At the point at which the system is ready to begin hardware/software integration, the first builds should have a minimum functionality established by checkout in the lab. Instruments such as waveform generators, scopes, and logic analyzers are typically used to ensure that the nodes toggle, there are no stuck-on faults on the board, the clocks and addresses reach their destinations, and so on. That is, the hardware has had some debugging but it is in a primitive state and is ready to take on intelligence to advance its own checkout.

PROM Monitor

The lowest level of intelligence on most machines is a basic monitor that is stored in programmable read-only memory. It usually contains some communication facilities to talk to boot devices (e.g., disk or tape communications link) and to the user. And, it contains sufficient power-on diagnostics so that it has confidence the CPU, memory, and I/O are working. But it does all of the above at the minimum possible level so that it can be used with minimal debugging.

MIPS provides PROM monitors for all of its M-series computer systems in the SPP source. These are typically used as models for the target system monitor. The PROM monitors assume that a reset forces the program counter to the start address. They then initialize the cache and reset the state of the machine before transferring control to either the stand-alone shell or to the user's image that has been booted from the communications link. The MIPS monitors store console baud rate, net address, and other variables in nonvolatile memory. Having the source for the monitors means that any of these actions can be modified by the user.

Another useful part of SPP is a set of power-on diagnostics that tests the basic functionality of the CPU, memory, and floating-point accelerator. These diagnostics are often incorporated into the PROM monitor for the target system.

The PROM monitors must be compiled, formatted, burned into PROMS or EPROMS, and placed into the target hardware. The standard C compiler described previously (sacc) can be used for compiling. SPP also contains utilities that understand the file structure and formatting requirements for data I/O PROM burners. Burning the devices and placing them in the system is left to the user.

Once the PROM monitor is on the target hardware, the user can begin debugging from a console. It is possible to finish the debugging completely on the target or to use a separate host to control the target remotely. The latter is the more typical scenario, since it gives the user more power in the full use of symbolic debugging. Some way to talk between the two machines is needed if this is the choice.

Host Communication

MIPS uses an RS-232 connection for remote debugging of its own computer systems. On the target system, one of the two serial ports is reserved for the console and the other is a port for the remote access. It is this second port that connects another system via the RS-232 cable. On the host system, it is common to use a separate serial I/O port in the I/O subsystem rather than the second port on the processor board. In any case, it is necessary to establish a physical connection and simple communication protocol to begin debugging. The SPP routines are provided in source format so that if the conventions above are not directly usable, it is possible to model a different protocol and integrate it with the rest of SPP.

Beyond the serial connection, MIPS computers all have Ethernet capability and, while optional, that has been found to greatly speed up the data transfers once the initial debugging is over. Again, the drivers are a part of SPP to be used, modified, modeled, or left out as appropriate for the user's configuration.

Most microprocessor developers are familiar with ICE as a software development tool and a way to connect to the target system for debugging. As a software development tool, the best ICE units cannot compare with the functionality of SPP. However, ICE is a convenient way to connect to the target and download programs. But the ICE probes affect the system timing even at low speed enough so that some problems cannot be debugged until the ICE is removed. This deficiency is accentuated at the high bandwidth of RISC processors (200 MB/s between the R3000 CPU and cache). When the hardware must be observed at these speeds, there is still no substitute for very high performance instruments (scopes, analyzers). ICE units and RAM/ROM emulators will find a place in low-speed systems for some time to come because of their familiarity and convenience (faster than burning a PROM). Still, a more useful debugging approach for high speeds is to rely on the system and integral communication links described here.

Monitor/Stand-alone Shell

The PROM monitor described above is deliberately kept simple so that it does not have to be debugged and so that the least amount necessary is placed in nonalterable memory. The full utility that a user would want is contained in a stand-alone shell, sash. The PROM monitor and sash are very similar, but sash contains many more commands, can control more device types, and comprehends all types of file formats including System V and

BSD. The partitioning between the PROM monitor and sash is simply a convenience in the MIPS systems to allocate the minimum functions to PROM and allow RAM to contain the full functionality. Between the PROM monitor and sash, the user has the capability to download programs over the communication links and to boot off of disk, tape, or Ethernet.

Debug Monitor

The next item to be loaded onto the target system is the debug monitor, dbgmon. This communicates to pdbx, which resides on the host. These two programs work together to allow the user full debugging capability accessing the target from the host. At this point, the user can set breakpoints, single-step, and examine the state of the machine and the instructions that are being executed. Dbgmon does more than disassemble the instructions. The debugger maps the machine instructions into the source statements from which they came and can print variables and addresses symbolically. It also allows the user to view these at assembly level or at machine code level.

With dbgmon loaded into the system, the user has the capability to debug his or her own stand-alone program or kernel in the same way users debug applications on top of UNIX. Figure 6-8 illustrates the fully configured system. Now besides the development environment, the user has all the facilities needed to download and debug software on the target hardware. Interactive sessions showing pdbx in two modes are contained in Figure 6-9. One mode shows it being used to download the user program automatically over Ethernet. The other mode shows the user interacting with the program over a serial line manually.

Cache Trade-offs

A separate utility contained in SPP is a memory system simulator containing models of the cache and main memory. The program is called cache 23. It is used to make trade-offs in memory size and speed. Cache 23 is driven by traces from UNIX profile output files, and it returns performance data on how the program executed. Sable is capable of being used for the same purpose, but cache 23 runs several orders of magnitudes faster. The value of cache 23, then, is that architectural trade-offs on realistically large benchmarks are possible early in the design cycle.

Simulating Motorola's 88000

Motorola offers a software simulator as part of their software tools package. This package can set and clear breakpoints, block-fill memory, change and display CMMU registers, display memory map, go to location and start execution, load a program, display/modify memory and registers, trace, and display program statistics. With these features it is possible to

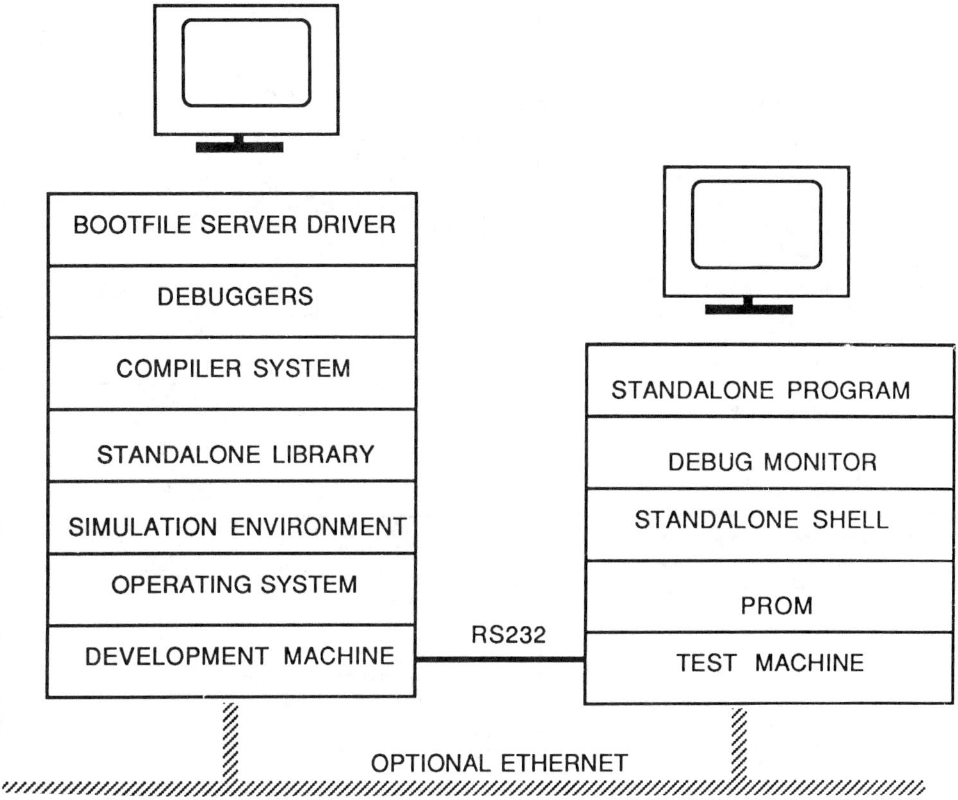

FIGURE 6-8
SPP development environment. Reprinted with permission of MIPS Computer Systems, Inc.

trace program execution and get information on how the processor would actually execute the program and interact with the CMMU.

Logic Analyzers

Logic analyzers provide the capability to monitor the hardware of the very high speed RISC systems. As mentioned before, in-circuit emulators have to work at much higher speeds than the processors they emulate. Logic analyzers present a much more cost-effective tool for investigating processor performance. Logic analyzers capture a large number of instruction cycles and store the captured results into memory. Logic analyzers, such as the DAS9200 system from Tektronix, are capable of capturing up to 32-KB samples. The code can be disassembled and then a software simulator can be used to deduce the results of the internal registers of the machine. The

```
% pdbx test_pgm
dbx version 1.31
Copyright 1986 MIPS Computer Systems Inc.
Type 'help' for help.
reading symbolic information ....
[using main.main]
main: 73 startup();
(pdbx) set $pdbxport= "test_machine"
(pdbx) set $use_sockets = 0
(pdbx) set $manual_load = 0
(pdbx) bp startup
[using startup.startup]
 main: 4 printf("Hello World!");
(pdbx) run
```

Development Machine

```
>> boot -n -f bfs()winchester: /test_pgm
   Obtaining dev_macnine: test_pgm from
    server dev_machine 672848+88624+567632 entry:
   0x80021000

>> boot -f bfs()dbgmon
   Obtaining dbgmon from server dev_machine
   21840 + 11408 +896 entry: 0xa0010000
   MIPS Standalone Debugger Version 1.5 MIPS OPT
   Wed Oct 28 13:58:12 PST 1987 root
   0xa0010000: move v0, gp <vo=0x0, gp = 0xa0332a60>
DBG: p pc 0x80021000
DBG: debug tty(1)

Hello World!
```

Test Machine

```
% convert -f s3rec test_pgm > test_pgm.srec
% convert -f s3rec dbgmon > dbgmon.srec
% sdownload /dev/tty < test_pgm.srec
Device: /dev/tty1
Done total line 4673
% sdownload /dev/tty < dbgmon.srec

Device: /dev/tty1
Done total line 2065

% pdbx test_pgm
dbx version 1.31
Copyright 1986 MIPS Computer Systems Inc.
Type 'help' for help.
reading symbolic information ....
[using main.main]
main: 4 printf("Hello World!");


(pdbx) set $pdbxport = "dev_machine"
(pdbx) set $use_sockets = 0
(pdbx) set $manual_load = 1
(pdbx) run
```

Development Machine

```
>> setrbaud 9600
>> sload tty(1)
Done: 4676 records, initial pc: 0x480020000
>> sloadtty(1)
Done 2065 records, initia pc: 0xa0010000
>> go 0xa0010000
DBG: p pc 0x802100
DBG: debug tty(1)

Hello World !
```

Test Machine

FIGURE 6-9
Pdbx sessions. Reprinted with permission of MIPS Computer Systems, Inc.

disassembly allows the tracing of the actual instruction-level activity of the 88100 Code P bus. Address and data values are displayed for each cycle along with the instruction and a time-stamp value. Instructions that were prefetched but not executed are indicated as not being executed. A time-stamp value, representing the time from the start of the acquisition to the time the cycle was clocked in, is stored with each bus cycle. This allows memory delay information to be monitored. Since address, data, and control signal information are highly interleaved on the 88100's Code and Data P buses, the time-stamp value stored represents the time the bus cycle was completed. All time-stamps of Code bus cycles indicate the time of prefetch rather than internal execution. The final version of disassembly will incorporate the data bus activity to the memory system over the Data Memory P bus. Thus the display can correlate when a memory access instruction was issued and when the access was actually completed by the data memory unit.

Once the logic analyzer has captured the required number of bus cycles on the instruction P bus, a software simulator can be used to deduce the values of the internal registers. This allows programmers to look at their code on a cycle-by-cycle basis to see what their programs are doing. Without the use of the software simulator, only the processor pins could be monitored.

This method of "program trace" is very effective. As an afterthought, it might be added that tracing through a program in an actual system would be very difficult. First of all, the clock cannot be stretched or stopped. This is because there is a phase lock loop on the processor that keeps the internal and external clocks synchronized. If the clock is stopped, a resynchronization sequence is involved that requires the assertion of reset. A second method of program trace is to put a trap instruction after each instruction to be executed. The trap instruction could direct program flow to a handler that would dump the internal registers. However, this methodology would disturb the program flow through the pipelines and would not give a true representation of when the instructions were executing with respect to each other. The logic analyzer solves this problem by letting the machine run in real time. Then, the internal register details can be figured out to give the hardware designer and the software writer a better insight into the workings of the system.

Debug Cards

Debug cards are very useful in monitoring system performance. Debug cards are generally complete systems that include the 88100 processor, the 88200 cache memory units, and a memory system. A monitor program is usually included that will put the system into a known state and allow the user to implement rudimentary functions. An important aspect of these cards is that they generally interface onto the back plane of a host computer. A card that sits on the Nu-bus of the Macintosh II is offered by Tektronix, and a VME bus card is offered by Tadpole Technologies. In both cases the processor has a cache memory management unit on both the Code P bus and the Data P bus. The Memory bus or M bus of the CMMU that connects to the memory system is common to both CMMUs. The M bus is also connected to the appropriate external bus, such as the Nu-bus or VME bus. The monitor program for the card can be contained in a ROM that resides on the M bus. ROMs are slow compared to M bus speeds, but the monitor would quickly be read into the cache where it would execute from zero wait-state memory.

Debug cards of this type are not generally intended to run the host system. They can interface to the host system over the back-plane and request data and programs. However, the general intent of the card is to have a program downloaded onto the card and then the card will run the program autonomously from the host system. The debug monitor can be used for such functions as download control, starting the program, setting breakpoints, and interrupting execution flow. Other functions might include memory and register modify and display. Debug cards play a very valuable role in allowing programmers to see how their software is going to run in an actual system.

Chapter 7

Evaluation and Development Boards

When development tools were mostly external to microprocessor chips, evaluation boards were used primarily as educational tools in the early stages of product or system development. They were used to show how a given processor could be applied in a working single-board computer configuration. As more development support is being built into the chips, evaluation boards or cards are becoming part of the development tool kit.

For instance, in the case of AT&T's DSP-32C digital signal processor, four cards are part of a complete development environment, when interfaced with a personal computer. These are a software development card, an emulator card, a DRAM (dynamic random access memory) expansion card, and a PC bus interface card. A fifth card provides a multiemulator interface for use when more than one target system is being developed simultaneously.

Another example is the evaluation board or card support provided by Motorola for its 68332 32-bit microcontroller. Here a low-cost cost development solution is provided by three boards. The first board, called a business card computer because of its size, $2\frac{1}{4} \times 3\frac{1}{2}$ inches, is the heart of an emulation system, with a powerful monitor/debugger. The second board is a development interface for the first board. The third board is a platform board with emulation memory for user code. This example is described in greater detail later in this chapter.

The above examples undoubtedly indicate a trend, but most of the evaluation boards being used for 32-bit CISC and RISC microprocessor development support are not smaller than a VME bus card and are like the AT&T and SPARC boards, whose descriptions follow.

Evaluation Board for 32-bit CISC Microprocessor

The evaluation board for AT&T's WE32100 32-bit microprocessor (Figure 7-1(a)) allowed the evaluation of the hardware and software capabilities and performance of the WE32100, the WE32101 memory management unit, and the WE32106 math acceleration unit in an application environment. However, if one wished to develop, test, and debug WE32100 microprocessor-based applications in an integrated hardware and software system, it was necessary to employ the WE321DS Microprocessor Development System described in Chapter 4.

The microprocessor operated at clock frequencies up to 18 MHz with accesses out of the RAM in memory bank 2 capable of being performed with zero wait-states. The WE32101 clock was used as the 18-MHz source. Other components included 64K of ROM containing the interactive monitor program and the system self-test, 96K of RAM, a programmable peripheral interface, three programmable 16-bit counter/timers, a programmable interrupt controller with seven available levels, a dual UART, and an eight-character LED display. Memory was physically divided into three banks. Bank 1 and bank 2 RAM were provided for program development and scratch pad space. Bank 0 and bank 1 were supplied with 64K of ROM and 32 bytes of RAM, respectively. However, they could support a wide variety of EPROM/RAM combinations, as follows:

Optional Bank 0 and Bank 1 Memory Configurations

	Memory Type	*Memory Size*
ROM	2764 EPROMs	32K
	27128 EPROMs	64K
	27256 EPROMs	128K
RAM	6264 RAMs	32K

Communication with the evaluation board was provided from two RS-232C serial I/O ports. One of these ports was configured for terminal communications, while the other was configured for communications with a host system running a UNIX operating system. This arrangement allowed for downloading of programs into the evaluation board for execution and debugging without reconfiguring the cabling. A typical cabling of the evaluation board with a terminal and a host system is shown in Figure 7-1(b).

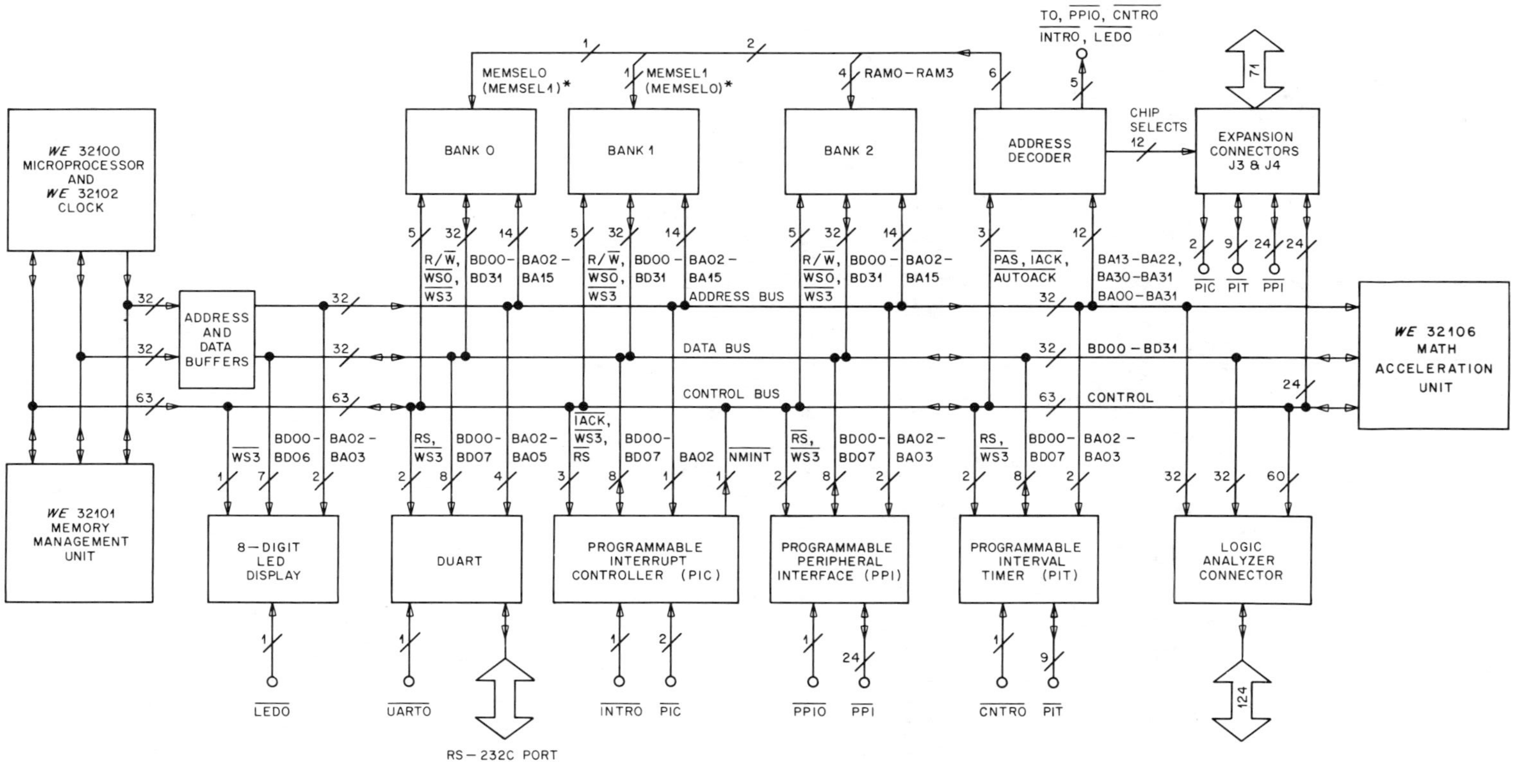

* MEMSELO AND MEMSEL1 ASSIGNMENTS ARE SWITCH SELECTABLE.

FIGURE 7-1(a) Evaluation board for AT&T's WE32100 32-bit microprocessor. Reproduced with permission © 1986 AT&T.

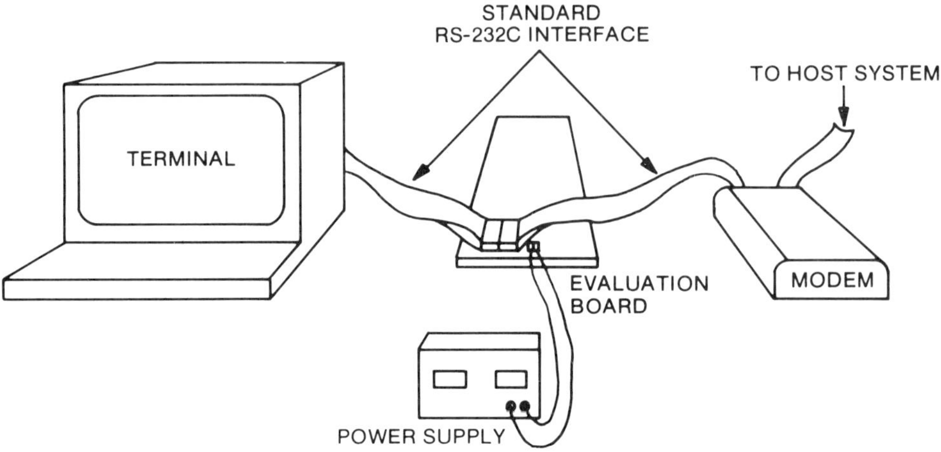

FIGURE 7-1(b)
Typical cabling of 32-bit evaluation board to terminal and host system. Reproduced with permission © 1986 AT&T.

In addition to being supplied with an interactive monitor program and a system self-test program in a 64K EPROM, the evaluation board was also supplied with the WE321SE Evaluation Software Programs for use on a host UNIX system. The interactive monitor program was an assembly-level, hex-oriented debugging tool designed to aid in the verification of hardware and the debugging of firmware. Some of its key features are as follows:

1. Download object files from host system
2. Initiate program execution
3. Execute program trace
4. Display and set software breakpoints
5. Select baud rates (300–9600 baud) for host/terminal communications
6. Support math acceleration software
7. Disassemble memory

The system self-test program provided for checking the minimal functional performance of the board. The results of the self-test were displayed on LEDs, which are on the board. The WE321SE Evaluation Software Programs provided an I/O library and supported the interactive monitor program. Cross software generation programs were also available. These programs, which were accessed while the interactive memory program was in the transparent mode, allowed users to create and debug programs before downloading to the evaluation board. When using the WE32106 math acceleration unit as part of the evaluation application, the MAU mode of the cross software generation program had to be used to provide the MAU support.

A SPARC Evaluation Board

The SPARC Evaluation Board described below is a version using the CY7C601 integer unit, with a CY7C608 floating-point controller and a CY7C609 floating-point processor. A later version using the CY7C601 with a CY7C602 floating-point unit with a CY7C604 cache controller and memory management unit is described in Appendix A.

The SPARC Evaluation Board is a single-board computer designed to show the SPARC chip set (integer unit, floating-point controller, and floating-point processor) operating at maximum speed. It contains high-speed static RAM, EPROM, and RS-232 channels. It is built on a VME card form factor and requires only a single +5-V supply for operation.

It is primarily intended to allow the downloading and running of small (less than 256K) benchmark programs. If the board is used with a Sun workstation, it can be plugged into an empty back-plane slot and receive the required power directly and, thus it will not require an external power supply. It can, in this situation, be connected to the Sun workstation through an RS-232 link to port B of the workstation.

The SPARC Evaluation Board contains a software program SPARC-Mon, written in Forth, that provides control over the board and its functions. Through SPARC-Mon you can download and run programs, read and write hardware ports, perform diagnostic tests, fill and dump memory, and so on.

SPARC Evaluation Board Features

Hardware

1. CY7C601 SPARC IU (integer unit) (20-Mips RISC processor)
2. Floating-point unit
   a. CY7C608 floating-point controller
   b. CY7C609 floating-point processor
3. SRAM, 256K, 17 ns (0 wait states)
4. EPROM, 256K/512K, 55 ns (2 wait states)
5. Dual RS-232 channels (12 wait states)
   a. 300 baud to 38.4Kbaud
   b. Polled or interrupt driven
6. Counter (for timing software, 64 wait states)
   a. 24-bit, binary

   b. 542.5 ns/count
   c. 1 to 16777215 counts
   d. Counter can be cleared, started, stopped, and read
7. LEDs (12 wait states)
   a. 7, addressable and writable
   b. 1, showing ERROR/output of IU
8. Wait-state logic for I/O and EPROM uses "stretched clock" method to generate 0–255 wait states
9. Valid address space logic checks that all accesses are valid and will generate an NMI (trap 1Fh) if not
10. Operates over a frequency range of 10–33 MHz at ambient temperature and nominal +5 V; accepts both 1x and 2x crystal oscillators
11. Requires approximately 3 A @ +5 V from a standard VME P1 and/or P2 connector
12. VME form factor card (9U) and connectors so that it will plug into the backplane of a Sun workstation; does not have a VME bus interface

Firmware

1. SPARC-Mon monitor
   a. Complete Forth interpreter
   b. Approximately 120K of RAM
   c. User applications in the range of 80000100h–8003E000h to avoid conflicts with SPARC-Mon memory usages
   d. Provides complete control over the board's hardware resources
   e. Built-in routines to access the floating-point registers and to perform floating-point arithmetic
   f. Handles communications over both serial ports with a terminal or with UNIX via the "tip" program
2. User can insert EPROMs/programs and run out of EPROM but will pay a two wait-state penalty/access

Basic Evaluation Board Timing

The CY7C601, running at 33 MHz, makes address and control signals available 26 ns after the clock rising edge (RE), which gives 4 ns setup time before the following clock RE to latch them. It also holds these signals until 7 ns after the next clock RE. This allows plenty of time for an external latch to latch these signals using the same clock RE as the IU. Using a fast, octal, D-Reg. such as a 74FCT374A more than meets the setup and hold parameters but, more important, gives a very fast clock-to-output of 1.5 to 6.5 ns.

A Data Store on the 7C601 is one of the few accesses that takes more than one clock cycle to complete. It takes two clock cycles and puts out the same address in both cycles. Normally, in a cache-based system, the first cycle is used to check if there is a hit, but in this design it is unnecessary. The 7C601 writes out data 15 ns after the first clock falling edge (FE) and holds the data until 4 ns after the second clock FE. Using an octal, transparent latch such as a 74FCT373A allows us to capture and hold the data throughout the second clock cycle. The latch-to-output delay is 2 to 8.5 ns.

During a Data Load cycle the IU requires that data be valid no later than 3 ns before the clock RE and that it be held for 5 ns after the clock RE. See Figure 7-4, which shows the timing requirements of the SPARC Evaluation Board.

Detailed Block Diagram Description

Figure 7-2 shows a basic block diagram of the SPARC Evaluation Board. Figure 7-3 is a detailed block diagram of the SPARC Evaluation Board.

The logic can be broken down into the following sections:

FIGURE 7-2
Basic block diagram of SPARC Evaluation Board. Reprinted courtesy of Cypress Semiconductor.

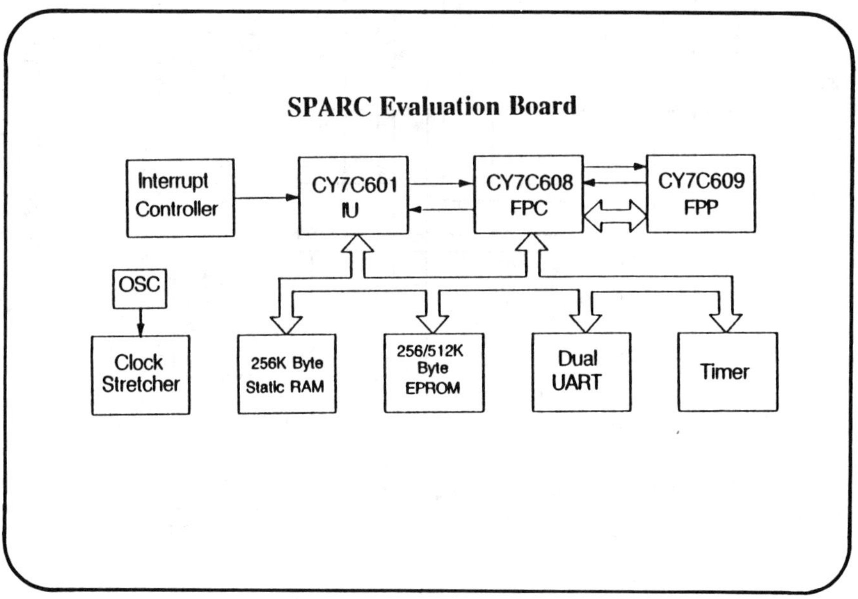

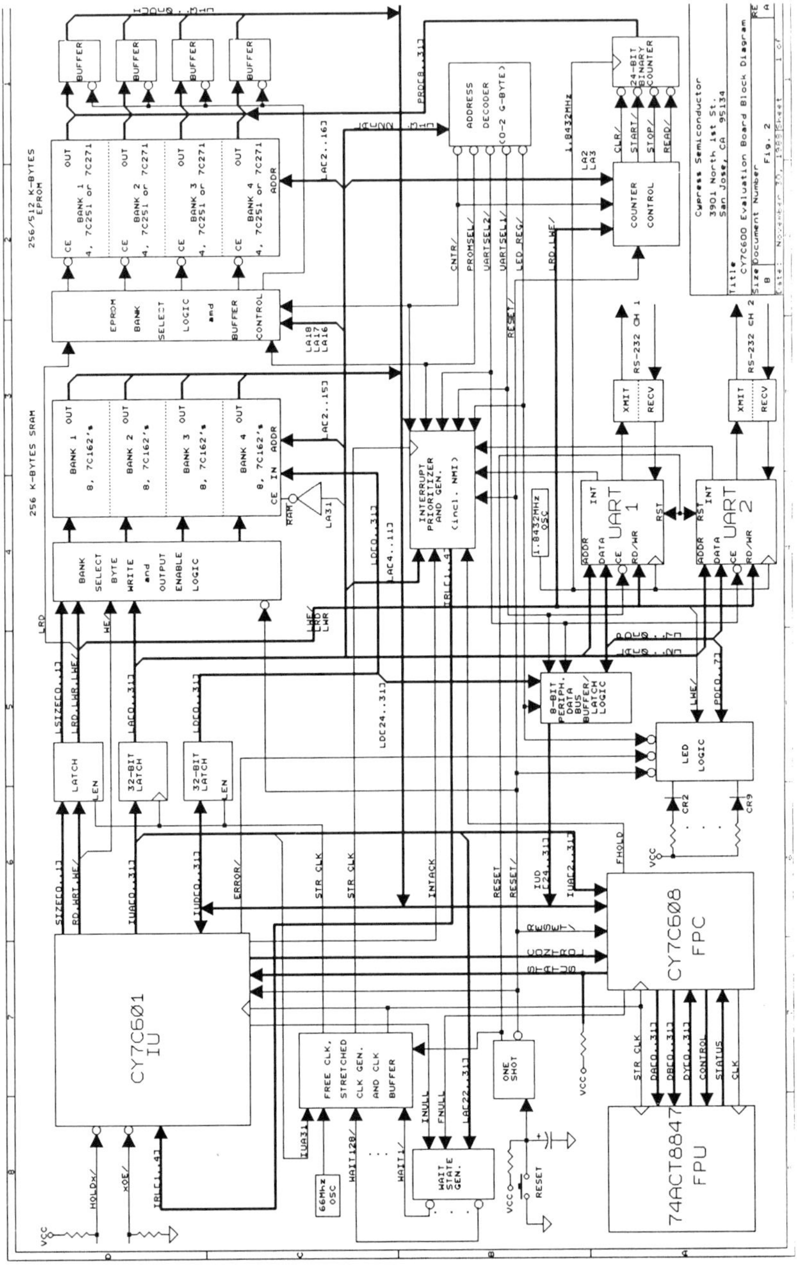

FIGURE 7-3 CY7C601 Evaluation Board timing. Reprinted courtesy of Cypress Semiconductor.

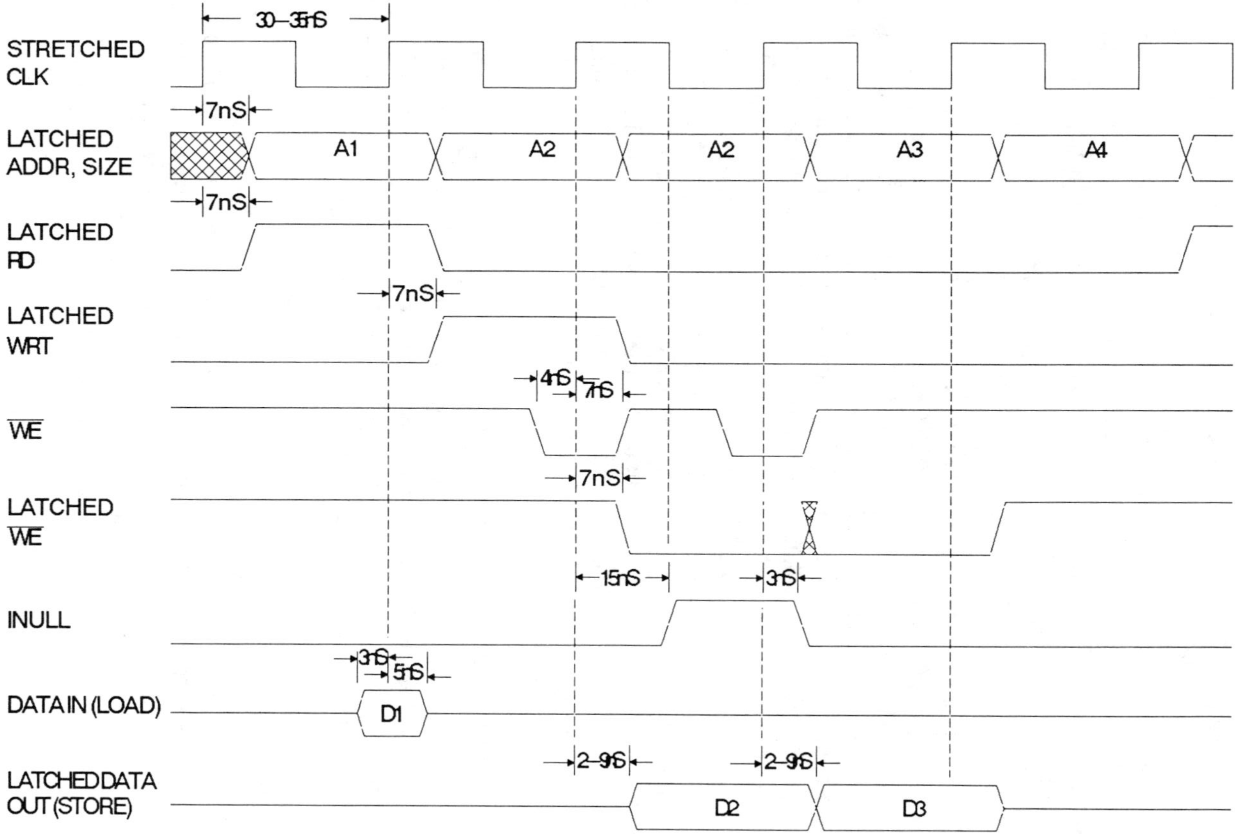

FIGURE 7-4 Detailed block diagram of SPARC Evaluation Board using CY7C600 components. Reprinted courtesy of Cypress Semiconductor.

Reset Logic

The SPARC Evaluation Board generates a Reset on power up and when the Reset button is depressed. Reset does not clear memory or IU/FPU registers. Therefore, after a Reset (non-power up) all information will be available for debugging purposes.

Reset will turn all of the LEDs on for diagnostic purposes. The trap error LED will only remain on for approximately 0.7 seconds and then will turn off. All of the other LEDs will stay on until turned off via software.

Computing Cluster (IU, FPC, and FPP)

The socketed computing cluster consists of the CY7C601 integer unit in a 207-pin PGA package, the CY7C608 floating-point controller in a 299-pin PGA package, and the CY7C609 floating-point processor in a 208-pin PGA package. These chips are connected in a tightly coupled configuration to provide both integer and floating-point capabilities. The board does not make use of the ASI (address space identifier) bits of the IU, any coprocessor signals, nor the MHOLD/BHOLD signals. The MHOLDA/, MHOLDB/ and BHOLD/ lines have pull-up resistors on them. There is a jumper, JP1, that should be set to the "NO FPU" position when a CY7C609 is not physically installed. There is a pull-down resistor on the FNULL line from the FPC so that if an FPC and FPP are not present, the IU will still operate correctly. There are pull-down resistors on the AOE/, COE/, DOE/, and TOE/ lines of the IU so that its outputs are always enabled.

Address/Data Latches

Since addresses from the IU and data to/from the IU are only valid around the IU clock rising edge, it is necessary to latch them in order to be used throughout the board. Fast octal D latches such as the 74FCT374A are used to latch "write" data during the positive portion of the clock.

Clock Circuitry

1. *Crystal Oscillator*. The SPARC Evaluation Board operates off of a socketed crystal oscillator so that its operating frequency can be easily changed. The board will accept an oscillator that operates at either the correct (1×) frequency or at twice (2×) the frequency. A jumper, JP7, chooses between the 1× frequency (OSC) and the 2× frequency (OSC/2). This is necessary, since the IU has certain minimum high and low times for its clock input and some crystal oscillators do not put out a "symmetrical"

square wave. In this case it is necessary to use a 2× frequency and divide it in half by a D F/F to "square it up."

2. *Wait State Generator.* This is a 7-ns TI PAL that decodes the upper latched addresses (A31–A22) from the IU as well as the INULL signal from the IU and a signal to disable INULL during the second clock of a write cycle. Its outputs are the WAIT1/ through WAIT128/ signals that are the parallel inputs to the fast binary wait state counters (74FCT163A).
3. *Stretched Clock (Wait State) Logic.* The SPARC Evaluation Board delays (holds) the IU not by the use of MHOLD and MDS signals but by "stretching" the clock. Since the IU operates on clock edges, this mechanism "fools" the IU by delaying the next clock RE the required number of wait states.

The decision of whether to stretch or not is based on address line A31. This scheme is used to simplify decoding but mainly to speed up letting the clock stretcher know whether to stretch or not. The address space of the SPARC evaluation board is divided in half by A31. If the address is in the upper 2-GB (FFFFFFFFh-80000000h) range then it is a RAM access and proceeds with no wait states. Otherwise in the range of 7fffffffh–00000000h, a variable number of wait states will be generated depending on what type of I/O or EPROM access it is. See Figure 7-5 for the timing of the stretched clock logic.

If the clock cycle is to be stretched, then the number of wait states (minus 1) is loaded into the 4-bit binary counters and at the same time the stretched clock F/F (JK) is set, which disables, or stretches, the clock. When the cascaded pair of counters reaches terminal count (TC), then the JK F/F toggles and the clock is enabled to the computing cluster and the rest of the board. The JK F/F has logic on its CLR input to prevent clock stretching during Reset and the first clock of a write cycle. As can be seen from the schematic of the clock stretcher logic, if clock stretching is started it cannot be stopped quickly enough to annul wait state insertion. Therefore, the minimum delay will be 1 cycle.

Actual clock stretching (or holding) is done in the clock buffer chip, which is a hex NOR gate. It is a 74AS1805 from TI that can sink and source 48 mA and, typically, has less than 0.5 ns skew between its outputs. Both stretched clocks and free-running clocks are output from this buffer.

The function of the stretched clock is to provide extra access time. It extends the low portion of the clock cycle as necessary to complete the required access. Figures 7-6a, b, and c show block diagrams of the clock stretcher, and 7-6d shows the stretching in a timing diagram.

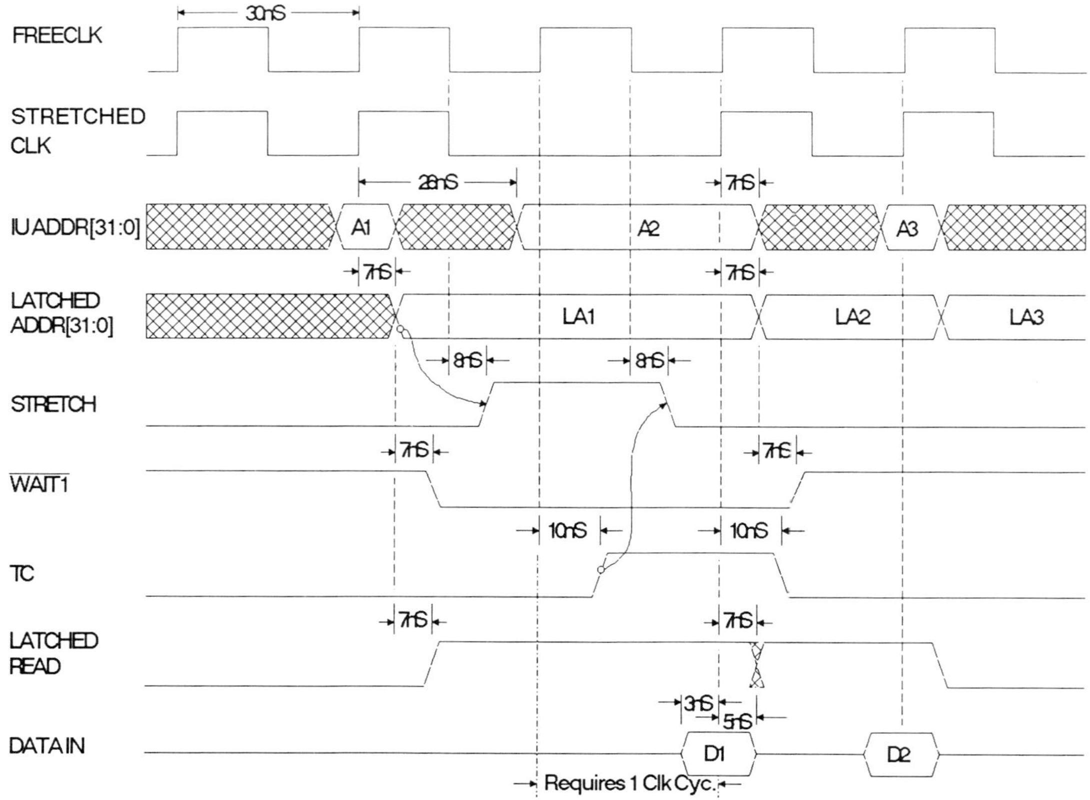

FIGURE 7-5 Timing of stretched clock logic (1 wait state hold timing at 33 MHz). Reprinted courtesy of Cypress Semiconductor.

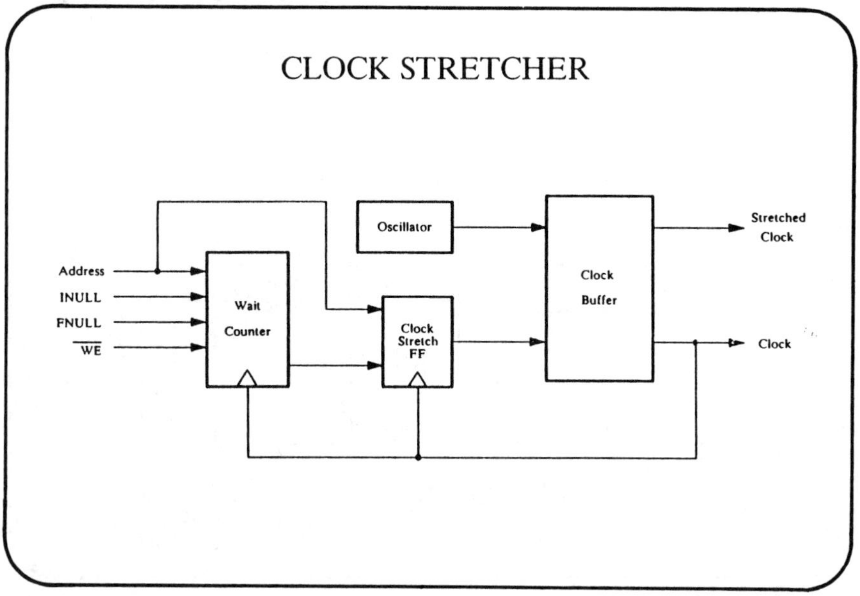

FIGURE 7-6(a)
Block diagram of stretched clock circuit. Reprinted courtesy of Cypress Semiconductor.

FIGURE 7-6(b)
Stretched clock waveforms versus original clock. Reprinted courtesy of Cypress Semiconductor.

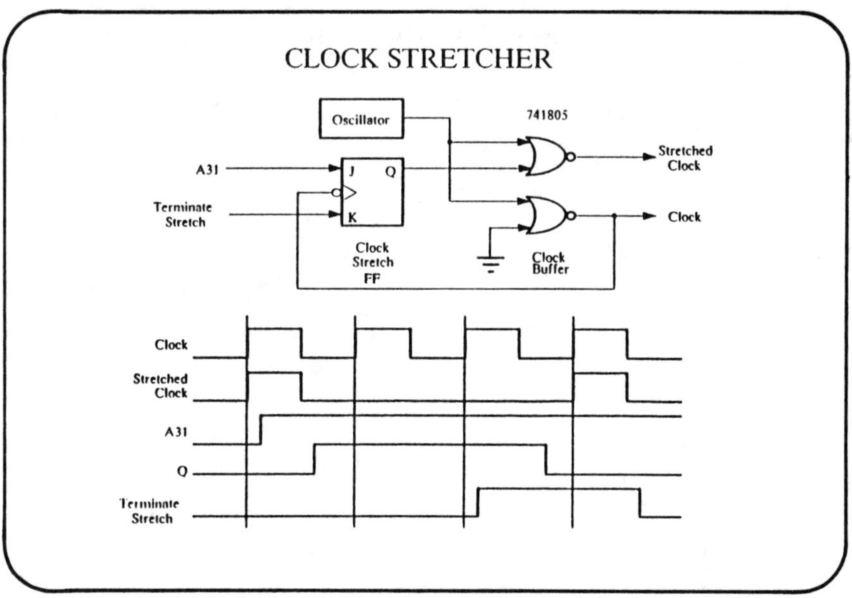

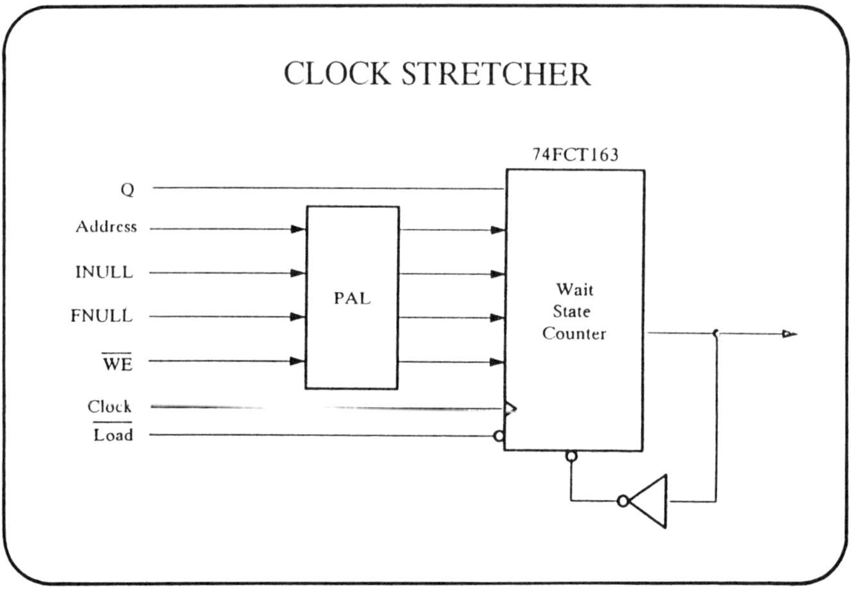

FIGURE 7-6(c)
Wait-state generator. Reprinted courtesy of Cypress Semiconductor.

FIGURE 7-6(d)
Stretched clock timing. Reprinted courtesy of Cypress Semiconductor.

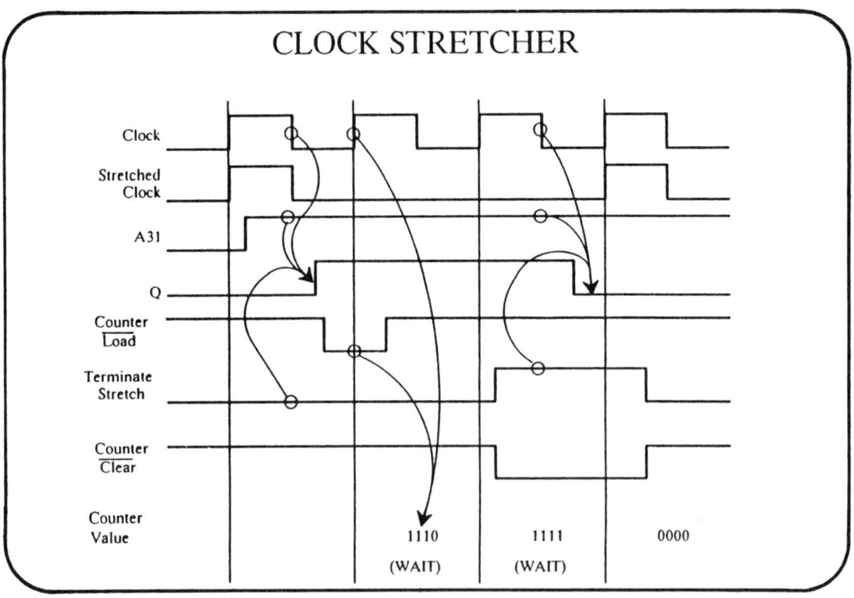

Address Decoder Logic

This is a 7-ns PAL from TI that decodes address lines A31–A22 in order to generate the chip select/enable signals for I/O and EPROM. As can be seen from the address range decoded, each address space is 4 MB.

Interrupt Logic

The SPARC Evaluation Board supports interrupts from two sources; each UART can generate a maskable interrupt to the IU, and there is logic to check address space validity that will generate a nonmaskable interrupt in the event of an invalid address. An invalid address means that the access is to an address not decoded by the address decoder logic.

If it is desired to disable all interrupts on the board, then it is only necessary to remove U125, a 74F04 which drives IRL3–IRL0 pins of the IU. The IU pins have pull down resistors so that no spurious interrupts will be received.

1. NMI Logic. This logic verifies an asserted output from the address decoder and, if not, will cause IRL3–IRL0 pins of the IU to be set to a logic "1," which is an NMI to the IU. This NMI will stay asserted until an INTACK from the IU is received coincident with 00011111b on address lines A11–A4 and provided that the FPC is not holding the IU.
2. MI Logic. Each UART can generate a maskable interrupt to the IU; UART 1 generates INT 5 to the IU and UART 2 generates INT 3 (lower priority) to the IU.

SPARC-Mon does not support interrupt-driven communications, that is, the UARTs are polled during operation. SPARC-Mon sets the interrupt level in the PSR to 0Fh so that all MI are ignored. However, on receiving an NMI, it will generate a "Trap 1F" message. If the interrupt level in the PSR is lowered to 02h, then SPARC-Mon will generate a "Trap 15" or "Trap 13" message for a UART 1 or 2 interrupt, respectively.

Static Random Access Memory

The SPARC Evaluation Board contains 256K of SRAM using 32 socketed 16K × 4 SRAMs, which can be used to store and run programs and which are used by SPARC-Mon for its required stack space. The RAM configuration allows byte, half-word, word, and double-word accesses.

It is necessary to latch the Address and Control signals as well as the

Write (Store) Data Signal. However, data being returned from the RAM array (Load) is connected directly to the data pins of the IU without intervening logic to delay the data. This is feasible due to the small size of the memory array, which does not place too heavy a load on the IU data bus. This also necessitates using separate I/O SRAMs so that the inputs can be connected to the latched data bus and the outputs can be connected directly to the IU pins.

1. *SRAM Address Timing*. Address signals LA15–LA2 will be latched and valid at the SRAMs no later than 7 ns after clock RE. See Figure 7-7 for a diagram of SPARC Evaluation Board memory timing and for the following timing analysis.
2. *SRAM Chip Enable (CE) Timing*. There are two CE pins on each 16K × 4 SRAM chip, and these are used as Bank Select signals for the SRAM array. Latched Address signals LA17–LA16 and their inverses are used to generate the CE/Bank Select signals. The delay through the inverters, 74AS1004As, is 1 to 4 ns. Therefore, worst-case SRAM selection is as follows:

   7 ns (Latch) + 4 ns (Inv) = 11 ns, max, after clock RE

3. *SRAM Output Enable (OE) Timing*. The SRAM OE signal is generated by the NANDing of Latched RD (LRD) and LA31 (since the SRAMs should only be driving during a RAM read) which, for a 74AS1000, adds 1 to 4 ns delay to the latched signals. Therefore, OE is valid as follows:

   7 ns (Latch) + 4 ns (NAND) = 11 ns, max, after clock RE

4. SRAM Byte Write Enable (BEW/) Logic and Timing. The 7C601, on a Store cycle, can write a byte, a half-word (16 bits), a word (32-bits), or a double-word (64-bits), so we have to generate the correct Write strobes depending on the type of access. Also, the IU can perform a Store Single access (32-bits), which requires two cycles, or a store Double (64 bits), which requires three clock cycles.

In order to generate the proper logic equations we need to use the Latched SIZE1–SIZE0 signals, the Latched Address signals LA31, LA1, and LA0, LRD, and Latched Write enable (LWE/). Signals with a / after them are low true. The following tables show the CY7C601 addressing conventions for how addresses and control signals are generated to access various sizes of data.

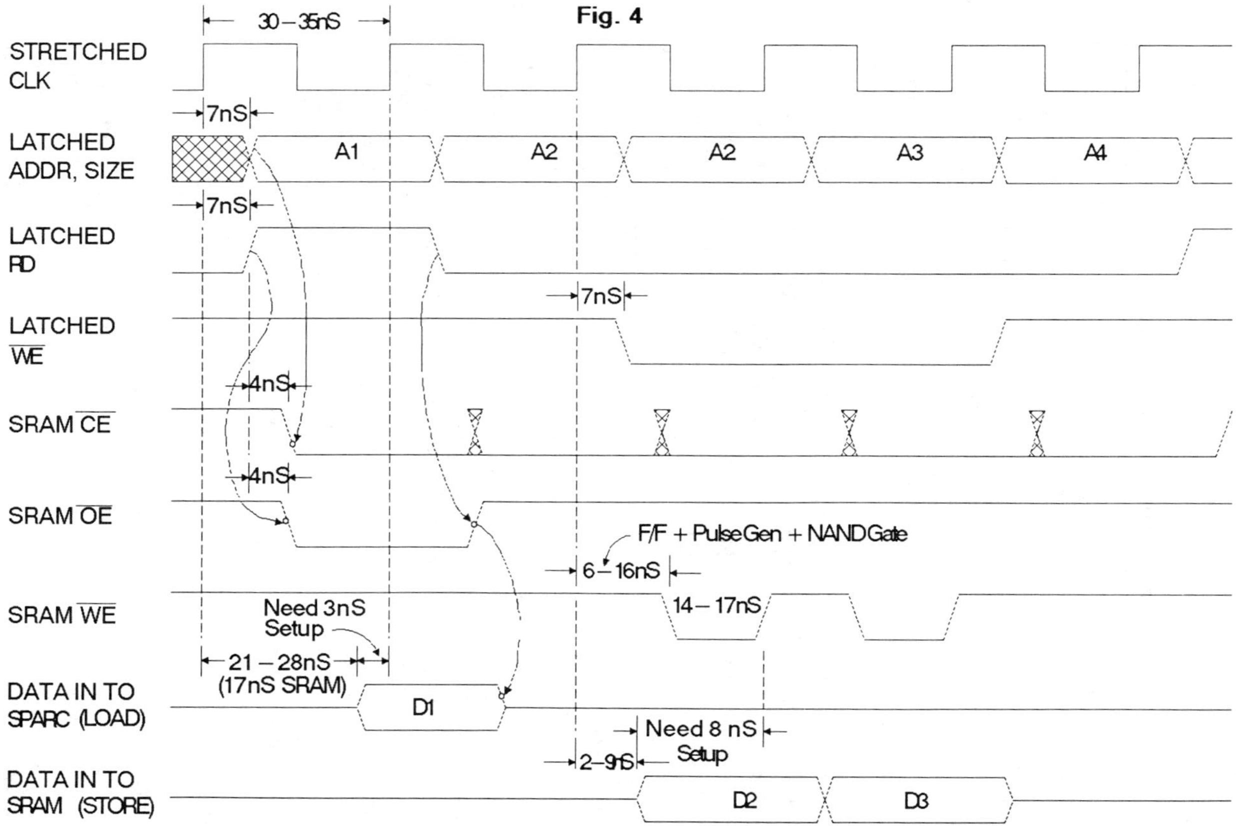

FIGURE 7-7 CY7C601 Evaluation Board memory timing. Reprinted courtesy of Cypress Semiconductor.

<table>
<tr><td colspan="8">63 Double Word 0</td></tr>
<tr><td colspan="4">31 Word 0 (MS, A2 = 0) 0</td><td colspan="4">31 Word 1 (LS, A2 = 1) 0</td></tr>
<tr><td colspan="2">15 Halfword0 A1 = 0 0</td><td colspan="2">15 Halfword1 A1 = 1 0</td><td colspan="2">15 Halfword2 A1 = 0 0</td><td colspan="2">15 Halfword3 A1 = 1 0</td></tr>
<tr><td>7 Byte0 0</td><td>7 Byte1 0</td><td>7 Byte2 0</td><td>7 Byte3 0</td><td>7 Byte4 0</td><td>7 Byte5 0</td><td>7 Byte6 0</td><td>7 Byte7 0</td></tr>
</table>

Size 1	*Size 0*	*Data Transfer Type*
0	0	Byte
0	1	Halfword
1	0	Word
1	1	Word (Load/Store Double)

A1	*A0*	*Byte*
0	0	0 (MS Byte)
0	1	1
1	0	2
1	1	3 (LS Byte)

As can be seen from the SIZE1–SIZE0 table, as long as SIZE1 = 1 then all BWE3–BWE0/ signals should be asserted low. When SIZE1 = 0 and SIZE0 = 1, then this is a halfword access and A1 is used to determine which halfword is being accessed. BWE1–BWE0/ (MS Halfword) will be asserted when A1 = 0 and BWE3–BEW2/ (LS Halfword) will be asserted when A1 = 1. Finally, when SIZE1–SIZE0 = 00, this is a byte access and A1–A0 are used to determine which BWEx/ lines will be asserted.

On the SPARC Evaluation Board a 16L8-15 PAL is used to generate the correct logic signals from the above signals. LA31 is used to determine whether this is a RAM access or not. As long as LA31 is high, then this is a RAM access and the rest of the equation can be processed. The four PAL outputs are used to enable separate NAND gates whose other inputs are the actual timing pulses used to generate the SRAM write strobes. Since the 16L8 PAL has low true outputs, the equations were De Morganized for proper polarity. The enabling signals from the PAL are available by the end of the first cycle of a Store Single access. They will remain asserted until the latched input signals such as Address and Size change after the end of the second cycle on the next clock RE. On a Store Double access they will remain asserted through both the second and third cycles, since Address, SIZE, and WE/ don't change throughout the access.

The actual write timing strobe is generated by a TTL pulse generator IC, which generates both positive and negative pulses, 14 to 17 ns wide, from an RE input. This component can be either a standard or custom part (delay line) to generate the pulse width required by the particular SRAM used. The signal to start a write is derived by clocking WE/ through a D-F/F with a stretched clock. Pulse timing is such that BWEx/ will be asserted at the SRAM 6 to 16 ns after clock RE and will be deasserted 14 to 17 ns later when data is actually written into the SRAM.

5. *SRAM Read Data Timing*. The longest path between the IU and the SRAMs is the chip enable path, since it takes 11 ns, maximum. Given a 30-ns cycle time, the following occurs:

   30 ns − 11 ns (CE) − 3 ns (IU set) = 16 ns, max, SRAM access time

6. *SRAM Write Data Timing*. As we stated earlier, a transparent latch adds 9 ns to the data from the IU. Therefore, Write data is valid at the IU 9 ns after clock RE in the second clock cycle. A typical fast SRAM such as the Cypress CY7C161 has a minimum Write pulse width and a minimum Write Data Setup time before Write pulse deassertion (RE). For a 17-ns access SRAM, the minimum Write pulse width is 14 ns and the minimum Write Data Setup time is 8 ns. Again, see Figure 7-7 for the timing analysis.

Erasable Programmable Read Only Memory

The SPARC Evaluation Board contains either 256K of socketed EPROM using 16 CY7C251 chips or 512K of socketed EPROM using 16 CY7C271 chips. SPARC-Mon occupies approximately the first 120K of EPROM but, if SPARC-Mon is removed, the entire amount of EPROM space is available for customer code, real-time operating systems, and the like. All EPROM access (code) by the IU are word (32 bits) wide.

Each EPROM has multiple chip enable (CE) inputs that are used to select the chip, enable its outputs, or put it in a low-power mode. When using the CY7C251 chips, address lines A17–A16 and their inverses are used to select the four banks of EPROMs. When using CY7C271 chips, address lines A18–A17 and their inverses are used to select the four banks of EPROMs. In addition, there is a master EPROM select line called PROMSEL/ from the address decoder that decodes the EPROM address space. The particular decoding used is controlled by jumpers JP2–JP6 on the board. These jumpers should all be set to the 251 position when using the 16K × 8 EPROMs and to the 271 position when using the 32K × 8 EPROMs.

Data from both the EPROM array and the 24-bit counter is buffered by a set of 74AS244 octal buffers before reaching the IU data pins in order to reduce IU data pin loading. This can be tolerated, since EPROM and I/O have wait states inserted for any access and the added layer of buffering does not increase the number of wait states.

1. *EPROM Address Timing*. Address signals LA15–LA2 will be latched and valid at the EPROMs no later than 7 ns after clock RE. See Figure 7-8

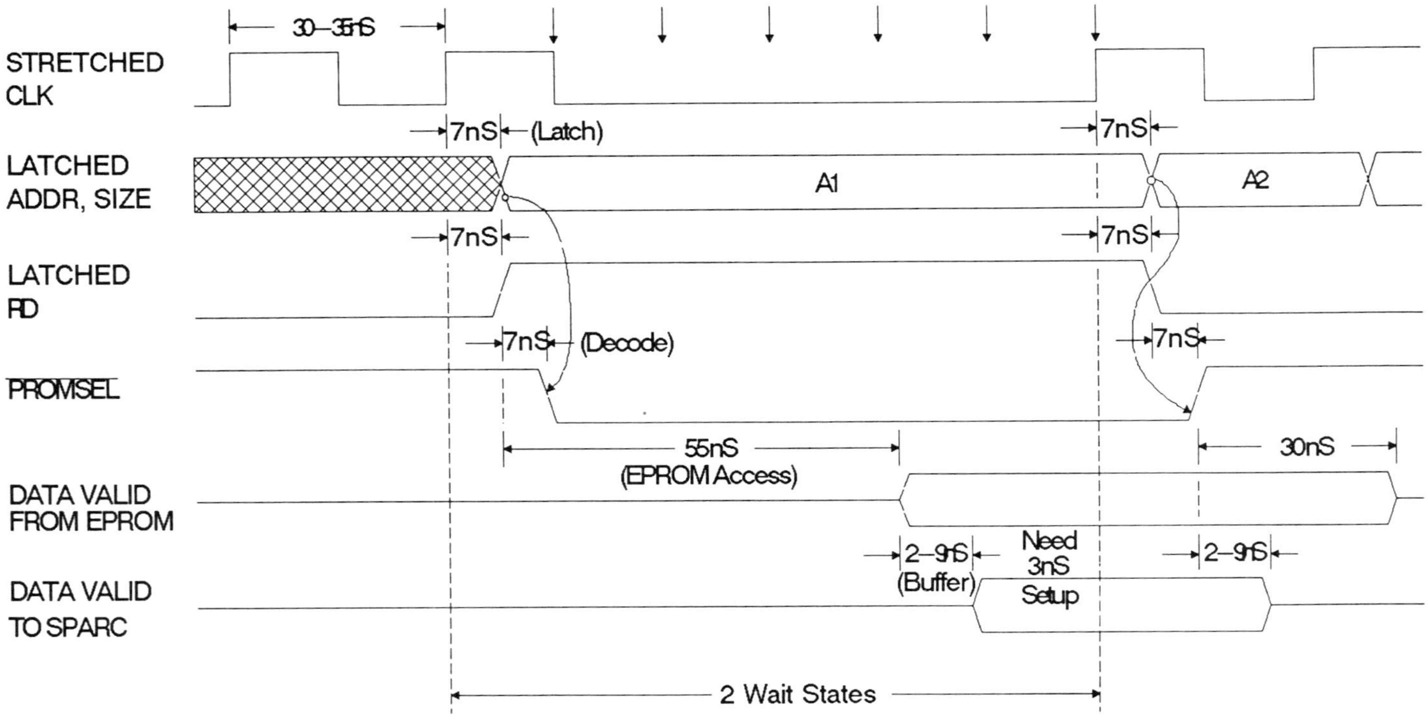

FIGURE 7-8 CY7C601 Evaluation Board EPROM timing. Reprinted courtesy of Cypress Semiconductor.

for a diagram of SPARC Evaluation Board EPROM timing and for the following timing analysis.

2. *EPROM Chip Enable (CE) Timing*. There are multiple CE pins on each EPROM chip, and these are used as Bank Select signals for the EPROM array. Latched Address signals LA17–LA16 (LA18–LA17 for CY7C271) and their inverses are used to generate some of the CE/bank select signals. The delay through the inverters, 74AS1004As, is 1 to 4 ns. Therefore,

   7 ns (Latch) + 4 ns (Inv) = 11 ns, max, after clock RE

   However, the worst timing path is from the address decoder to one of the CE pins. This timing is as follows:

   7 ns (Latch) + 7 ns (Decode) = 14 ns, max, after clock RE

3. Using 55-ns access EPROMs says that data will be available at the IU as follows:

   14 ns (Select) + 55 ns (Access) + 9 ns (Buffer) = 78 ns, max, after the clock RE

which, for a 30-ns clock cycle, indicates that we need two wait states.

RS-232 Serial Communication Channels

The SPARC Evaluation Board contains two RS-232 serial channels. Each channel conforms to the RS-232 definition of data terminal equipment (DTE). This means that if the board is to communicate with other DTE, such as another computer, then a "Null Modem" cable must be used. The connector type is a DB25 male connector. SPARC-Mon initializes both channels to 9600 baud, 8 data bits, no parity, 1 stop bit and tries to communicate using Serial Channel 1 (also called Channel A).

Each channel uses a National Semiconductor NS16550A UART operating off of a 1.8432-MHz crystal oscillator and Maxim interface chips so that +/− 12-V supplies are not required. Reading and writing the UART is accomplished via byte 0, data bits 24–31.

The base address of UART 1 is 01000000h, and the base address of UART 2 is 01400000h. Their registers can be accessed by reading/writing byte 0, data bits 24–31, at the following addresses:

Receiver Buffer (Read)—01000000h (01400000h) (only when DLAB bit (MSB) of the Line Control register is cleared to "0")

Transmit Holding Register (Write)—01000000h (0140000h) (only when DLAB bit (MSB) of the Line Control register is cleared to "0")

Interrupt Enable—01000004h (01400004h)

Note that UART 1 generates INT 5 to the IU and UART 2 generates INT 3 to the IU

Interrupt Identification (Read)—01000008h (01400008h)

FIFO Control (Write)—01000008h (01400008h)

Line Control—0100000Ch (0140000Ch)

Modem Control—01000010h (01400010h)

Line Status—01000014h (01400014h)

Modem Status—01000018h (01400018h)

Scratch—0100001Ch (0140001Ch)

Divisor Latch (LSB)—01000000h (01400000h) (only when DLAB bit (MSB) of the Line Control register is set to "1")

Divisor Latch (MSB)—01000004h (01400004h) (only when DLAB bit (MSB) of the Line Control register is set to "1")

Note: The output frequency of the internal baud rate generator is 16 × the baud rate. The 16-bit divisor is equal to 1843200/(baud rate × 16). For instance, 9600 baud requires a divisor of 12.

For further information concerning programming and using the NS16550A, see the National Semiconductor data sheet. The following data, control, and status signals are supported:

- XMIT Data (Pin 2)
- RECV Data (Pin 3)
- Data Terminal Ready (Pin 20)
- Data Set Ready (Pin 6)
- Request to Send (Pin 4)
- Clear to Send (Pin 5)
- Data Carrier Detect (Pin 8)
- Signal Return (Pin 7)

Note that the Ring Indicator signal is not supported.

24-bit Binary Counter

The 24-bit binary counter is a discrete counter implementation consisting of a 22V10 PAL and three 74LS590 counters that are concatenated to produce the 24-bit count. The counters operate off of the 1.8432-MHz clock (542.5 ns/count) that drives the UARTs. The 22V10 implements the control of the counters by accepting a chip select signal (CNTR/) from the address decoder as well as address lines A3 and A2, the latched RD and WE/ signals as well as RESET/. The PAL also synchronizes these IU signals (from the stretched clock) to the 1.8432-MHz clock used by the counters. These two stages of synchronization are what cause the large number of wait states, 64, to access the counter.

The counter does not receive any commands from the data bus but, instead, decodes its commands from the address bus. The following commands are supported:

1. *Clear* the counter. Depressing the Reset switch or writing anything to address 00400000h will clear the counter to all zeros. It is necessary to clear the counter before timing anything, as it is the only way to load the counter.
2. *Start* the counter. Writing anything to address 00400004h will cause the counter to start counting.
3. *Stop* the counter. Writing anything to address 00400008h will cause the counter to stop counting. The count will be held until the next Clear or Start command.
4. *Read* the counter. A read of address 0040000Ch will place the contents of the counter on the data bus. Counter bits 0–7, the least-significant bits of the count, will be placed on byte 0 of the bus, data bits 24–31. Counter bits 8–15 will be placed on byte 1 of the bus, data bits 16–23. Counter bits 16–23, the most significant bits of the count, will be placed on byte 2 of the data bus, data bits 8–15. Data byte 3, bits 0–7, will always read FFh.

In general, it is necessary to strip off byte 3, shift the rest of the bytes to their proper order, convert the number to decimal, and multiply by 1/1843200 to obtain 542.5 ns/count to calculate the delay between starting and stopping the counter. Also, one count should be added to compensate for hardware implementation.

LED Register/Display

The LED circuitry consists of a data latch (74ALS273) connected to byte 0, data bits 24–31 at address 01800000h, and the 8 LEDs, CR2–CR9. The register is accessed when the address decoder asserts LED _ REG/. Data bit 24 controls LED CR2. When bit 24 is cleared to a logic "0," then the LED is turned on. When data bit 24 is set to a logic "1," then the LED is turned off. Similarly, data bits 24–30 control LEDs CR3–CR8. However, CR9 (TRAP ERROR) is controlled differently. It gets turned on when the IU asserts an ERROR/signal. It can be turned off only by writing a "0" and then a "1" to data bit 31.

During a Reset all LEDs will be turned on for diagnostic purposes. LED CR9 will turn back off after about 0.7 seconds, but all the rest will remain on until turned off via a software write to address 01800000h.

SPARC-Mon Firmwave Monitor

Memory Access

SPARC-Mon can read and write memory locations as either bytes, halfwords (16-bit), words (32-bit), or double words (64-bit). Memory locations may be accessed using any of the SPARC address spaces (ASI numbers). Commands are also provided for displaying (dumping) the contents of a range of memory locations and for filling ranges of memory with various patterns.

Device Access

Since I/O devices are accessed as memory locations, SPARC-Mon may access any I/O device. Optionally, device drivers may be provided for certain devices, allowing a higher-level interface to those devices. Convenient access to serial lines is always provided.

Arithmetic

SPARC-Mon can calculate arithmetic and logical expressions, either interactively or as part of a user-defined program. The set of integer operators includes addition, subtraction, multiplication, division, shifts, and, or, exclusive-or, and others. If floating-point hardware is present, SPARC-Mon may be configured with floating-point operators.

Arithmetic is entered using Reverse Polish Notation, similar to an HP scientific calculator. This allows arithmetic to be easily combined with other SPARC-Mon commands, including memory access. The operands and results of calculations may be entered and displayed in binary, hex, octal, or decimal. SPARC-Mon uses IEEE floating-point format in its operations.

Looping and Conditionals

The SPARC-Mon commands may be combined with control structures, such as subroutines, loops, and conditional statements. Programs created in this fashion may be typed directly at the keyboard. Loops may also be conditional.

Assembler/Disassembler

SPARC-Mon contains a resident interactive assembler/disassembler. Short assembly-language subroutines may be entered directly from the keyboard or downloaded from a host machine over a serial line. The subroutines may be assembled either on the host machine or on the SPARC Evaluation board running SPARC-Mon. Once loaded, assembly-language subroutines located anywhere in memory may be disassembled via the adr dis command. SPARC-Mon can even disassemble itself. To start disassembling at the beginning of SPARC-Mon, address 0, type "0 dis" and it will start disassembling. The assembler has macro capability and structured conditionals (e.g., IF statements), which often eliminate the need for labels in assembly code.

C Program Compiling and Execution Environment

A sample shell script as listed below will compile and link programs for execution on the evaluation board:

```
cc -04 -N -Bstatic -t 80008000 $1.c -o $1.out -1m
cp $1.out $1
strip $1
```

-04;	turns on the optimizer to level 4, i.e., the highest level
-N;	allows the text segment to be writable and places the correct "magic number" in the object module to allow it to be downloadable by SPARC-Mon
-Bstatic;	links all libraries in statically
-T;	locates the text segment at absolute address 80008000h
$1.c;	substitute parameter for C program name
$1.out;	substitute parameter for C executable with symbol information
-lm;	link math libraries as needed

Note 1: The cp command creates a duplicate executable file and the strip command removes the symbol table for downloading by SPARC-Mon. The $1.out file contains symbol information that may be useful when using the adb debugger.

Note 2: A similar shell script could be used with the FORTRAN compiler, e.g., substituting F77 for CC

C programs compiled with the above options can then be downloaded to a SPARC-Mon machine for execution. SPARC-Mon provides an execution environment suitable for such programs, including window overflow and window underflow trap handlers and simulation of some common UNIX system calls, for example, print{. The simulated system calls are sufficient to allow many simple C programs to execute as though they were running on a UNIX machine. Programs in this category include those that use only standard input and standard output for I/O (SPARC-Mon does not require the existence of a disk drive, therefore the C execution environment does not include disk file I/O).

Software Debugging

Programs running under the control of SPARC-Mon may be debugged with conditional breakpoints, single-stepping, tracing, disassembly, and register display. The user may specify an arbitrary sequence of SPARC-Mon commands to execute every time a breakpoint is reached. Those commands may do something as simple as displaying the contents of one or more registers or as complicated as decoding and symbolically displaying a software data structure and possibly restarting based on fields within that data structure.

If the program results in an exception (e.g., Memory Address Not Aligned), SPARC-Mon handles the exception and saves the registers so that the user may examine the conditions that caused the problem.

Any time a program is suspended (i.e., at a breakpoint or after an exception), SPARC-Mon can display a C stack back-trace, showing the nested list of active subroutines and their arguments.

Command Line Editing

For typing command lines, SPARC-Mon has a powerful command line editor, with intraline editing, command completion, and interactive history. The intraline editing features implement a subset of the EMACS (Editing Macros) text editor commands, allowing movement back and forth within the line being typed and replacement of a character or word anywhere in the line. Command completion allows the user to type just part of a command name, and SPARC-Mon will finish the name. History (^p) allows any one of the eight most recent commands to be recalled and reissued, perhaps with some editing first.

Forth Programming Language

SPARC-Mon is based on the Forth programming language. SPARC-Mon contains a complete implementation of Forth 83, which is available to the user if she or he wishes to use it. It is not necessary to know Forth in order to use SPARC-Mon, however. The SPARC-Mon commands may be used in isolation, requiring only the simplest of Forth knowledge (e.g., the number is typed first, then a space, then the operator).

The SPARC-Mon Forth implementation contains many extensions beyond the Forth language as specified in the Forth 83 Standard. Included are such useful programming tools as a structured decompiler, a call finder, and a word finder. These make it very easy to use SPARC-Mon to develop and debug Forth programs for applications such as hardware diagnostics.

Downloading/Uploading

Portions of memory may be transferred to and from a host machine. The transfer uses a serial line, with either the XMODEM protocol or a fast binary protocol. Motorola S-Records are optionally supported, but they are not recommended because the S-Record format is very inefficient.

Either binary code (using dlbin), assembly language source code (using dl), or Forth language source code (using dl) may be transferred.

Configurability

The source code for SPARC-Mon is provided, so the customer may add any custom features desired. SPARC-Mon source code may be recompiled on a Sun-4 workstation. The source code is written in a combination of Forth and assembler. The Forth cross compiler, which runs on a Sun-4, and a native Forth for the Sun-4 are provided as part of the SPARC-Mon package. This allows you to test Forth code under UNIX if you desire.

A MIPS R3000 Evaluation Board

Integrated Device Technology, Inc., has developed their 7RS382 board as a design of a typical embedded application of their IDT79R3000 RISC microprocessor. This board is a complete RISC system for use as a low-cost evaluation system. It consists of an R3000 CPU, IDTR3010 FPA (optional), 16K instruction and 16K data cache, 128K RAM, 128K EPROM, a real-time counter, and a two-channel UART. An expansion bus on the board provides all of the signals that a designer needs to add a DRAM system or other specific I/O. Except for the expansion bus, this board is typical of the heart of an embedded microprocessor-based control for LANs, laser printers, avionics, and the like. The differences would center on which peripherals are used and the memory requirements.

This board was designed as an example for designers of the construction of a system using their IDT79R3000 in a low parts count, cost-effective solution. It was not meant to show a design for the highest performance achievable. For example, it does not include cache block refill. Higher levels of performance can be achieved through further optimizations and tradeoffs. This design uses 4K × 16 cache memories, a generic single-level read/write buffer, and SRAM for the main memory. The cache memory has been optimized to accommodate caching up to 8 MB of main memory, while minimizing parts count, PAL states, and system cost. Although the board is designed with SRAM for simplicity, the width of the cache and the read/write control logic has been designed with DRAM-size memory in mind. A basic block diagram of the board is shown in Figure 7-9.

The Generic R3000 CPU System

A generic R3000 CPU subsystem consists of CPU/FPA, cache memory, read/write buffers, read/write control logic, and byte decode logic. The system memory is responsible for decoding addresses and responding with an

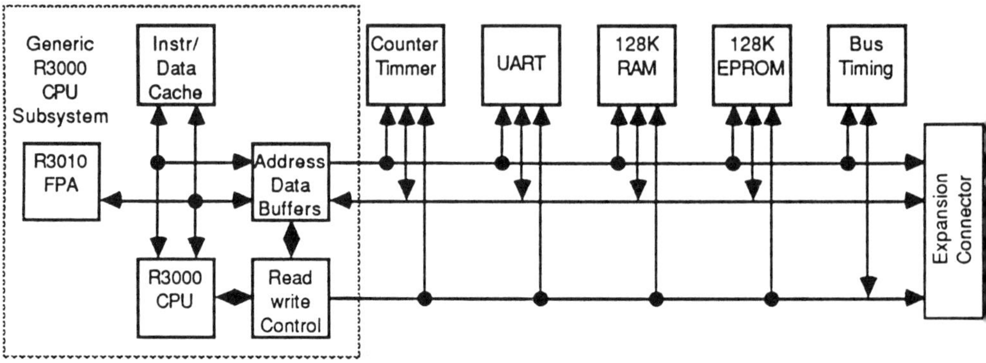

FIGURE 7-9
Block diagram of IDT's 7RS382 Evaluation Board. Reprinted courtesy of Integrated Device Technology.

acknowledge (similar to DTACK) when enough time has passed for the memory or I/O function. A block diagram of a generic R3000 CPU with memory is shown in Figure 7-10.

The cache memory of the R3000 uses standard architecture SRAM to store recently accessed data and instructions. The R3000 generates signals that drive the SRAM directly, connecting to the OE/ and WE/ pins. For instructions and data reads, the R3000 outputs an address on each clock phase to the cache memory and tests the results for a match (hit). If the data is not in the cache (miss), the R3000 drops from synchronous cache access mode to use the asynchronous handshake signals to talk to the system memory. The R3000 can be programmed at initialization time to bring back either the word needed or a block of surrounding words. The block refill choice raises the probability of a hit on the next cache access. If a block of instructions is being fetched from system memory, the R3000 can be initialized to execute each word as it is placed into the cache (streaming) rather than having to reread the cache to proceed.

The interface between the R3000 CPU and the R3010 FPA is the cache data bus and several control signals. The FPA synchronizes with the CPU by monitoring the control signals. On each instruction fetch the FPA decodes the operation code and determines whether it is a floating-point operation. In the event that it is, appropriate action is taken. Examples are transferal of data to and from the data bus and internal registers, or performing arithmetic operations. The FPA also incorporates an internal phase lock loop to match its clock-to-data timing with that of the CPU. The interconnect is straightforward pin to pin. The length of the clock lines between the FPA and the CPU must be matched as noted in the schematic and data sheets.

In general the read/write buffers of a typical system match the speed

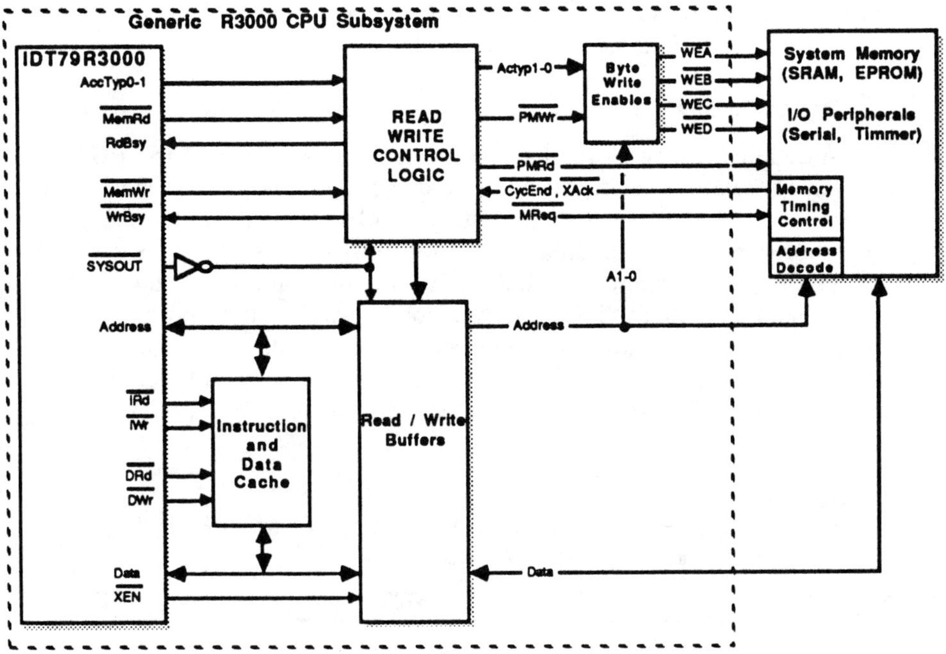

FIGURE 7-10
Block diagram of Generic R3000 CPU with memory. Reprinted courtesy of Integrated Device Technology.

of the R3000 CPU to the bus and memory subsystem. When a Write (Data Store) is performed, the data and address are deposited into the write buffer without stalling the CPU. Thus the R3000 can continue to execute code from the cache even though the system memory has not completed the write operation. The R3000 will continue to write into the write buffer on subsequent Store operations as long as there is room in the internal write buffer FIFO. When full, a Write Busy must be asserted such that the R3000 will stall on the next Write operation to avoid losing data through FIFO overrun. When using a one-level deep write buffer like the one employed on the RS382 board, the control logic must assert the Write Busy immediately such that further Writes will stall the R3000.

The read buffer is the path back from the system memory to the R3000. Together with the write buffer, it isolates the cache memory from the system memory, allowing code to be executed from the cache while Writes are retired to system memory. In more complex systems this isolation allows devices like DMA to access system memory with minimal impact on execution throughput. Read/write control logic takes requests from the R3000 and translates them into strobes for the system memory, coordinates the read/write buffers, and asserts the busy signals of the R3000.

Design Review

A detailed review of the function of each block in a generic CPU subsystem as it relates specifically to an R3000 design follows. A generic subsystem includes: 1 PAL for read/write control, 1 PAL for byte enables, 4 bidirectional registers for data, 4 registers for address, 4 74F280s and 1 register for parity generation, and 1 PAL for rest time initialization. The system memory utilizes: 1 PAL for memory timing control and 1 PAL and decoder for address decoding. The remainder of the design is dedicated to memory and I/O peripherals generic to most CPU designs. A block diagram of the R3000 CPU to system memory path is shown in Figure 7-11.

FIGURE 7-11

R3000 CPU to system memory path. Reprinted courtesy of Integrated Device Technology.

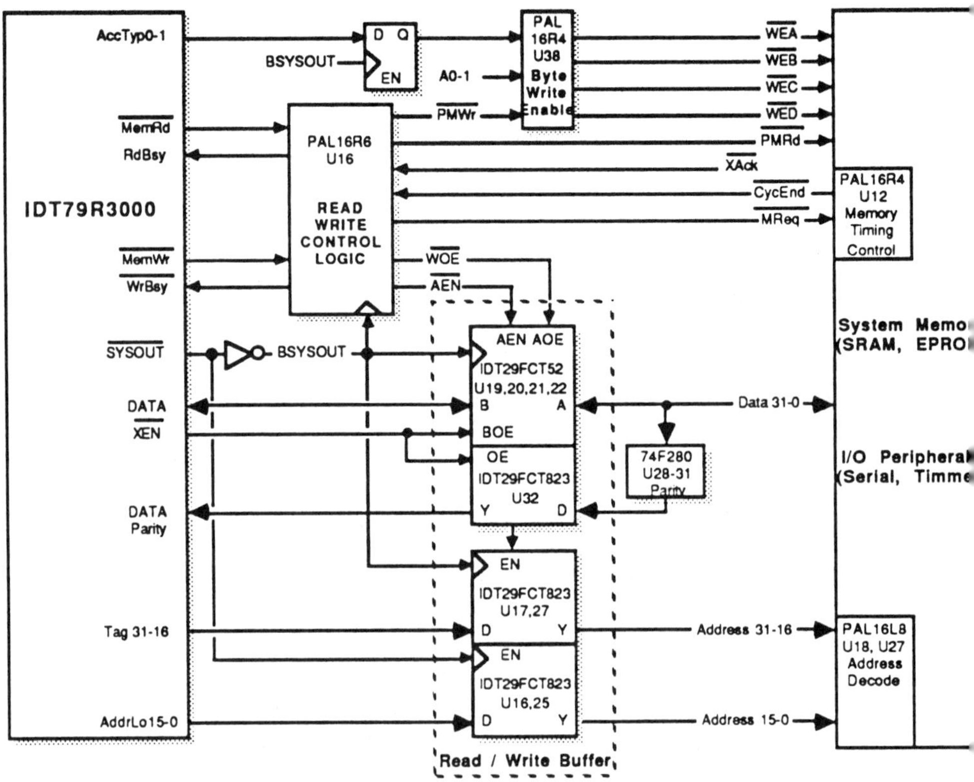

Cache Memory

Cache memory lowers the average access time of memory by the processor. Thus a system can be designed with slower main memory and still use a microprocessor with higher clock rate to provide higher performance. There is in effect a trade-off between performance and the cost of the cache. In the 7RS382 board design, a balance was chosen to reduce the cost of the cache memory and yet maintain reasonable performance as an embedded controller. The goal of the design was to realize system memory access latency of less than 4 to 5 CPU clock cycles. The R3000 uses standard architecture SRAM to implement a direct-mapped cache. Such a direct-mapped cache operates as follows: the low-order address bits of the desired word are used as an index (address) into the cache, thus specifying a single cache entry, and the processor matches the upper address bits of the desired word with an address in the cache word. If there is a match (referred to as a hit) the corresponding data is the desired data.

The depth of the cache has a direct influence on the probability of a cache hit, which is referred to as the hit rate. The higher the hit rate, the lower is the average access time for a given system memory speed. When the actual main memory access time is low, the size of the cache has less effect on reducing latency. The R3000 cache uses two banks of cache memory, one for instructions and the other for data. In order to save on pins, the R3000 multiplexes the address bus with instruction and data addresses. The data bus is also multiplexed between data words and instructions. In order to give the SRAM the full clock period for cache access, latches are placed on the address bus for each bank (instruction and data). Figure 7-12 shows how the R3000 is connected directly to the memory to provide a streamlined cache solution.

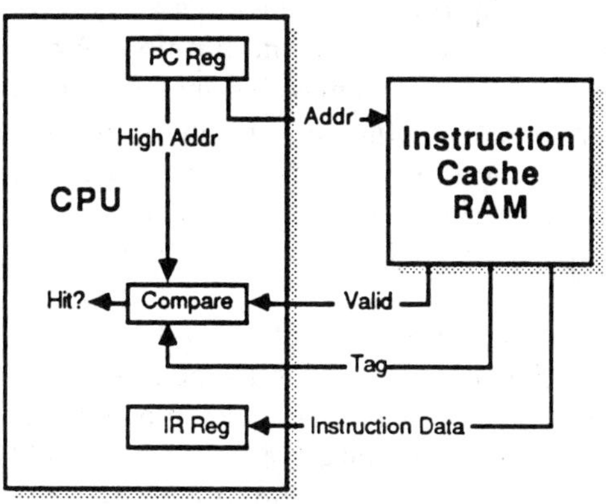

FIGURE 7-12
R3000 with direct mapped cache. Reprinted courtesy of Integrated Device Technology.

Read/Write Control

As in all such designs, a block of logic controls the read/write data path between the CPU and the system memory. On the RS382 board this is accomplished by a PAL. Two main functions are performed by the read/write control. The first function is to respond to the R3000 with Read Busy (RdBsy) and Write Busy (WrBsy) in response to read and write requests from the R3000. The second function is to generate read and write pulses to the rest of the memory and I/O system as well as the control address and data bus buffers. In performing this function the control logic receives a read or write request and then generates a system memory request asserting the appropriate read and write strobe. The system must then respond to timing control with a Cycle End signal, which tells the control logic to deassert the Read or Write Busy to the R3000.

In order to design the read and write control logic, the designer must define the system bus architecture and corresponding rules. The rules are either requirements of the R3000 or the system memory. Some of the following rules were used in the design of the RS382 board:

The CPU will continue to execute after the first write, which is allowed by the R3000 and a one-level-deep write buffer.

The CPU must stall on a write if a write is already in progress to the system memory, as required by a one-level-deep write buffer.

The CPU must stall on system memory if a write is in progress to the system memory because the memory has only one bus.

The CPU must stall on a write if streaming during a read because of a shared address register between the read and write path.

The basic read and write cycles are shown in Figures 7-13 and 7-14, respectively.

The write control is centered on a flip-flop that generates a Write in Progress signal (WIP/). The WIP/ controls the input to the flip-flop. When there is no write in progress, the Memory Write (MemWr/) signal is used to set the flip-flop to indicate a write in progress. Once a write in progress is registered, the Cycle End (CycEnd/) and the Transfer Acknowledge (XAck/) signals are monitored for the end of the write in progress. Either of these signals will end the write cycle. The write control flip-flop is shown in Figure 7-15.

The read control logic is centered on a flip-flop that generates a Read Busy signal (RdBsy). The Read Busy signal is active high to the R3000 and is always asserted until a CycEnd/ or XAck/ signal is received. When either of these signals goes low, Read Busy is cleared to inactive (low) and is held until MemRd/ goes away. The CycEnd/ and XAck/ are monitored only when

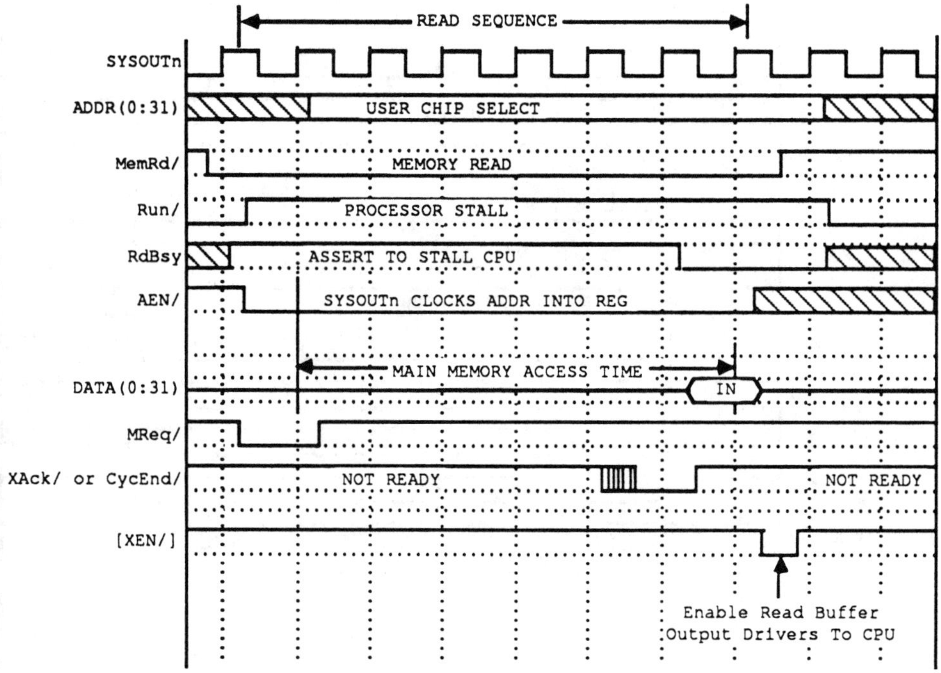

FIGURE 7-13
Basic read cycle. Reprinted courtesy of Integrated Device Technology.

there is a write in progress and during a memory read; otherwise they are ignored. The memory read strobe (PMRd) is asserted as soon as there is a MemRd/, when there is no write in progress. It is held as long as there is a Read Busy signal. RdBsy is asserted when Mem/Rd is deasserted because during block refill RdBsy must be held off. Early assertion of RdBsy will signal a retry. The read control flip-flop is shown in Figure 7-16.

The Memory Request signal (MReq/) is an active-low strobe asserted for one clock cycle. It can be triggered by either a memory write (MemWr/) when there is no write in progress or a memory read (Mem/Rd/) when there is no read in progress. Required timing control logic generates the cycle end from memory request. Normally the timing control is a counter that takes into account the request and the address and then counts an appropriate number of clock cycles. The memory request flip-flop is shown in Figure 7-17.

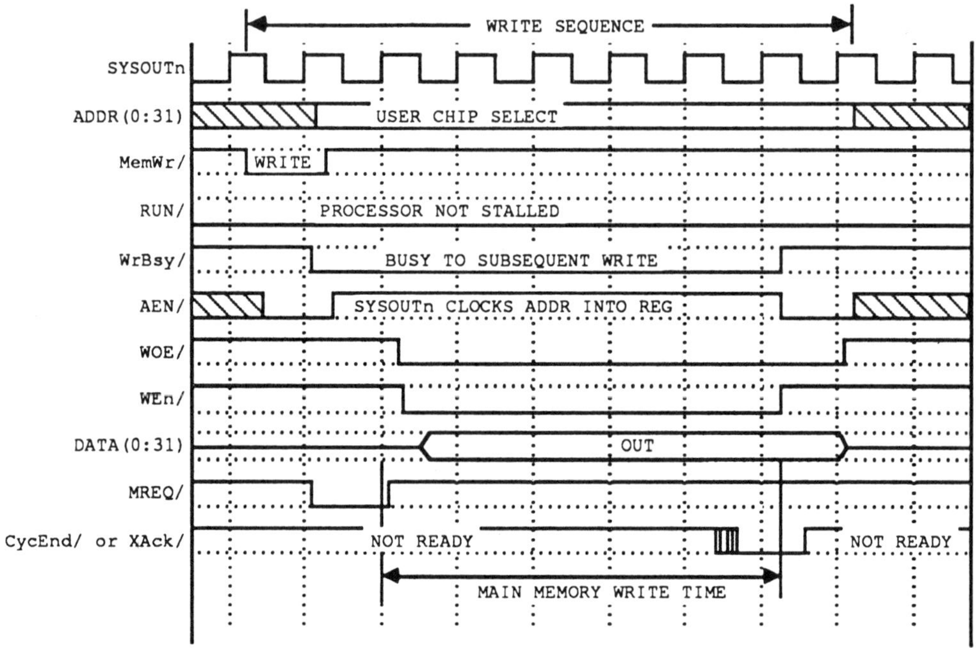

FIGURE 7-14
Basic write cycle. Reprinted courtesy of Integrated Device Technology.

FIGURE 7-15
Write control flip-flop. Reprinted courtesy of Integrated Device Technology.

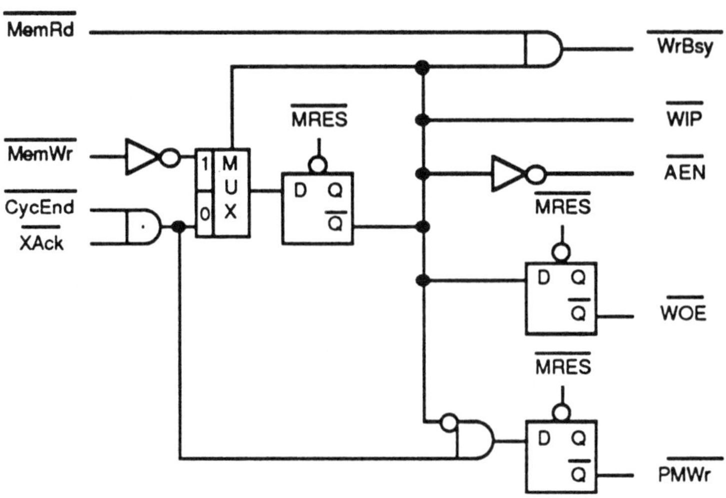

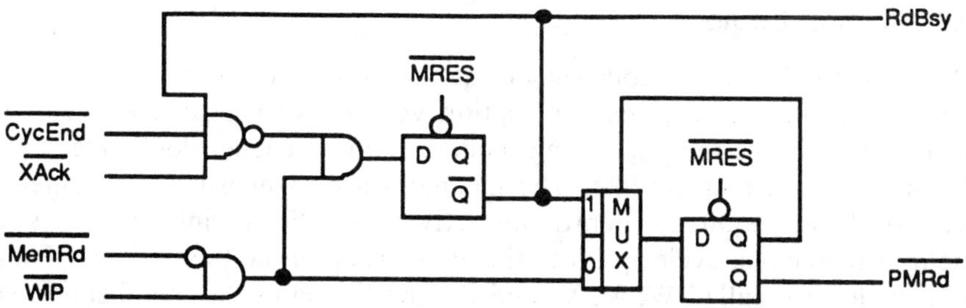

FIGURE 7-16
Read control flip-flop. Reprinted courtesy of Integrated Device Technology.

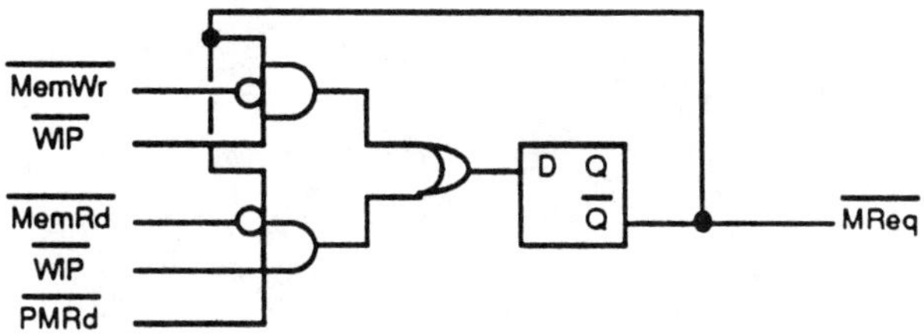

FIGURE 7-17
Memory request flip-flop. Reprinted courtesy of Integrated Device Technology.

FIGURE 7-18
Timing of initialization sequence. Reprinted courtesy of Integrated Device Technology.

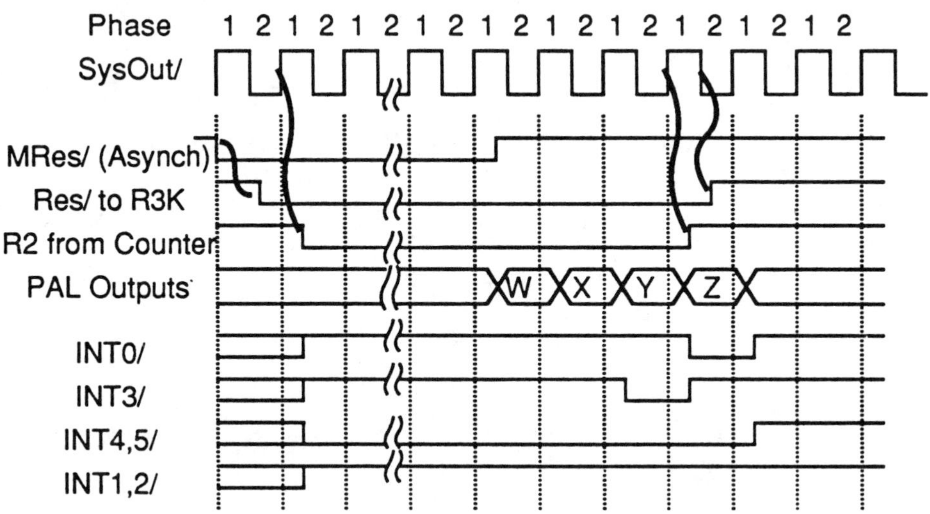

Initialization Logic

The R3000 CPU Reset/input signal not only forces the processor to commence execution at the reset exception vector, but it also allows for the introduction of various operating modes, such as cache block size, "Endianness," test, and the like. The mechanism by which these parameters are introduced requires synchronous activation of the six interrupt lines for the four final clock cycles prior to the deassertion of the Reset signal. These four cycles are called W, X, Y, and Z. The timing of the initialization sequence is shown in Figure 7-18.

Chapter 8

In-Circuit Emulators

In-circuit emulators have been key development tools since they were introduced by Intel in the 1970s for use in the application of the 8080 microprocessors. The emulator for that CPU had a maximum bus bandwidth of 5 MHz. Today in-circuit emulators are being used quite effectively at clock frequencies at least as high as 33 MHz in the application of microprocessors like Intel's 80386 and Motorola's 68030. As the clocks are increased to higher frequencies for both RISC and CISC CPUs, there has been some question about how well they can compete with software simulators. However, companies such as Hewlett-Packard have indicated that they expect to develop emulators that will compete at higher frequencies than 33 MHz. Also, built-in test capabilities, which are being provided in some high-speed microprocessors, have reduced the amount of external equipment required in emulator-based development and consequently reduced its cost. There is every indication that emulation will survive in the development picture. At least in competition with simulation, engineers find emulation easier to use.

Today emulators provide support to microprocessor applications in development, design verification, and test roles. In development and verification, emulation is applied to the CPU. In test equipment, emulation is applied to the CPU, memories, and bus cycles. In code development, memory emulation is sometimes applied, but generally emulation in development systems refers to the CPU microprocessor.

When reduced to the fundamentals, an emulator provides the following three basic functions:

> First, it provides run controls over the target processor, which itself is monolithic with no access to the internal registers or control functions that are typically available to the operator of a mainframe or minicomputer. Examples are: starting and stopping the processor, initializing registers, and starting on a specified address.
>
> Second, emulation provides a window into the inner workings of the target processor. It allows one to examine the program execution at the register level.
>
> Third, emulation provides an experimentation environment where trial programs can be evaluated as they run on the target processor. Ideally they are a light link back to the software design environment, making it possible to make corrections efficiently.

There is a wide range of in-circuit emulators on the market. They provide development support for 4-bit to 32-bit microprocessors and operate at increasingly higher frequencies. There are many well-designed emulators, among which is the family of emulators in the Hewlett-Packard 64700 series. Here the emulator is combined with the logic analyzer to provide analysis as well as emulation.

The Hewlett-Packard 64700 Series

Figure 8-1 is a block diagram of the major components of the 64700 emulation system. The emulation processor resides on the run control board within the emulator and is connected to the processor socket in the target system via line driver buffers and a target system cable. The emulation processor, typically a test version of the target system processor, forms the core of the run control board. Surrounding the emulation processor are three basic functional circuit blocks: run control, memory and memory mapping, and the analysis bus interface.

The run control circuitry controls the operation of the emulation processor so that the user can always maintain control of the emulator. Since each microprocessor has a different internal architecture, the run control circuitry must be tailored to provide the interface to the host control system. Run control affects operations such as single-stepping, sharing interrupt signals with the target system, signaling the analyzer, keeping the host controller aware of the status of the emulation processor, and so on.

The emulator contains its own internal emulation memory, which can

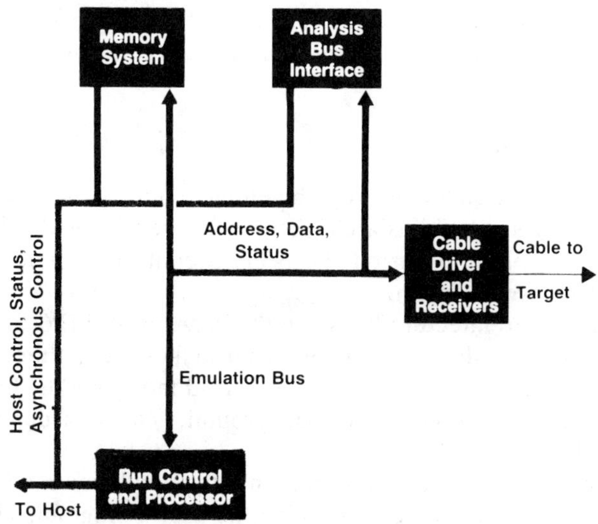

FIGURE 8-1
Major components of Hewlett-Packard 64700 series emulators. Reproduced with permission. © Copyright 1988 Hewlett-Packard Company.

be used as a substitute for the main memory, RAM, and ROM that will eventually be in the target system. This memory can be located anywhere in the address space of the emulation processors. Mapping of the memory and its designation as RAM, ROM, or guarded are handled by the mapping circuitry, which provides a translation function between the address being output by the emulation processor and the address being supplied to the emulation memory array.

The analysis bus interface provides the link between the signals traversing the address, data, and status buses of the emulation processor, the emulation bus analyzer, and the state/timing analyzer. The analyzers will be discussed more in the following sections.

Two other system blocks, the host controller and the analysis system, are not shown in Figure 8-1. The host controller works with the run control system, memory subsystem, and analysis system to provide the coordination of all the complex tasks going on within the emulator. It interfaces with the user to convert commands to proper operational sequences. The host controller also interfaces independently with the state/timing analyzer, emulation bus analyzer, and memory subsystems. This enables the host controller to communicate with emulation memory and read and write to it while the emulation processor continues to run the user's code. Likewise, the state/timing analyzer or the emulation bus analyzer can be set up or unloaded without stopping the emulator. This type of architecture is called dual-bus, referring to the two independent processor buses (emulation and host control) within the emulator.

System Architecture

Many possible architectures can be employed in emulator design. The HP64700 series uses a dual-bus architecture (Figure 8-2), with a foreground or background monitor to permit the user to control the microprocessor emulation in the target system.

The dual-bus architecture has been the heart of the HP64000 family of products since its introduction in 1979. What makes the architecture of the HP64700 series new is the addition of the hybrid monitor. The hybrid monitor, residing either in foreground or background, is a major departure from the original HP64000 family design.

The monitor is the control program that the emulation microprocessor is running when it is not explicitly running the user's program. The monitor program is very similar to the small programs often supplied with single-board computers. Typically 2K to 3K in size, the monitor handles reading and modifying memory, internal registers, I/O ports, and so on.

Single-stepping and software breakpoints are also handled by the monitor program. While the use of the monitor is easily understandable, its integration within the emulator plays an important role in the emulator's range of useful applications.

Complex applications typically require the use of a foreground monitor. This monitor is a block of code that runs in the same address space (fore-

FIGURE 8-2
Hewlett-Packard 64700 emulator system architecture. Reproduced with permission. © Copyright 1988 Hewlett-Packard Company.

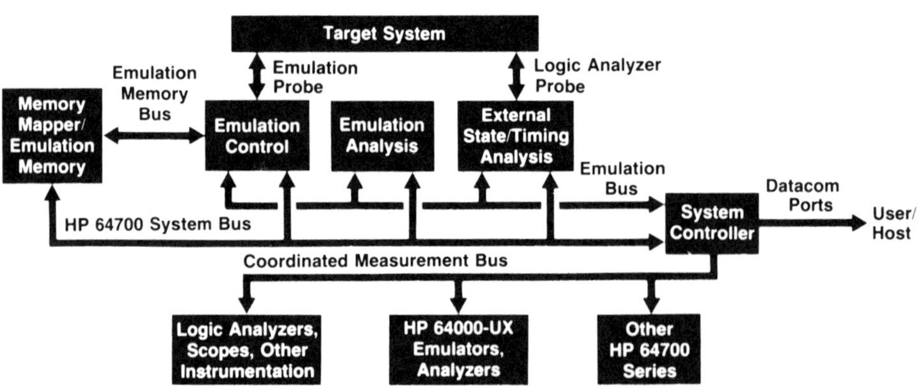

ground) as the user's own program. It is linked with the user's run-time code at assembly time so that when control of the emulation processor is switched from the user program to the monitor program, real-time system events can still be serviced. These events, such as multiple interrupts and watchdog timers, are very difficult to handle properly with other monitor implementations. The foreground monitor is often the architecture of choice for complex multitasking environments.

Background monitors are easier to use when trying to get a new design off the ground. The monitor is present in a background or shadow memory, and control is transferred to this monitor by remapping (called "jamming") the monitor instructions from the background address space directly to the emulation processor. This type of monitor makes no assumptions about the user's environment. The emulation processor can always enter the monitor when instructed to do so. Problems arise from this architecture when it becomes necessary to keep a user's target system alive while the emulator is running in background. The new HP64700 emulators solve this problem by offering the best of both worlds. Once the user answers a simple configuration question, either the emulation system will operate with the monitor in the background, or, by assembling and linking the monitor with the run-time code, the user will have the advantages of a foreground monitor.

The Effects of Emulator Cables

With so much emphasis on the operation of high-speed microprocessors at higher and higher clock frequencies, Hewlett-Packard made a study on the electrical effects of emulation probes. They dealt with four factors that influence probe transparency. They are overshoot or ringing in signal lines, probe loading effects on the target system, interference from signal lines on one another, and distortion caused by interconnect techniques. It is said that the results of the study led to a dramatic improvement in emulation cable performance. Cable fidelity could be made relatively independent of cable length, but not without increasing its cost. In fact, the emulation cable is often the most costly component of a high-performance emulator.

Overshoot and ringing were found to be due to transmission line effects. Several techniques to control this problem were explored. The most direct one was to terminate the line in its characteristic impedance. This solution was found to be practical in only a few cases, because the target system could not drive the resulting load. Each design appeared to require its own special solution, which in some cases involved using a small capacitor of about 100 pF in series with a resistor equal to Z (see Figure 8-3).

In cases where the loading effect of the emulation probe is critical, active buffers are incorporated in the probe assemblies and clock signals are terminated in the emulator using the technique described above. With

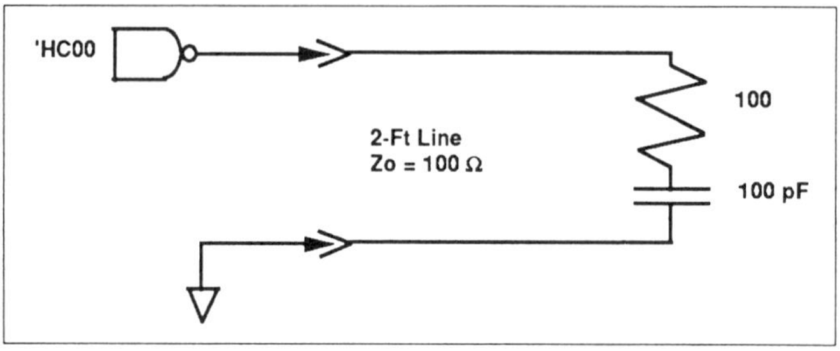

FIGURE 8-3
Termination to reduce overshoot and ringing. Reprinted from Steven S. Peurifoy and Arnold S. Berger, "Cabling technology vital to high-end emulator performance." Courtesy of *Personal Engineering & Instrumentation News* and Hewlett-Packard.

this approach, target devices are isolated from the cable with a slight increase in signal delay, but a good-quality signal is provided to the emulation circuitry.

Signal interference within a probe cable was found to be due to capacitive or inductive coupling. In the case of capacitive coupling a solution can be found by interposing a ground line between the two signal lines (Figure 8-4). Further improvement can be realized by enclosing the lines in a grounded conductor such as a coaxial cable. The effects of inductive coupling can also be alleviated by placing a ground wire between each pair of signal lines (Figure 8-5). Most of the current associated with a signal returns on

FIGURE 8-4
A ground path between lines A and B reduces capacitive coupling. Reprinted from Steven S. Peurifoy and Arnold S. Berger, "Cabling technology vital to high-end emulator performance." Courtesy of *Personal Engineering & Instrumentation News* and Hewlett-Packard.

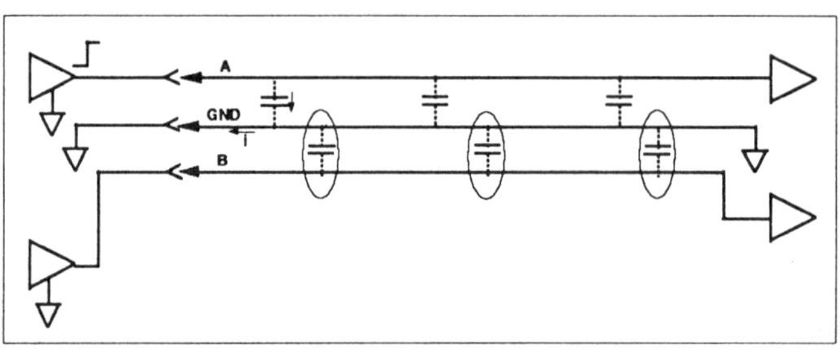

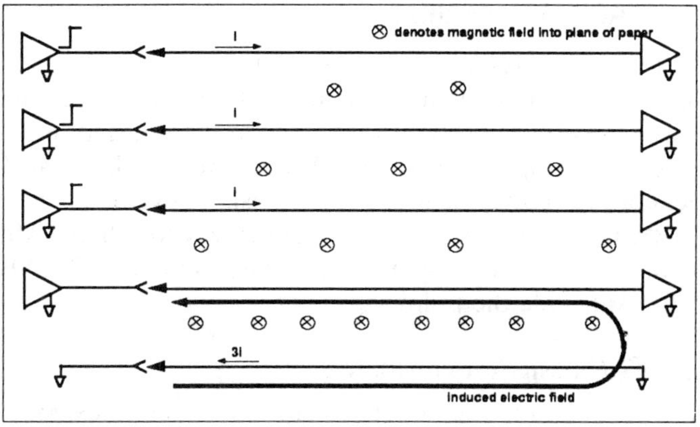

FIGURE 8-5
A ground line between pairs of signal lines reduces inductive coupling. Reprinted from Steven S. Peurifoy and Arnold S. Berger, "Cabling technology vital to high-end emulator performance." Courtesy of *Personal Engineering & Instrumentation News* and Hewlett-Packard.

the ground wires immediately adjacent to it because that path has minimum inductance. The magnetic field is thus concentrated between the signal wire and its ground wires, and its intensity drops rapidly with distance from the signal wire. Also, the loop area is small and the flux linking the loop due to a given magnetic field strength is small. Hewlett-Packard engineers concluded that the key to reducing a circuit's generation of inductive interference and its susceptibility lies in reducing circuit loop areas. It was found that cable performance could be improved keeping the ratio of layer-to-layer conductor spacing to signal conductor diameter larger, resulting in lower capacitance from one signal wire to another. Also, the spacing between signal wires and ground wires should be smaller compared to the layer-to-layer spacing to more closely confine the fields to the vicinity of the signal wire.

Advanced probe designs pay considerable attention to interconnect design. At the emulator end, a pin-grid array connection scheme can provide a short direct connection between the cable ground wires and the emulator ground plane. Every signal pin is adjacent to multiple ground pins in the connector itself. At the target plug end, of course, the processor being emulated dictates the pin arrangement. However, it is possible to make modifications to reduce the effects of such factors. For instance, a ground plane can be placed on the lower side of the plug board to reduce the area of the loops in the pin-out. Also, there should be proper ground and power planes on the development boards.

Emulation in ASIC Verification

The RPM™ Emulation System is the first ASIC design verification tool that assures that ASICs work in the system the first time. ASIC netlists are automatically converted into hardware that can be plugged into the target system for real-time operation of the design. Built-in logic analyzer and stimulus generator features facilitate rapid design development and debugging. The RPM Emulation System is a practical way to achieve ASIC design verification in the context of system hardware and software. RPM Emulation System features include the following:

- Supports ASIC designs up to 100K gates
- Real-time emulation in the target system with clock rates up to 10 MHz
- In-circuit emulation with up to 1260 I/O signal lines
- Integral 128-channel logic analyzer and stimulus generator
- Convenient connection of existing ICs to emulation logic
- Powerful graphic user interface
- Accepts popular ASIC netlists and libraries
- Integrates into existing ASIC design environments

System Overview

The RPM Emulation System combines emulation setup and control software with reprogrammable emulation hardware and debug tools, providing a complete environment for real-time hardware emulation of ASIC designs (Figure 8-6).

RPM emulation software includes a netlist parser/expander, test vector converter, fully automatic design partition and place and route software, stimulus generator/logic analyzer software, and the RPM control panel user interface (Figure 8-7).

The RPM Emulation System integrates directly into the existing ASIC design environment. The emulation software accepts and converts ASIC designs in popular netlist format.

The RPM emulation hardware is centered on emulation modules, which contain a matrix of RAM-based reprogrammable logic arrays (Figure 8-8). The logic arrays contain gate and register primitive functions that are connected together to implement a hardware emulation of the ASIC design. Netlists are automatically converted into connection data that is used by the control processor to program the logic arrays.

The in-circuit interface provides a physical connection between the emulated design and the target system. Active in-circuit interface pods contain logic that buffers signals to the target system and minimizes delays of selected time-critical logic paths.

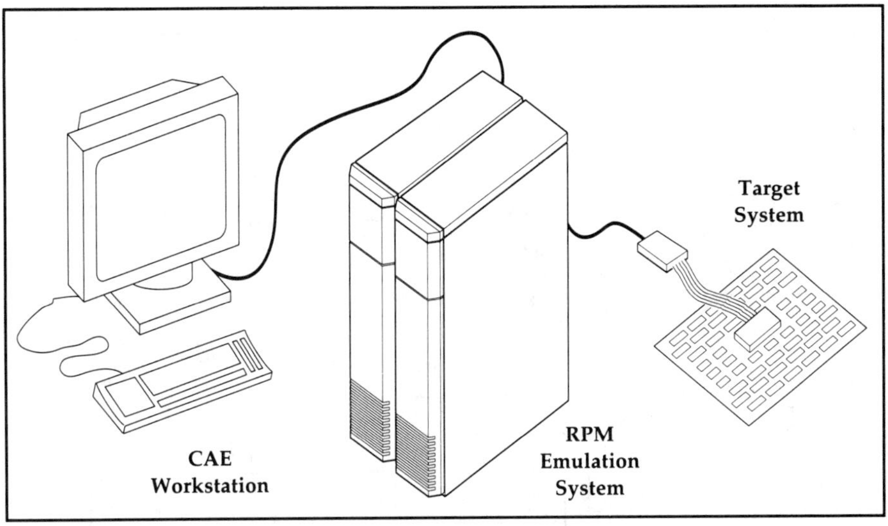

FIGURE 8-6
The RPM emulation system. Reprinted courtesy of Quickturn Systems, Inc.

FIGURE 8-7
The RPM emulation software. Reprinted courtesy of Quickturn Systems, Inc.

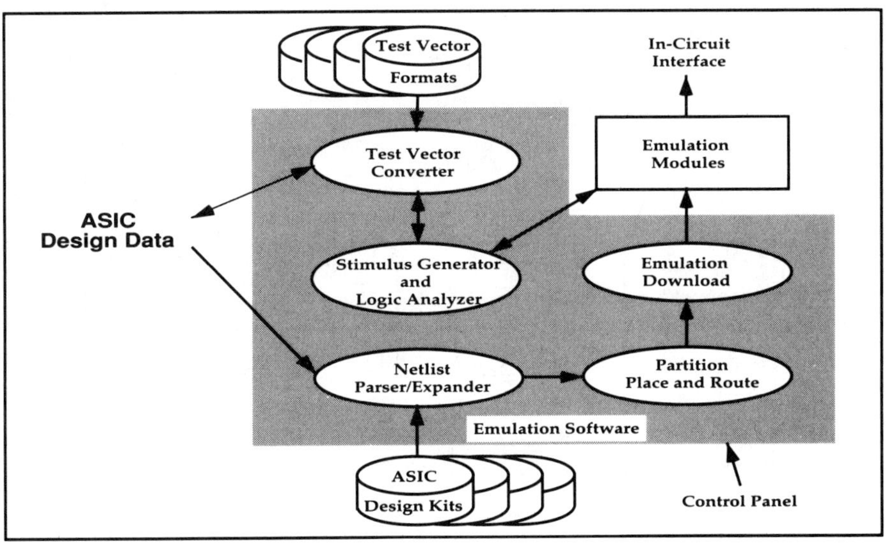

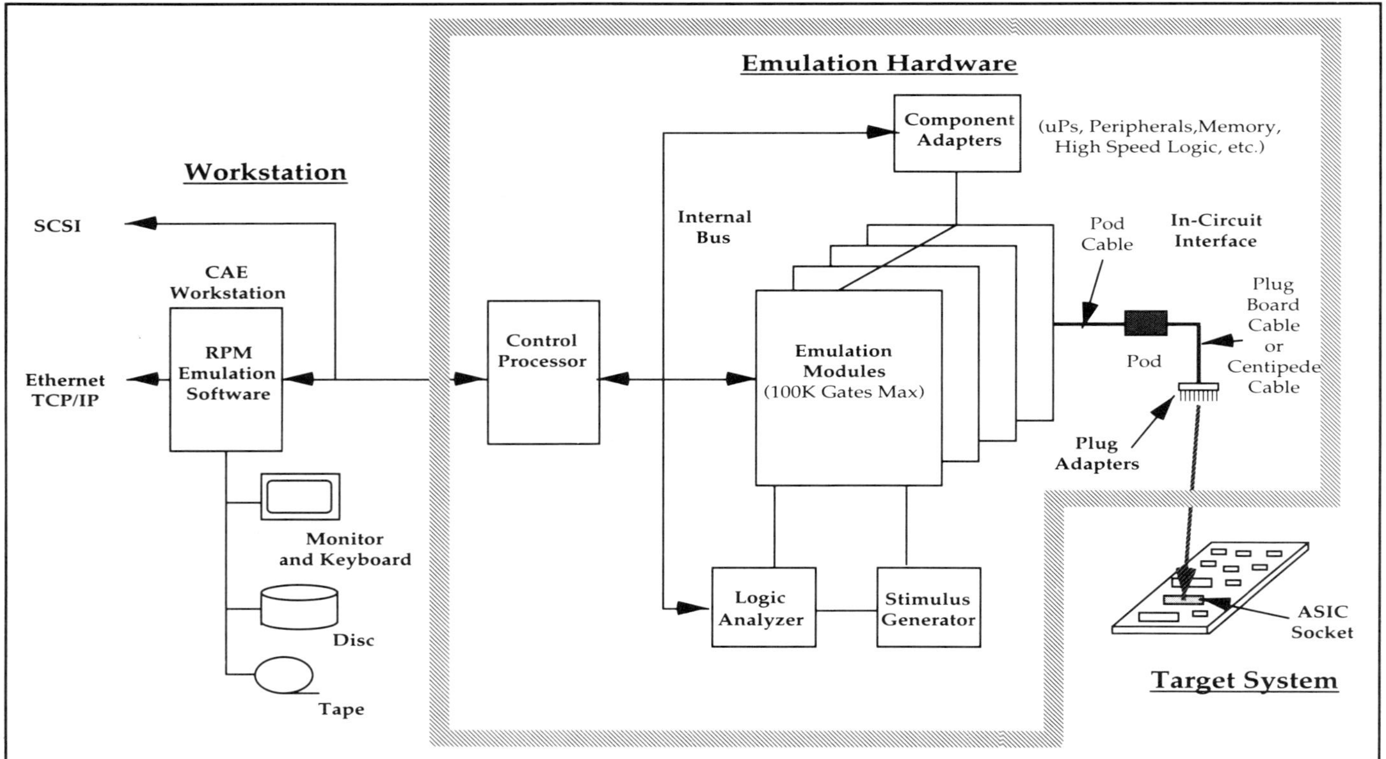

FIGURE 8-8 The RPM emulation hardware. Reprinted courtesy of Quickturn Systems, Inc.

Plug adapters allow the in-circuit interface to plug directly into sockets for popular ASIC package configurations.

Component adapter cards allow existing ICs to be connected to the emulation modules inside of the RPM Emulation System cabinet. Microprocessors, peripherals, memory, or any user-definable logic can be included in the emulation without taking up emulation gates, and the RPM emulation system operates as if they were part of the ASIC emulation.

The RPM Emulation System can be used as a very fast simulator and as an ASIC in-circuit emulator. As a fast simulator, integral stimulus generator and logic analyzer tools allow simulation vectors to be applied to the design in real time and responses to be captured and reviewed. As an in-circuit emulator, the ASIC emulation can be operated in the target system in real time.

In-Circuit Emulation

The RPM Emulation System provides a design verification capability not available in the simulation environment—in-circuit operation of the ASIC design. Operating the design in the system solves two fundamental limitations of simulation: the functional accuracy of the simulation vectors, and the ability to generate enough vectors to thoroughly verify an ASIC design.

A complete and accurate simulation requires the development of a set of simulation vectors that perfectly models the operation of the ASIC in the target system. However, it is virtually impossible to either develop an absolutely complete and accurate specification for the ASIC design, or to create a set of simulation vectors for the design that assures 100 percent conformance to the specification.

Plugging the design into the target system and operating it for extended periods of time is the only way to confirm that the design functions as required. The tedious and error-prone task of vector development is replaced by operation of the ASIC design with the complete system. Millions of vectors are effectively applied to the design as it operates in real time. When emulation of the design is functionally correct in the system, the silicon version will be functionally correct.

Emulation Performance

While current simulation technology is effective for behavioral simulation, and best-case and worst-case timing simulations, it is too slow to simulate enough real-time operation of ASIC designs to assure correct functionality in the system. Emulation of ASIC designs with the RPM system is implemented with real gates connected with real wires and operates in real time. Emulation with stimulus vectors is at least three orders of magnitude faster than simulating with a simulation accelerator. Although the performance is design dependent, the emulation modules support emulation of designs using stimulus vectors with clock rates up to 5 MHz. When ASIC designs are emulated with an in-circuit interface, some designs can be operated with 10-MHz clocks using the reprogrammable logic in the pods. When the target system is designed to operate much faster than attainable in-circuit emulation speeds, target system clocks can be slowed to allow in-circuit emulation within an order of magnitude of full speed.

The time required to complete the emulation process is determined by three steps:

1. netlist conversion
2. loading the design into the emulation hardware
3. emulation

ASIC netlist conversion on the RPM Emulation System is comparable to the place and route process used to lay out a gate array. The time required for the complete conversion process depends on design size and can range from a few hours for a few thousand gates to a full day for the largest designs. After the initial conversion of a netlist, an incremental conversion process enables changes to be made in only a few minutes.

Once an ASIC design is converted, loading it into the emulation hardware is much like loading a RAM. Loading a design takes only a few minutes, and the design is active in the emulation hardware until the next design is loaded.

The time it takes to complete an emulation with the RPM Emulation System depends on how much real-time operation is required for thorough functional verification. Since emulations take place in real time, minutes of real-time emulation take only minutes to complete.

Control Panel

The RPM Emulation System user interface, called the control panel, is menu-driven with pop-up forms and runs on Sun 3/60 workstations (Figure 8-9). All emulation features are accessed from the control panel. The pull-down command menus are organized into four categories: design, instrumentation, vectors, and timing.

Design—Designs are opened with this menu by identifying the ASIC netlist and the ASIC libraries used for each netlist.

Instrumentation—Stimulus generator and logic analyzer functions are specified with this menu, including commands to specify clocks, probes, triggers, and events. In-circuit interfaces and component adapters are also specified with this menu.

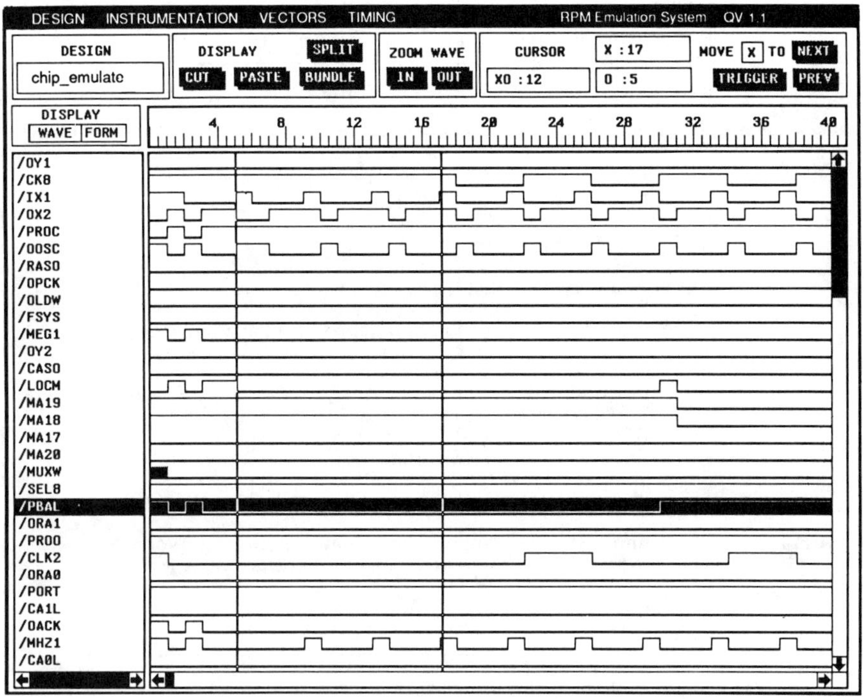

FIGURE 8-9
Displays on the control panel (user interface). Reprinted courtesy of Quickturn Systems, Inc.

Vectors—When stimulus vectors are used in the emulation, this menu controls the import and export of stimulus vectors in the selected format.

Timing—The timing specifications for correct operation of the emulation are specified with this menu.

The commands are summarized in Table 8-1.

The control panel displays emulation data in the logic analyzer window as waveforms or as a tabular listing of vectors. Powerful display and cursor control features facilitate rapid design analysis and debugging.

ASIC Design Data

There are three types of ASIC design data used by the RPM Emulation System: netlists, libraries, and stimulus and response vectors. The same netlists and stimulus and response vectors that are developed for the ASIC supplier are automatically converted by the emulation software for emula-

TABLE 8-1
Control Panel Command Summary

Command Menu	Command	Operation
Design	Open	Open a Design
	Configure	Convert Netlist for Emulation
	Emulate	Start Emulation
	Clocks	Set-Up Design Clocks
	Net Weight	Set Path Weights for Timing
	Pods	Set-Up In-Circuit Interface Pods
	Adapters	Set-Up Component Adapters
	Errors	Error Messages
	Session	Record and Replay an Emulation Session
	Exit	Exit an Emulation Session Without Saving
Instrumentation	Probe Set-Up	Select Logic Analyzer & Stimulus Generator Probes
	Variable Clock	Set-Up Internal Clock For Logic Analyzer
	Events	Set-Up Logic Analyzer Events
	Trigger & Acquire	Set-Up Logic Analyzer Trigger and Acquire Conditions
	BNC Out	Assign Event-Driven Clock to BNC Out
Vectors	Import	Import Simulation Vectors From Workstation
	Export	Export Vectors to Workstation
Timing	External	Specify ASIC I/O Timing
	Internal	Specify Internal ASIC Timing

Reprinted courtesy of Quickturn Systems, Inc.

tion. The supported formats for netlists, ASIC libraries, and stimulus and response vectors are described in the RPM design interface support schedule.

Emulation Modules

Emulation modules contain a matrix of high-density, high-performance, reprogrammable CMOS gate arrays. Each emulation module is capable of emulating designs up to 25K gates. When a maximum of four emulation modules are installed in the system, ASICs with up to 100K gates can be emulated.

In addition to implementing gate-level functionality, the reprogrammable logic arrays are used to connect stimulus generator, logic analyzer, component adapter, and in-circuit interface functions to any node in the ASIC design. Multiple ASIC designs may be emulated simultaneously with the same emulation hardware.

Logic Analyzer

The logic analyzer has features similar to a bench-top logic analyzer including selectable clocks, trigger, and events. Up to 128 channels are available for capturing emulation response data with acquisition clock rates up to 6 MHz. Acquisition memory is 8K vectors deep and is expandable to 32K vectors. Emulation vectors are acquired when the conditions are met for trigger and acquire events during emulation. Trigger specifications include a user-definable 8K vector offset that determines the number of vectors captured before and after the trigger.

Stimulus Generator

Vectors may be imported from the workstation and applied to the ASIC emulation with the stimulus generator. Up to 128 stimulus generator channels are available for driving the emulation at vector clock rates up to 6 MHz. Stimulus vectors are stored in an 8K vector stimulus generator memory that is expandable to 32K vectors.

A total of 128 stimulus generator and logic analyzer channels are available for an emulation, and they can be synchronized to start applying vectors to the design and recording responses at the same time. In this way emulation responses can be compared with expected responses imported from the simulation environment.

In-Circuit Interface

The in-circuit interface feature allows an ASIC design to be operated directly in the target system. Each in-circuit interface includes a flat-woven cable (the pod cable), and an active pod. Two in-circuit interface options are available for connecting pods to the plug adapters—a plug board cable and plug board (Figure 8-10) and a "centipede" cable. The plug board cable and plug board are used when automatic connection of a large number of signals is preferred, such as with high-pin-count PGAs. With the flexibility of individually connecting pins into the target system, and for low-pin-count ASICs, the centipede cable provides another connection alternative.

There are five in-circuit interface connectors on each emulation module, each with a maximum of 68 signals. Five pod cables allow 340 signals to be connected to the target system. An RPM emulation system with four emulation modules can support up to 1360 signals to the target system. In-circuit interface signals are specified with control panel forms and are automatically connected from the emulation module to the in-circuit interface by the emulation software.

The pods provide current limit protection, buffer signals, and facilitate automatic connection of signals to the plug adapter pins. The pods contain a reprogrammable logic array for implementing a small number of time-critical logic paths to minimize the path delays and improve the performance of the emulation.

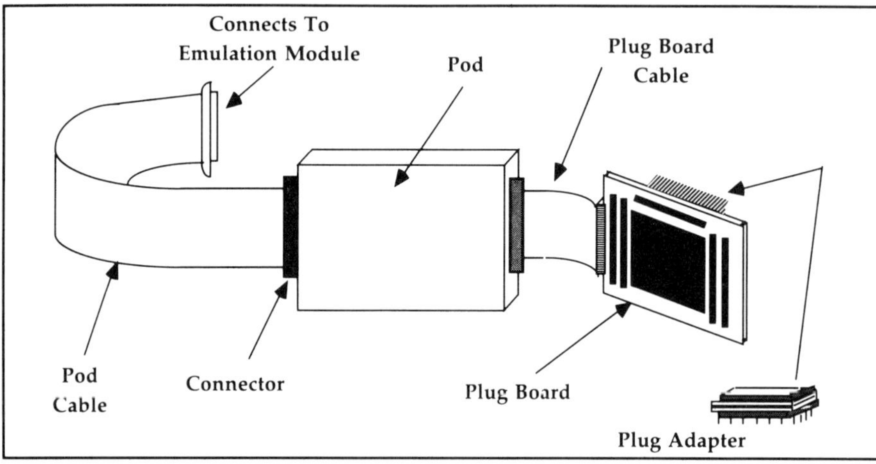

FIGURE 8-10
RPM in-circuit interface with plug board and PGA plug adapter. Reprinted courtesy of Quickturn Systems, Inc.

Emulation in Functional Testing

Emulation in Functional Testing as applied with the TS 6000 digital test system, manufactured by Talon Instruments, is described below.

There are at least three ways of controlling microprocessor operations:

1. Microprocessor emulation
2. Memory emulation
3. Bus cycle emulation

Microprocessor Emulation

With microprocessor emulation the UUT's microprocessor is physically or electrically removed and replaced with a connection to a "microprocessor pod" that contains the equivalent microprocessor. This allows microprocessor operations to be controlled by the tester, as depicted in Figure 8-11.

Memory Emulation

With memory emulation the UUT's microprocessor is left in place, but the instructions it normally executes are removed and replaced with those developed by the test engineer. Again control of the microprocessor operations is achieved. This method is accomplished by downloading special programs to RAM, installing special test ROMs, or connecting "memory

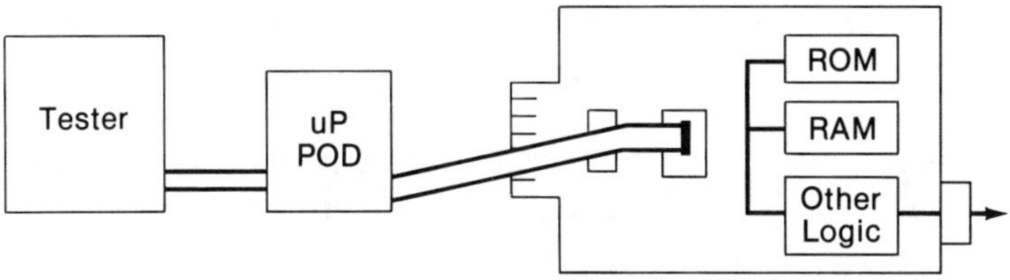

FIGURE 8-11
Tester using microprocessor emulation. Reprinted courtesy of Talon Instruments.

emulation pods'' to the tester's memory emulation system, as depicted in Figure 8-12.

Bus Cycle Emulation

A microprocessor communicates to peripheral chips by fixed cycles such as the Read cycle, Write cycle, Read-Modify-Write cycle, Interrupt cycle, and so on. The bus cycle emulation technique emulates the cycles as required. As with the microprocessor emulation technique, the UUT's microprocessor is disabled and replaced by an interconnection to the bus cycle emulator. The bus cycle emulator is hardware or software configured to emulate the required cycles of the UUT's microprocessor.

The Microprocessor Pod

The purpose of a microprocessor pod is to provide an interface through which the TS6000 may assume the role of the UUT's microprocessor. To accomplish this role, a microprocessor pod's basic functions are to execute at speed bus cycles (e.g., Write, Read, Interrupt Acknowledge) using the

FIGURE 8-12
Tester using memory emulation. Reprinted courtesy of Talon Instruments.

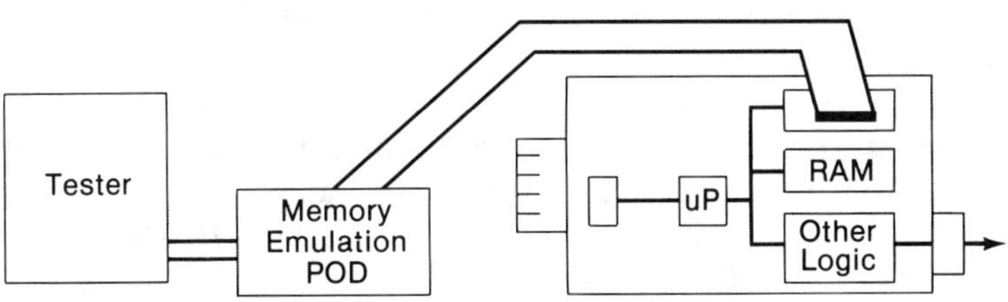

UUT's microprocessor clock and to set and test microprocessor control and status lines. In addition, a pod automatically monitors the UUT microprocessor's power and clock status. Using these basic functions, higher-level functions such as the Write and Read statements are provided for unique user tests and the bus, ROM, and RAM tests are provided for testing the microprocessor's "kernel."

Physically, the microprocessor pod consists of a free-standing case with ribbon cables to connect to the TS6000 and ribbon cables to connect to the UUT. Microprocessor pods may be connected directly to the UUT by plugging into the UUT's microprocessor socket. Dependent upon the microprocessor type, some pod types may be connected to the UUT by simply clipping over the UUT's microprocessor.

Microprocessor Pod User Tests

The following paragraphs summarize the preparation of user tests that use a microprocessor pod to test a microprocessor-based UUT's functions. These tests not only include the internal functions of a UUT but also include the external UUT functions, which require the board-edge stimulus and response capabilities of a TS6000's port simulator.

User tests are prepared using the Test Definition menu which is accessed via DEVELOP TESTS. Locating the display cursor to the right of a test step number will cause the TS6000's instruction set window to be displayed.

Bus Test

The Bus Test is a built-in standard test which tests the condition of the microprocessor bus. In general, output and bidirectional signals of the microprocessor are driven both high and low to detect "stuck-at" and "tied-in-common" conditions, while microprocessor input signals are sampled to detect irregular conditions.

It is generally recommended to execute the Bus Test as the first operation when initially using a microprocessor pod to ensure the integrity of the UUT's microprocessor interface. The command for Bus Test is: EXECUTE BUS TEST.

The Bus Test should initially be "learned," that is, RUN CYCLE = EXECUTE AND LEARN. Figure 8-13 illustrates the Bus Test results display for the M68020 microprocessor pod following execution in the learn mode. The status of the data and address buses is displayed in the top portion, and the condition of the control signals with the signal names listed vertically is displayed in the lower portion. Figure 8-13 indicates that data bus bit 00 is stuck low ("L"), data bits 03 and 04 are tied in common ("C"), address bit 07 is stuck low, and the control lines DACK0-, DACK1-, and AS-did not respond as expected ("*").

The learn feature of Bus Test does not apply to the address or data

```
                    TEST DEFINITION
SYNC EVENT : NONE
RUN  CYCLE : EXECUTE AND LEARN
RUN/EXECUTE: ALL
                                          CYCLES:0001
COMPLETE:     D1       @ STEP 100 # FAILURES:0000

 RESULTS:      33222222 22221111 11111100 00000000
               10987654 32109876 54321098 76543210
 DATA BUS:     -------- -------- -------- ---CC--L
 ADDRESS BUS:  -------- -------- -------- L-------

 CONTROLS:     IACR BH DD BBB III OERDRDA FFF SS
               PVDS EA AA GAR PPP CCMB/SS CCC II
               NEIE RL KK .C. LLL SSCEW.. 210 ZZ
               DCST RT 10 .K. 210 ...N... ... 10
 STATUS:       ---- -- ** --- --- ------* --- --

SIGNATURES P1=     P2=     P3=     P4=     UP=

   F1      F2      F3      F4      F5      F6
                                         RETURN
```

FIGURE 8-13
Learn bus test example. Reprinted courtesy of Talon Instruments.

buses; that is, an unexpected result is always reported as an error and should be corrected before proceeding. The data and address buses are analyzed to detect any "stuck-at" or "tied-in-common" lines with errors indicated by an "H," "L," or "C."

Following execution of the learn mode, the Bus Test will display an asterisk ("*") beneath any control signal that was not free to function. The user should determine whether the noted condition is normal for the UUT and, if not, correct the condition before proceeding. The status of the "learned" control signals is used as the expected response when executing the Bus Test in any of the test modes. An asterisk will be displayed beneath a control signal when its tested response does not equal its learned response.

ROM Test

The second standard built-in test of the microprocessor pods is ROM Test, which accomplishes fault detection by comparing the checksums of a UUT's ROM to that of a known good card.

The checksum for a specific ROM Test must first be learned similar to that of Bus Test. The command for a ROM Test is:

EXECUTE ROM TEST FROM $____ TO $____.

The preferred practice is to define a series of individual ROM Tests of a large ROM array rather than one single test in order to isolate potential faults. Since most ROMs are byte wide, this corresponds to a separate ROM Test for each individual ROM chip for microprocessors which have an 8-bit-wide, byte-oriented data bus. Similarly, for a 16-bit, word-oriented data bus microprocessor, a separate ROM Test would be defined for each pair of ROM chips (note that the address limits for a 16-bit-wide microprocessor ROM Test are constrained to even values). Finally, since most 32-bit-wide microprocessors provide some means of automatic data bus sizing capability, a ROM Test may apply to one, two, or four ROM chips dependent upon available UUT memory.

Figures 8-14 and 8-15 illustrate the resultant display following learn and test mode execution, respectively. The ROM Test display contains three types of information. The first is the checksum data, which is the basis for fault detection and is self-explanatory.

The second item is designated as the "1's AND 0's TEST" and provides diagnostic information. A failure of either the "1's TEST" or the "0's TEST" indicates that the data was either stuck high or stuck low throughout

FIGURE 8-14
Learn ROM test example. Reprinted courtesy of Talon Instruments.

```
                 TEST DEFINITION
SYNC EVENT : NONE
RUN  CYCLE : EXECUTE AND LEARN
RUN/EXECUTE: ALL
                                          CYCLES:0001
COMPLETE:     D1       @ STEP 002 # FAILURES:0000

 ROM TEST RESULTS:
   CHECKSUM IS: FB54
   1'S AND 0'S TEST (PASS/PASS)
   ADDR     DATA    ADDR     DATA    ADDR     DATA
  000000 0F0F0000 000004 00000000 000008 00000000
  000010 00000000 000020 00000000 000040 00000000
  000080 FCFC4646 000100 3C3C2F2F 000200 0F0F0000
  000400 DADA6D6D 000800 00000000 001000 00000000
  002000 00000000 004000 00000000 008000 00000000
  010000 00000000 020000 00000000 040000 00000000

SIGNATURES P1=      P2=      P3=      P4=      UP=

    F1       F2       F3       F4       F5       F6
                                              RETURN
```

```
                    TEST DEFINITION
SYNC EVENT : NONE
RUN  CYCLE : SINGLE PASS
RUN/EXECUTE: ALL
                                          CYCLES:0001
 HALTED :      D1        @ STEP 001 # FAILURES:0001

 ROM TEST RESULTS:
   CHECKSUM IS: DE40  S/B: FB54
   1'S AND 0'S TEST (PASS/PASS)
   ADDR     DATA     ADDR     DATA     ADDR     DATA
  000000 0F0F0000 000004 00000000 000008 00000000
  000010 00000000 000020 00000000 000040 00000000
  000080 0F0F0000 000100 3C3C2F2F 000200 0F0F0000
  000400 DADA6D6D 000800 00000000 001000 00000000
  002000 00000000 004000 00000000 008000 00000000
  010000 00000000 020000 00000000 040000 00000000
SIGNATURES P1=      P2=      P3=      P4=      UP=

   F1      F2       F3        F4       F5       F6
 RESUME  TEST                                 RETURN
         CHANNLS
```

FIGURE 8-15

ROM test example. Reprinted courtesy of Talon Instruments.

the entire ROM Test address space. Such an error would likely indicate unprogrammed ROMs or a fault in the basic ROM enabling circuitry.

The third type of information on the ROM Test display is referred to as the "address boundary" data and provides information to diagnose possible addressing errors. The displayed addresses are relative to the ROM Test address space; that is, address "000000" refers to the first address designated by the ROM Test. Since the ROM is initially read at relative address zero and then at consecutive individual address bits, addressing errors will cause a pattern of data values to result. Assuming that a printed copy of the learned ROM Test address boundary data has been retained, the user may compare it to the faulty address boundary data and determine the faulty address bit. A "stuck low" address bit would cause the data read at that address bit to be the same as that read at address zero. Conversely, if an address bit is "stuck high," the data read at address zero would be the same as that read at the address bit boundary that is "stuck high."

RAM Test

The RAM Test is the third standard built-in test of the microprocessor pods, differing from the Bus Test and ROM Test in that it need not be learned.

A RAM Test may be defined by the user using the Test Definition menu and selecting the Execute statement, RAM Test, and the first and last address boundaries of the RAM, as follows:

EXECUTE RAM TEST FROM $____ TO $____

As in the ROM Test, the preferred practice is to define a series of RAM Tests of a large array rather than one single test in order to isolate potential faults and to keep the duration of an individual RAM Test comparable to other tests. As before, the address limits must be set consistent with the addressing capabilities of the UUT's microprocessor.

The RAM Test is composed of two distinct parts: (1) a test of the RAM's data bus, and (2) a test of the RAM chip array and the RAM's address bus. Figure 8-16 is an example of the results that are displayed following execution of a RAM Test, and it illustrates the test results for the data bus, the RAM chips, and the address bus.

A RAM array's data bus is generally separated from the microprocessor's data bus by bidirectional buffers. Assuming that the Bus Test had been executed and detected no faults, any faults detected by the RAM Test's data

FIGURE 8-16
RAM test example. Reprinted courtesy of Talon Instruments.

```
                    TEST DEFINITION
SYNC EVENT : NONE
RUN  CYCLE : SINGLE PASS
RUN/EXECUTE: ALL
                                          CYCLES:0001
 HALTED :     D1          ∂ STEP 001 # FAILURES:0001
+--------------------------------------------------------+
|                                                        |
|                                                        |
| RESULTS:        33222222 22221111 11111100 00000000    |
|                 10987654 32109876 54321098 76543210    |
|                                                        |
| DATA BUS:       -------- -------- -------- --------    |
| CHIP TEST:      -------- ******** -------- --------    |
| ADDRESS BUS:    -------- -------- -------- -----*--    |
|                                                        |
| FAILED ADDRESS (HEX):    80000004                      |
| DATA IS (HEX):           55AA5555                      |
| DATA S/B (HEX):          55555555                      |
+--------------------------------------------------------+
SIGNATURES P1=      P2=      P3=      P4=      UP=

  F1       F2       F3       F4       F5       F6
RESUME   TEST                                 RETURN
         CHANNLS
```

bus test would normally be isolated to the RAM's data bus. The RAM Test data bus test consists of verifying the data bus to and from the RAM array and the basic RAM write and read controls by writing and reading data patterns to the initial RAM Test address and comparing the response data. A data bus error is indicated for any bit that fails to read back correctly. A RAM Test data bus error will terminate the RAM Test without attempting to execute the RAM chip and address bus portion of the test.

The test of the RAM chip array and the RAM's address bus is accomplished by writing and reading a series of data patterns throughout the address space. The results of the test may be categorized as follows:

1. A RAM chip error with no address errors indicates a faulty cell within the identified RAM chip.
2. A RAM address bus error with no chip errors indicates an address bus bit error common to the entire array.
3. A RAM chip error and an address error normally indicate an address bit error in a limited portion of the array.

If the RAM chip array and address data bus test successfully passes, a delay of approximately one second is inserted, which is then followed by an attempt to read the last pattern. An incorrect read of the pattern results in an indication of a memory refresh error.

Figure 8-16 illustrates an example of an address bit 02 error for RAM chips of data bits 16–23. This particular UUT had a RAM array consisting of byte-wide static chips with a failure symptom indicating that the trace for address bit 02, which was connected to the data bit 16–23 RAM chip, was open.

Microprocessor Setup

Figure 8-17 illustrates the TS6000 Microprocessor Setup menu for the M68020 microprocessor pod. The Microprocessor Setup menu is used to define the operating configuration of the microprocessor pod.

The Pod Timeout function effectively serves as a watchdog timer for all pod functions. Should the pod be unable to complete a function within the period specified by the Pod Timeout value, execution will be aborted. Typically, the Pod Timeout value should be set to twice the period required by the longest RAM Test.

A microprocessor pod executes bus cycles using the UUT microprocessor's clock. The UUT Clock parameter allows the user to adjust the pod's resultant clock phasing to match the UUT microprocessor's clock. Typically, the UUT Clock value is set to the value closest to the frequency of the clock that the UUT microprocessor receives.

The Bus Timeout function serves as a watchdog timer for individual

```
                    MICROPROCESSOR SETUP
 UUT µP TYPE : MC68020        POD TIMEOUT: 35 SECS
 UUT CLOCK   : 10 MHZ
 BUS TIMEOUT : 050 CLK CYCLES
 OVERLAY ADDR: 00000000             UPCTL: 0103

——————— CONTROL WORD (UPCTL) BIT ASSIGNMENT ———————
 BIT 00 -BUS CYCLE TIMEOUT                   ENABLED
 BIT 01 -TERMINATION ERROR FLAG              ENABLED
 BIT 02 -**NOT ASSIGNED**                    DISABLED
_BIT 03 -**NOT ASSIGNED**                    DISABLED
 BIT 04 -BUS CYCLE RE-RUN                    DISABLED
 BIT 05 -DMA OPERATION                       DISABLED
 BIT 06 -**NOT ASSIGNED**                    DISABLED
_BIT 07 -**NOT ASSIGNED**                    DISABLED
 BIT 08 -FUNCTION CODE 0                     ENABLED
 BIT 09 -FUNCTION CODE 1                     DISABLED
 BIT 10 -FUNCTION CODE 2                     DISABLED
_BIT 11 -**NOT ASSIGNED**                    DISABLED
 BIT 12 -**NOT ASSIGNED**                    DISABLED
 BIT 13 -**NOT ASSIGNED**                    DISABLED
 BIT 14 -**NOT ASSIGNED**                    DISABLED
_BIT 15 -**NOT ASSIGNED**                    DISABLED

  F1       F2       F3        F4       F5       F6
LEARN                                         RETURN
```

FIGURE 8-17
Microprocessor setup menu. Reprinted courtesy of Talon Instruments.

microprocessor bus cycles. Most microprocessor types either require that the slave function acknowledge a bus cycle or permit the slave function to extend a bus cycle until it is ready to complete the cycle. The Bus Timeout function permits the user to specify the maximum number of microprocessor clock cycles for an individual bus cycle. The Bus Timeout function may be enabled or disabled using the pod's microprocessor control word UPCTL described in a subsequent paragraph. When the Bus Timeout function is enabled, any bus cycle that is not completed within the number of clock cycles specified by the Bus Timeout value will be artificially terminated by the pod. If the Termination Error Flag or UPCTL is enabled, the test will be aborted; otherwise, execution of the test will be continued.

The microprocessor control word UPCTL provides the means to enable and disable various microprocessor unique functions. UPCTL may be configured either from the Microprocessor Setup menu or by the SET test statement. For the M68020 example of Figure 8-17, UPCTL bit 00 permits enabling/disabling the Bus Timeout function. The Termination Error Flag permits enabling or disabling all sources of termination errors, that is, Bus Timeout, reset line transitions, and the BUSERR condition. UTPCTL bit 04 enables or disables the bus cycle rerun feature of the M68020 bus. Bit 05 enables or disables the microprocessor DMA function. Bits 08–10 permit the user to set the value of the bus function codes.

Closely related to the UPCTL control word is a microprocessor unique register referred to as UBUS. Effectively, UBUS permits the user to test the status of individual microprocessor control signals, microprocessor unique error conditions, and the interrupt vectors using the IF and READ test statements. Refer to the specific microprocessor pod reference manual for additional information.

Learn Address Map

The TS6000 microprocessor pods permit the user to automatically learn the address map of the UUT's microprocessor using a function referred to as Learn Map. The Learn Map function is provided as a softkey selection on the Microprocessor Setup menu and is not often used when troubleshooting a UUT for which no documentation is available. It is generally recommended that the user first execute the microprocessor pod Bus Test prior to learning the address map in order to test the integrity of the UUT's microprocessor interface.

The process of learning a microprocessor's address space by executing writes and reads should be considered no more than an estimate and an aid to the user. Detecting RAM and ROM is relatively simple. However, because of the design features of a specific UUT such as undriven buses, a partially decoded address map, and nonuniform read and write properties of peripheral chips, categorizing the entire address map is not an exact process.

In order to put an upper limit on the duration of the process, the Learn Map routine divides the entire address space into fixed segments. The size of the segments is dependent upon the type of microprocessor, for example, 64 bytes for 8-bit microprocessors and up to 64K bytes for 32-bit microprocessors. The Learn Map routine than examines and classifies each segment. Classification designations vary by the type of microprocessor pod. Typical examples are RAM, ROM, ROM/IO, BERR, and DATA BUS ALL LOGIC HIGH.

Figure 8-18 illustrates an example of the Learn Map results for an M68020-based UUT.

Read/Write Commands

The primary test statements provided by a microprocessor pod for user tests are the WRITE and READ statements. Use of these microprocessor-independent statements eliminates the need to prepare tests using the microprocessor's assembly language.

To support the WRITE and READ statements, a microprocessor pod is allocated 65,536 bytes of RAM for storage of 99 microprocessor data tables, referred to as MTBLs. The width of each microprocessor data table

```
                 LEARN MICROPROCESSOR ADDRESS MAP

 START      END                    TYPE

00000000 00001FFC   BERR-(READ)
00002000 00002FFC   BERR-(READ/WRITE)
00003000 00003FFC   BUS CYCLE TIMEOUT (READ)
00004000 00004FFC   BUS CYCLE TIMEOUT (WRITE)
00005000 00005FFC   BUS CYCLE TIMEOUT (READ/WRITE)
00006000 00006FFC   RAM (BYTE)
00007000 00007FFC   RAM (WORD)
00008000 00008FFC   RAM (LONG)
00009000 00009FFC   ROM/IO (BYTE)
0000A000 0000AFFC   ROM/IO (WORD)
0000B000 0000BFFC   ROM/IO (LONG)
0000C000 0000CFFC   ROM/IO (BYTE)-BERR/WRITE
0000D000 0000DFFC   ROM/IO (WORD)-BERR/WRITE
0000E000 0000EFFC   ROM/IO (LONG)-BERR/WRITE
0000F000 8FFFFFFC   RAM (LONG)
90000000 FFFFFFFC   ROM/IO (LONG)


  F1        F2        F3        F4        F5        F6
                                                  RETURN
```

FIGURE 8-18
Learn address map example. Reprinted courtesy of Talon Instruments.

may be specified as byte, word, or longword. Each table may be assigned a six-character name and may contain up to 8192 elements. In addition, the user may specify a mask value for each bit of the data table.

Microprocessor data tables may be used as the source for the data transmitted by the WRITE statement and as the destination for the data received by the READ statement. In addition, microprocessor data tables may be used to specify the addresses of the WRITE statement's destination and the READ statement's source. The width of a microprocessor data table used to specify address information must be sufficient to define the specific microprocessor's address bus.

In summary, the format and parameter choices of the WRITE statement are as follows:

WRITE	[width]	[source]	@[address]	[cycle type]
WRITE	Byte	ASCII STNG	@HEX CONST	MEM CYCLE
	Word	HEX CONST	MTBL	I/O CYCLE
	Long	MTBL		VARIABLE
		VARIABLE		

Examples of the WRITE statement are as follows:

(a) WRITE.B $A5 @ $C3D2 I/O
Writes the hex byte A5 to the hex address C3D2 using an I/O cycle.
(b) WRITE.B $A5 @ $C3D2
The same as example (a) except that a memory cycle is implied.
(c) WRITE.W V1 @ $C3D2
Writes the contents of variable V1 to hex address C3D2.
(d) WRITE.L V2 @ V3
Writes the contents of variable V2 to the address specified by the contents of variable V3.
(e) WRITE.W MTBL88(BLOCK) @ MTBL99(BLOCK)
Writes the first value of table 88 to the address specified by the first value of table 99 and repeats the process for all values of the two tables. In this case, the two tables must be of equal length.
(f) WRITE.W MTBL88(V6) @ MTBL99(V7)
Same as example (e) except that the numbers of the memory data table values are specified by variables. This format is useful in FOR/NEXT loops.

The format and parameter choices for the READ statement are as follows:

READ	[width]	[destination]	@[address]	[cycle type]
READ	Byte	MUBL	@HEX CONST	MEM CYCLE
	Word	USTATUS	MTBL	I/O CYCLE
	Long	VARIABLE		UBUS
				VARIABLE

Examples of the READ statement are as follows:

(a) READ.B V1 @ $C3D2
Reads a byte from hex address C3D2 and stores it in a variable.
(b) READ.W V1 @ V2
Reads a word from the address specified by variable V2 and stores it in variable V1.
(c) READ.L METBL66(BLOCK) @ MTBL77(BLOCK)
Reads a 32-bit longword from the address specified by the first value of table number 77 and stores it in the first value of table 66. It then repeats the process for all values of the two tables. As before, the two tables must be of equal length.
(d) READ.W MTBL66(BLOCK) @ $FF00 STEP 2
Reads a block of data words into table 66 starting with the address

$FF00 and incrementing the address by 2 on consecutive read cycles.

The remaining paragraphs describe an example of a user test that uses a microprocessor pod's WRITE and READ statements.

The example tests the ability of the microprocessor to communicate with an on-board peripheral component having eight consecutively addressed write and read configuration registers. Admittedly this is a somewhat simplified case, but it does demonstrate the basic principle of board-edge functional testing; that is, a functional test consists of exchanging data between two interfaces or back and forth on a single interface. This example was also selected to permit a first-time user to use a microprocessor pod on a circuit card having the necessary equivalent RAM.

Figure 8-19 illustrates the test program named EXAMPLE1. Test step 001 writes the data of MTBL1 to the addresses specified by MTBL2.

Step 002 reads the data from the addresses specified by MTBL2 into MTBL3. Step 003 compares the mask-enabled data of MTBL3 acquired while testing a UUT to that which was acquired while learning a known

FIGURE 8-19
Test example 1. Reprinted courtesy of Talon Instruments.

```
                TEST DEFINITION
SYNC EVENT : NONE
RUN  CYCLE : SINGLE (PASS/FAIL)
RUN/EXECUTE: ALL



  NAME: EXAMPLE1   TITLE: INTERNAL FUNCTION
STEP
001   WRITE.B MTBL1(BLOCK) @ MTBL2(BLOCK)
002   READ.B MTBL3(BLOCK) @ MTBL2(BLOCK)
003   TRAPTO L80 IF MICRO <> EXPECTED
004   PRINT'PASSED TEST:EXAMPLE1'@ CRT
005   END
006
007 L80:
008   PRINT'FAILED TEST:EXAMPLE1'@ CRT
009   HALT
010


KEYPAD --------
  F1      F2      F3       F4      F5      F6
 RUN     TEST    PORT     MEMORY  MORE    RETURN
 TEST    CHANNLS TABLES   TABLES  OPTIONS
```

good card. If the data does not compare, execution jumps to L80 at step 008; otherwise, execution continues with step 005 and ends at step 006.

The EXAMPLE1 test requires only eight steps and has been intentionally simplified for discussion purposes. Additional features may be added to increase the comprehensiveness of the test (e.g., increased data table values) and to aid the operator (e.g., data comparison routines and print messages).

Chapter 9

Software Development Tools

The principal software development tools are the following:

Operating System
Editor
Compiler
Assembler
Linker
Simulator
Debugger

There are various versions of these basic tools as well as ancillary tools with a variety of functions.

The execution of the user's commands is the responsibility of the operating system, which may be UNIX, CP/M, SunOS, or the like.

Source code is entered into the development system through the editor, which may be either line-oriented or character-oriented. The latter, which is also referred to as a screen editor, is most widely used in current microprocessor development systems. A dynamic display of the test is automatically updated as the user edits the code.

The source code generated on the editor cannot be executed by a microprocessor. It is stored as ASCII characters and is translated into object code (machine language) by a compiler and assembler. Assemblers in various forms have been widely used in the past to make the translation. They continue to be used following a compiler or alone where assembly language is applied. They include one-pass and two-pass types, crossassemblers, macroassemblers, and metaassemblers. However, with the trend toward the more extensive use of high-level languages, the compiler has become the principal tool in the conversion of source code to object code or machine language.

The development cycle environment for the SPARC RISC microprocessors is shown in Figure 9-1. Initially, engineers could choose from two environments when developing the code for these processors. The native environment was based on the Sun-4 workstation, but code could also be developed on a Sun-3 workstation in a cross-development environment. The SPARC software development environment was built upon the operating system SunOS, a converged version of UNIX System V and Berkeley 4.3bsd. The environment consists of text editors, compilers, assemblers, a linker, libraries, and debuggers. To further assist the developer, there is a Network Software Environment and the SPARC Architectural Simulator, which simulates the user's SPARC-based target system in software. The Network Software Environment coordinates large software development projects by automatically tracking and maintaining consistent versions of related development objects, including source code, project requirement documents, drawings, and test plans that are used during different project development phases.

The set of language-based tools includes a code generator, global optimizer, assembler, peephole optimizer, linker, and libraries. All of the compilers permit designers to mix code written in different programming languages. For example, C programs can call FORTRAN mathematics libraries, and FORTRAN programs can call C routines. Sun's compiler technology also supports position-independent code and shared libraries, a feature of SunOS 4.0. Sharing libraries reduces the amount of code by up to 50 percent, saving both disk space and real memory.

Optimizing Compilers

Compilers are optimized to reduce program size and to increase the execution speed of the program. Techniques for optimization depend on a detailed analysis of program structure to remove redundancies and to exploit concurrency. Compilers have generally been optimized for CISC microproc-

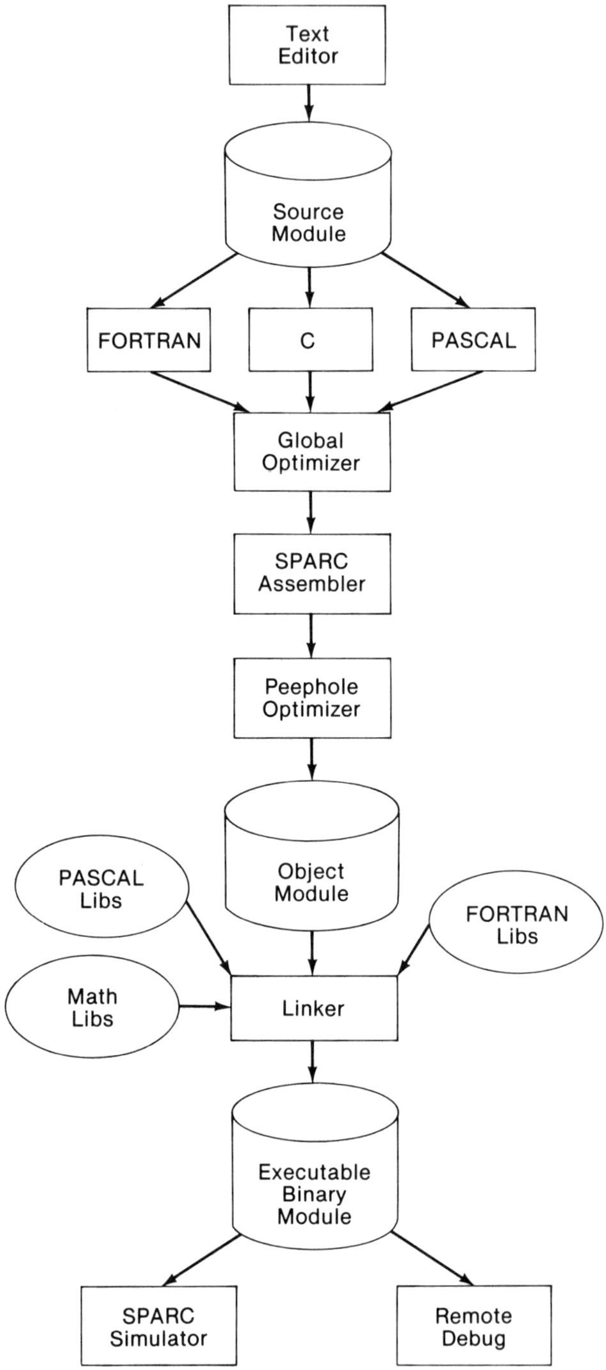

FIGURE 9-1
The software development cycle for SPARC RISC

essor applications, but the greatest need for optimization has arisen in the application of RISC microprocessors.

Much of the development of optimizing compilers has gone hand in hand with RISC microprocessor development. By compiling significant programs while chips were still on paper in the design phase, designers identified bottlenecks and were able to decide how best to eliminate them at an early stage. IBM used this approach with their 801 computer and their PL.8 compiler. The effort at the University of California at Berkeley, which led to SPARC, was combined with intensive work leading to the Portable C compiler. Likewise at Stanford University the development of the MIPS chips was done simultaneously with the development of an optimizing Pascal compiler. The evolution of the MIPS compiler development from 1980 to 1988 is shown in Figure 9-2.

As microprocessors have become more powerful and faster, data movement has increased in importance as a critical system bottleneck. Traditional compilers have not done much relative to data movement. They have accepted the data layout established by programming language conventions. It became clear from studies that accompanied the early RISC computers that compiler control of data layout and register allocation could lead to much faster programs. During the 1980s compiler optimizations have brought data into registers, used it, and returned it only when other parts of the computation need it. New tools with sophisticated features such as

FIGURE 9-2
The evolution of MIPS compiler development. Reprinted with permission of MIPS Computer Systems, Inc.

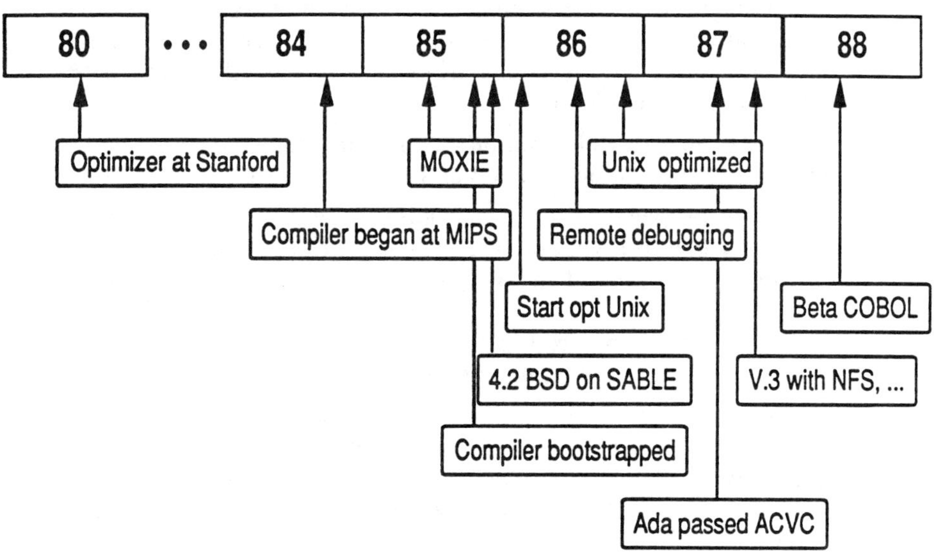

dependency analysis permit data references to be analyzed, leading to better register allocation, cache utilization, and the use of special-purpose instructions.

One can better understand the details of optimizing compilers by looking at the compilers associated with some of the principal RISC architectures.

SPARC Optimizing Compiler

RISC microprocessors employing the SPARC architecture depend on intelligent compilers developed for them to utilize their technology and achieve their highest performance. Sun Microsystems provides compilers for several languages including C, FORTRAN 77 with VMS extensions, Pascal, LISP, and Modula-2. All of these compilers, except the one for LISP, are based on the same code generation and optimization technology. In addition, many other language products are available from third parties, including common LISP, Ada, Prolog, and Mainsail.

The global optimizer performs a series of data flow analyses and transformations, which are for the most part machine independent. It performs the following optimizations:

Loop-invariant code motion—finding those computations within a loop that yield the same results for each iteration of the loop and moving them out of the loop.

Common subexpression elimination (local and global)—saving expression values and reusing them, instead of recomputing them.

Induction-variable strength reduction—after the strength reduction, if any induction variable would be dead except for its use in a loop termination test, rewriting the test and deleting the induction variable by dead code elimination.

Tail recursion elimination—converting some recursive subroutine calls into iteration, thus potentially saving call overhead.

Copy propagation (local and global)—Copy operations that assign a simple value to a variable are of interest to copy propagation if, at run time, the source of the assignment can be referenced faster than its target. For each such copy, all local uses of the target that can be reached by this copy are replaced by the source, if the source is not redefined between the copy operation and the uses.

Register allocation (modified graph coloring)—deciding which objects are worth putting in registers and which objects can share a register with others within a region of code. All scalar objects that are referenced and can fit in machine registers are candidates for allocation to registers.

Loop unrolling—replacing the body of a loop with several copies of the body, adjusting the loop control code accordingly. This optimization reduces the run-time looping overhead by reducing the number of loop iterations taken. By increasing the size of the loop body, the opportunities for instruction scheduling are increased.

Dead code elimination—An expression computation is dead if there is no execution path along which the computation can reach it. A variable definition is dead if it cannot reach any uses or it can reach computations that are used only in that definition itself. This optimization deletes all reachable dead computations and updates reaching definition information. Unreachable code is deleted by a postpass optimizer.

After the SPARC assembler there is a machine-dependent phase, referred to as peephole optimization, that involves instruction scheduling, leaf routine optimization, subroutine in-linking, elimination of unnecessary jumps, deletion of unreachable code, loop inversion, utilization of machine idioms, register coalescing, and constant propagation.

Several levels of optimization are supported by SPARC compilers. Each level requires various amounts of compilation time and produces various qualities of code, resulting in important performance gains. The amount of compilation time required and the speed of the resulting code generally increase as the optimization level decreases.

MIPS Computer Compilers for the R3000 RISC Processor

The MIPS compiler system was designed to satisfy the following objectives:

To support a reduced instruction set by providing for functions not present in the hardware

To produce code and data that achieve maximum run-time efficiency

To be compile-time efficient

To be tested and bootstrapped before hardware is available

The structure of the MIPS compiler is shown in Figure 9-3. Multiple front ends are connected to a common optimizing generator and a full set of common execution facilities. With this structure the optimizations and execution facilities can be leveraged across many applications from multiple languages. Six programming languages are supported. They are C, Pascal, FORTRAN 77, Ada, PL1, and COBOL. The dates at which these languages were accommodated by the MIPS compiler are shown in Figure 9-4. The

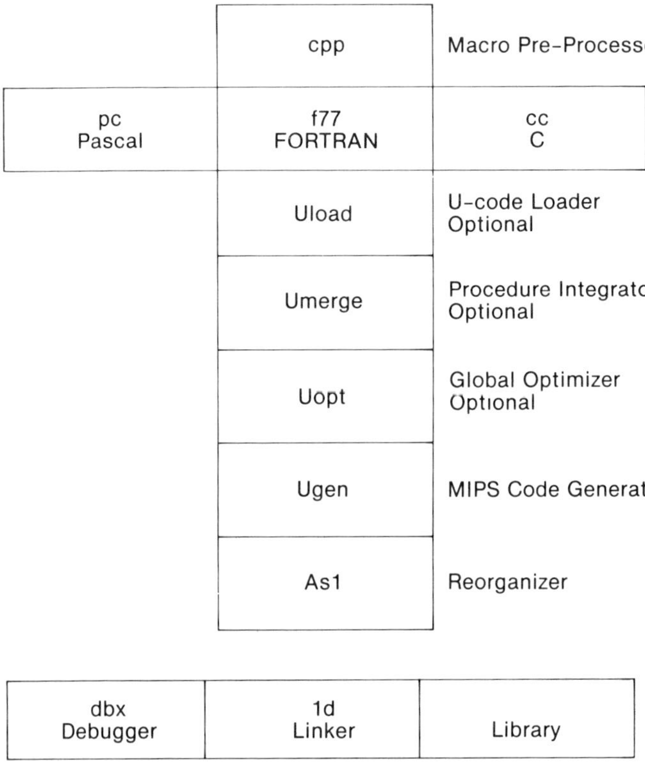

FIGURE 9-3
Structure of the MIPS compiler. F. Chow, M. Himelstein, E. Killian, and L. Weber, "Engineering a RISC Compiler System," COMPCON Proceedings, Sept. 12–14, 1988 © 1988 IEEE.

compiler has separate front ends to translate each language and a common backend to generate machine code. Run-time libraries provide language-dependent functions for each language.

MIPS Computer Systems put their emphasis on the optimizing code generator and on the execution environment. Various levels of optimization can be invoked by a switch in the compiler. The defaults level does pipeline scheduling and local optimizations. Pipeline scheduling takes care of the load and branch delay slots, rearranging the object code and finding useful instructions to place in these locations. These optimizations can be turned off for debugging purposes. Also, it is both automatic and transparent to the programmer because it is done at the assembly level.

The front ends translate the semantics of each language into an intermediate representation, called U-Code, used by several of the common components. U-Code is a stack-oriented pseudo-machine language, descended

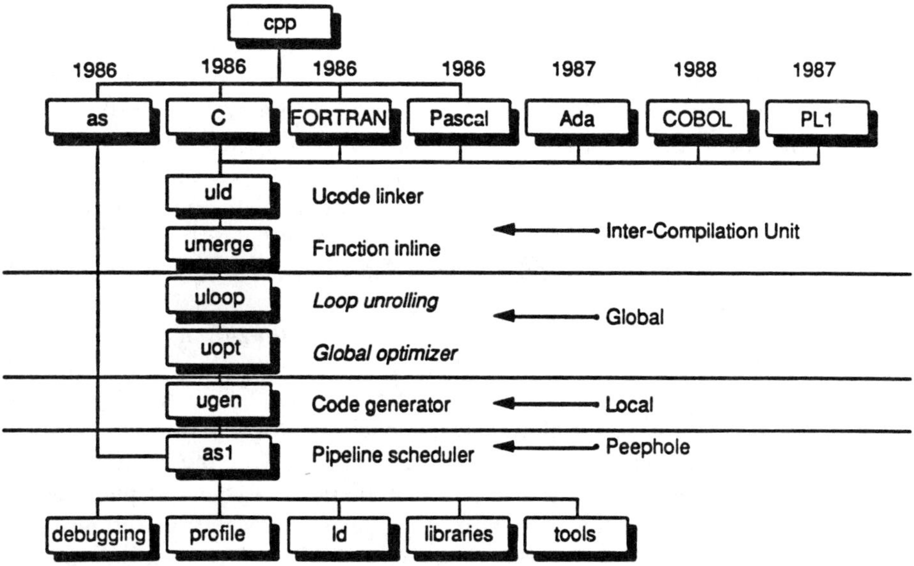

FIGURE 9-4
The evolution of front-end development for various languages. Reprinted with permission of MIPS Computer Systems, Inc.

from the version in use at Stanford University. With U-Code as the intermediate representation, the system provides uniform compilation and optimization support to high-level languages.

Each front end produces a symbol table file that contains loader and debugger information. Each compiler phase may add to or modify the symbol table to provide information for the loader or extend debugger information. The loader uses the symbol table information for relocation and merges it into the final object module. In doing so, it uses the symbol table to provide a robust debugging environment even with the optimizations.

Every compiler phase performs optimizations. An optional phase, called Uopt, performs global optimization and register allocation. The code generator, called Ugen, does local optimizations and produces MIPS assembly-language output. The final assembly phase, called Asl, does peephole optimizations and architectural-dependent pipeline scheduling as well as producing a MIPS machine-language object module.

Optimization has different levels, depending on how fast the optimizations can be performed and the nature and extent of the work. The default level consists of both peephole and local optimization. The next level, global optimization, transforms code within individual procedures. Copy propagation, constant-propagation loop invariant code motion, global common subexpression elimination, strength reduction for induction expressions, linear function test replacement, loop induction variable elimination, redundant assignment elimination, and register allocation are included. These global optimizations require extensive work to achieve their objectives. Special techniques such as data-flow analysis and graph coloring are involved.

The highest levels of optimization are provided across procedure and compilation unit boundaries. MIPS compilers provide these levels of optimization by linking an entire load module before applying the optimation transformations. The optimizations provided under these levels include interprocedural register allocation and procedure integration (in-line expansion of called procedures).

A range of profiling tools lets users adapt the behavior of their programs to the instruction level. The programmer can identify "hot spots" in the program where the source code can be turned to further increase performance. Also, the compilers allow the feedback of profile information to the compiler phases when the program is recompiled, giving the compiler more data in performing optimizations. Close interaction between compiler design and hardware architecture results in compiler-synthesized operations with efficiency that matches and sometimes exceeds their hardward-only implementations. If a third-party, multitargeted compiler had been used, this customized compilation approach could not have been pursued.

Compilation Techniques

The key to machine-specific optimization is instruction scheduling. The MIPS processor is designed to make all processor units visible to software, which permits the compiler to emit instructions in an order that utilizes each unit to its maximum efficiency. In the machine's pipeline, load and branch instructions are delayed operations. Instead of providing expensive scheduling and interlocking in hardware, the compiler is relied upon to ensure consistency between instructions. This optimization improves performance by an average of 20 percent. A similar optimization is performed for floating-point operations in which add, multiply, divide, and memory operations can execute at the same time.

Procedure call, entry, and return represent an area where the compiler must synthesize the proper functionality using the primitive operations provided in hardware. A good design allows effective subsetting of the complete procedures call mechanism according to the needs of the specific call. In the MIPS design, a procedure call, entry, and return may be as fast and simple as executing two instructions in two cycles. The protocol specifies how to pass parameters and what registers to preserve across the procedure call. The called procedure is allowed to organize its stack frame (if any) as appropriate.

For each procedure, the code generator allocates space for the largest parameter list used in the body. The code allocates this space once on entry to the procedure. Each point of call requires no further stack manipulation, so the stack pointer may double as the frame pointer. The first four words of parameters are placed in specific registers (the parameter computation is often done directly into the appropriate register) and other parameters are placed in the stack. The bulk of the calls (80 to 90 percent) require no memory references because they pass four or fewer parameters. The compiler divides registers into saved and unsaved. The protocol specifies that the callee may freely use the unsaved registers but must preserve the saved registers. Variables set before a call and used after the point of call will be allocated to saved registers because their contents will be preserved across the call.

A procedure that does not call any other procedure is called a leaf procedure. Leaf procedures do not need to save parameters and return address registers for further calls, and the parameters and return address are left in the input registers. No data movement is necessary. A leaf procedure can allocate register values to unsaved registers, eliminating the need to save and restore registers. When there are no registers to save and no calls to make, a leaf often needs no stack frame because everything resides in registers. Thus all the procedure call overhead is eliminated, except for the call instruction and the return. An extreme example of the efficiency that can be obtained by these techniques is shown in a program that makes 10 million calls to a leaf procedure within a loop (100 calls for each iteration). A MIPS M/120 RISC computer, rated at 12 times the speed of a VAX 11/780, runs this particular program in 2.4 seconds; a VAX 11/780 runs the same program in 153 seconds.

Global Optimization

The MIPS global optimizer performs all of the common global optimizations mentioned above. Although these optimizations are not machine specific, they are more effective on RISC computers than on conventional ones. First, the low level of machine functions exposes more candidates for global optimization. The effects of these global optimizations are likely to be beneficial, since they are not replaceable by more complex machine instructions in a RISC architecture. MIPS has refined the U-Code intermediate language to make it expose most MIPS machine primitives to the optimizer. Also, most of the optimizations are specifically tuned to the MIPS environment. For example, the optimizer knows which constants and uses of constants are appropriate to assign to registers.

The global optimizer also provides global register allocation. The load/store nature of RISC machines makes register allocation especially effective. By globally analyzing usage patterns, the register allocator decides which variables best reside in registers. A priority-based global-coloring algorithm is used, which often achieves optimal results with reasonable compile-time cost. Register allocation helps realize the full benefits of the other global optimizations, since intermediate computational results are best saved in registers.

Interprocedural register allocation is another capability built into the optimizer. Register allocation is normally performed globally on a per-procedure basis. With this optimization, procedures are processed before their callers and register usage information for each procedure can be propagated to its callers. By propagating register usage information upward via the call graph, the register allocator can avoid reuses of the registers already used by the callees. Thus MIPS compilers can make full use of the large and uniform register set provided by the processor.

The MIPS compiler provides a procedure integrator that selectively copies procedures in-line at their points of call. Apart from saving the cost of procedure calls, procedure integration also enhances the global optimizer's impact due to the enlarged context in the caller and the opportunity to customize the procedure body to the caller's environment.

Optimizing compilers for the R3000 RISC microprocessor serve to hide the details of the machine from programmers. The compilers not only compensate for the lack of certain functions in the machine, but by doing a good job of optimization, they can actually improve performance. Thus the programmer need not worry about the idiosyncrasies of the RISC computer on which the software is run.

One frequent problem is porting of programs that rely on special machine-dependent features. The MIPS compiler provides compile-time flags that generate extra code to support unaligned data. Cost is minimized by the processor providing instructions to load misaligned data, and the initial porting process is eased. Eventually, the program can be tuned by removing occurrences of unaligned data.

Dependence on static data in some FORTRAN programs is another example of how a proper compiler design can aid program portability. Although this dependence is outlawed by FORTRAN standards, many programs display such dependence often unknown to programmers. The MIPS compiler design provides a flag to force static definition of variables, which enables the programmer to easily ignore whether the program is dependent on static variable allocation.

MIPS compilers also assist in program porting. Assigning optimization levels based on compilation time is consistent with the programmer's practice of using more optimization only in the late stages of the project. Also, the user can easily characterize a program's performance with the sophisticated profiling tools found in these compilers. Default encoding of compressed line-number information in load modules represents another important feature, which permits some degree of program debugging without compiling especially for that ability. The development environment is improved by all of these features.

C Compiler, Assembler, and Linker for Intel's 80960

Project development with the 80960 microcontroller involves the main components shown in Figure 9-5. They include the iC-960 C compiler, the asm960 assembler, the LNK960 linker, and debugging tools. The macro preprocessor facilitates expansion of assembly macros. A user-transparent C preprocessor is included in the C compiler. Maintenance tools support further development, such as decreasing code density and making it possible to maintain often-used routines in libraries.

The iC-960 C compiler makes is possible to define the nature of the compilation process. As an example, operation controls can name and locate source text input files. Operation controls are also available for defining macros for use at compile time, as well as controlling the compiler's operation. Other controls select the proper instruction set for the particular version of the 80960 being used in the project development. Also, there are output file controls that can be used to suppress generation of an object file, while the programmer searches for syntax errors.

With control over the compiler's operation, the user can preselect from three optimization levels. Level 0 provides minimal optimization, but compile time is fast. It is typically used for debugging. In level 1 the compiler optimizes for code length using constant propagation and arithmetic identity techniques. Level 2 optimizes for processor speed.

FIGURE 9-5
Software tools for Intel's 80960 microcontroller

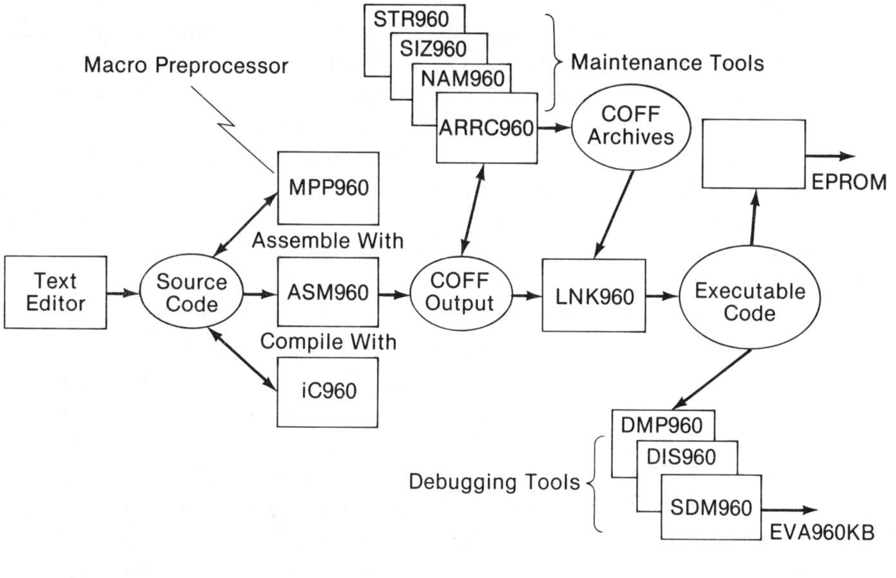

One way in which the compiler optimizes for processor speed is through procedure incorporation. The compiler recognizes nonrecursive procedures that do not call other procedures. Then it evaluates whether they can be incorporated as in-line code, or it marks them for future branch and link conversion by the linker. Other optimization routines evaluate both program loops for possible unrolling and recursive procedures for potential branching to the procedure, rather than continuous calls to the procedure.

Machine-dependent optimizations occur at all three levels. As an example, the hardware register scoreboarding capability allows the processor to continue executing instructions when it encounters a LOAD instruction. When the LOAD begins, the processor sets the scoreboard bit on the target register. Then, during the time consumed for information retrieval, the processor executes any subsequent instructions that do not reference the target register by transparently checking the scoreboard bit. Once the target register loads, the scoreboard bit resets.

To facilitate code development for embedded systems, the C compiler provides volatile and constant type-specifiers. The volatile type-specifier supports memory-mapped I/O. Such data is updated so frequently that having the processor cache it in internal registers is risky. Thus the volatile declaration directs the processor to look to the system memory for the latest value.

In some embedded systems, a value must be initialized and remain constant throughout the entire application. The constant type-specifier supports this requirement by allowing the programmer to declare a variable and initialize it with a value. In addition the constant declaration ensures that the value can never be changed during program execution.

The C compiler also provides two ways to insert in-line assembly code instruction. The simplest method is by prefixing them with the reserved word "asm" followed by the assembly language command in parentheses and quotes, such as (asm "flushreg"). A second and more powerful method relies on the compiler's support of assembler functions. Assembler functions are a special form of the normal C function declaration and are used to ensure that speed-critical sections of code are executed as fast as possible. They resemble the C case statement, but instead of using case 1, case 2, and so on, templates are defined to specify a sequence of assembler instructions to be generated when the template conditions are met. Including assembly code within the function thus avoids the overhead of a function call.

The most notable feature of the 80960 assembler is that it maintains the symbolics that the C compiler creates, as well as generating its own. It also incorporates in-debugger directives. This combination of symbolics and directives simplifies the software developer's debugging task.

If a programmer declares a leaf procedure, the assembler can force the linker to perform some optimizing. The linker attempts a BRANCH and LINK rather than a call to procedure. (Each time the code calls a procedure, a new set of on-chip registers is allocated. Only after the fourth procedure call does the processor begin writing the oldest set of cached registers to external stack memory. This process is fast, but it does not incur additional overhead in setting up new register sets.) BRANCH AND LINK is a better alternative for procedures that do not call other procedures and have few parameters to pass. If the processor has enough available registers, it branches and links instead of allocating a new set of registers. When a programmer declares that a procedure is to be a leaf procedure, it tells the assembler and linker that it is permissible to branch and link.

Typically, the linker operates transparently though the C compiler system, producing either executable or relocatable object code. Executable code can go to the target system or to the evaluation board. Relocatable code can be linked to other programmers' modules or libararies as the project development progresses.

Directed by its own UNIX-like controls, the linker not only generates code for the target system, but it also configures target memory and performs some of its own machine-dependent optimizations. For instance, when possible it changes certain procedure calls to BRANCH AND LINK or CALL SYSTEM instructions.

Since there is no operating system upon which to run the object code, as there would be if the programmer were writing a PC application, object code is of less interest to the developer than it would be in a process involving a chip like the 80386. However, the Common Object File Format (COFF) standard for object files enables third-party vendors to develop their own tools.

To find the more elusive bugs, programmers can use the disassembler, which produces assembly listings of object or executable files. Symbol tables can be examined via the name lister tools, and the dumper outputs selected parts of an object or executable file. An archiver stores object file libraries, and a sizer returns the size of each section in an object file. Where the code size is critical, the stripper removes nonessential information from the object file. EPROM programming is aided by a ROM image builder utility included in the assembler. The ROM960 generates code suitable for most PROM programmers. It operates on executable code to generate hex format files to fit the size and width of the target system's memory.

Debuggers

Debugging involves understanding the dynamic behavior of a program to find the causes of program errors. The user interactively controls program execution and analyzes the program's data flow and control flow information to understand its dynamic behavior. Initially, only an overview of flow of the program is usually wanted to provide a general understanding of its behavior. After a problem has been narrowed down to a particular area within a program, as much information as possible is wanted. It should be formatted in a manner that is easily understood.

To meet this need a window-oriented display is essential, since a debugging environment needs to present a large amount of detailed information simultaneously. At different stages of debugging, detailed information about different aspects of a program is needed.

For any kind of software development it is essential that the software engineer be able to debug at the source language level. This debugging must occur at both high level and low level (assembly language level). Thus, the capability of a debugger to adapt to a variety of source languages is very important. C is the most commonly used high-level language, but others such as Pascal and Ada are likely to be encountered.

In the case of embedded systems the normal execution environment is both real-time and event-driven, and developers are concerned about compactness and the execution speed of their object code. For example, interrupt routines and drivers may not meet response time requirements if compiled without full optimizations. It is generally not acceptable to debug nonoptimized versions of object code. Timing problems apparent with optimized object code may not appear with less optimized equivalents. The capture and use of optimization information for debugging is a requirement for an embedded systems debugging environment.

Debugging becomes more difficult when multitasking kernels are employed in embedded systems to better manage the control flow of a program. Interactions between various tasks are the source of difficult problems. To make debugging and testing more efficient, the debugger must be able to obtain the task status information from the kernel. Because several parts of an application execute concurrently, it is essential to be able to examine and control all of the tasks. If only one task can be examined at the same time, it is not possible to follow the time-dependent relationships with other tasks and their environments.

The XRAY Debugger Family

The XRAY debugger family satisfies the requirements stated above. Its key features include a rehostable/retargetable design, flexible viewpoints and user-defined screens, a macro definition facility, use of compiler optimization information, and multitasking debugging. Portable design techniques were used in the debugger design in order to readily adapt to changing target processors, execution environments, host systems, and object module formats.

A large portion of the debugging environment is common for all target processors and execution environments. The core components include: a command processor; an expression analyzer, which evaluates source-language expressions specified in debugger commands; a macro processor, which interprets source-language macro procedures at debug time; a symbol table handler; and control routines for interactive execution, breakpoints, memory control, and I/O handling. The host-dependent components of this debugger design are the window package and the virtual memory package. Rehosting the debugger environment requires changes to only some parts of the host-dependent components.

A well-defined procedural interface (Absolute Environment Interface) is used to connect the debugger core to target-dependent components. The target-dependent components are: the object module reader; disassembler; target processor address arithmetic and display routines; compiler-specific, run-time environment routines for stack frame layout, stack walking, frame pointer manipulation, and access to saved registers; and tasking and extended addressing support routines. Retargeting the debugging environment involves modifying these components.

The Execution Environment Driver uses the Execution Environment Interface to connect the debugging execution environment to the debugger core. The following components, which are dependent on the execution environment, are included: target memory access routines, register access routines, target execution and breakpoint routines, and I/O handling. The execution environment can be any of the following: an in-circuit emulator, a simulator, a monitor in target environment, and a real-time operating system.

Support Software Tools for a CISC Microprocessor

A collection of software tools allowed the use of a UNIX system to support the microprocessor development system described in Chapter 4. AT&T used their Model 3B2/400 computer as the host to develop source code for their 32100 microprocessor. As mentioned previously the 3B2/400 consisted of the 32100 plus other components to form the UNIX Microsystem, as AT&T called it. The program for a 32100 in a user's product or system was developed through the use of the 3B2/400 computer and a group of software generation tools. This support software provided utility programs that enabled the user to write applications software in C language or assembly language and then translate it into machine code.

The software development tools consisted of the following:

Archiver and Library Maintainer

Assembler

Cc Command

Optimizing C Compiler

C Preprocessor
Optimizer
Object File Host Converter
Object and Archive File Converter
Object File Compressor
Disassembler
Object File Dumper
Link Editor

These tools provided programs, the use of which required familiarity with UNIX system operation, particularly with regard to use of the system text editor and the shell (command interpreter) programs. The functions of these programs were as follows:

Cc Command. Called the optimizing C compiler, translated C source files into assembly language files, assembled the resulting files and assembly files into relocatable object files, and linked the files with other object files and/or libraries to form an absolute object file for the processor.

Assembler. Constructed a relocatable machine code object file from an assembly-language source file.

Disassembler. Produced a listing of each specified object file. The listing included the assembly-language statements and the binary code associated with those statements.

Link Editor. Collected relocatable object files produced by the assembler and reconstructed them into a single load file for execution or as input for a subsequent link editor.

Object Code Optimizer. Optimized the text of an unstripped link-edited load module.

Object File Dumper. Dumped selected parts of each of the specified object file arguments. It accepted both object files and archives of object files. The options allowed selective processing of each file according to the needs of the programmer.

Symbol Table Printer. Printed the symbol table of each specified object file. Files specified could be relocatable or absolute object files, or archives of these files. Each symbol table entry contained the name of the symbol, its value expressing an offset or an address (depending on its storage class), its type and derived type, its size in bytes (if available), the source line number (if available), and the object file section in which this symbol was defined.

Object File Size Printer. Printed the section's name and size information in bytes for each section in the specified object file.

Object File Stripper. Stripped the symbol table, relocation, and line number information from object files, including archives. It was normally run only on production object files (modules) that had been debugged and tested, since no symbolic debugging information is available from a file once it has been stripped. Options controlled the amount of information stripped from the symbol table.

C Program Beautifier. By use of the cb command, read C programs either from its arguments or from standard input and wrote them on the standard output with spacing and indentation that displayed the structure of the code.

C Flowgraph Generator. By use of the cflow command, analyzed a collection of C, yacc, lex, assembler, and object files and attempted to build a graph charting the external references.

C Program Trace Generator. By use of the ctrace command, allowed one to follow the execution of a C program, statement by statement. Ctrace read the C program in file (or from standard input if one did not specify file), inserted statements to the text of each executable statement and the values of all variables referenced or modified, and wrote the modified program to the standard output.

C Program Cross Referencer. By use of the cxref command, analyzed a collection of files and attempted to build a cross-reference table. It produced a listing on standard output of all symbols (auto, static, and global) in each file separately or in combination.

Lexical Analyzer Generator. By use of the lex command, generated programs to be used in simple lexical analysis of text. The input files (standard input default) contained strings and expressions to be searched for and C text to be executed when strings were found.

C Program Checker. By use of the line command, attempted to detect features of the C program files that are likely to be bugs, nonportable, or wasteful. It also checked type usage more strictly than the compilers. Among the things that have been detected were unreachable statements, loops not entered at the top, automatic variables declared and not used, and logical expressions whose values were constant. Moreover, the usage of functions was checked to find functions that return values in some places and not in others, functions whose values are not used or whose values are used but not returned.

Macro Processor. By use of the m4 command, acted as a macro processor intended as a front end for Ratfor, C, and other languages. Each of the argument files was processed in order; if there were no files, or if a file name was -, the standard input was read. The processed text was written on the standard output.

Object File Comment Section By use of the mcs command, manipulated the comment section, normally the .comment section, in an object file. It was used to add to, delete from, print, and compress the contents of the comment section in a UNIX system object file.

Profile Data Printer. By use of the prof command, interpreted a profile produced by the monitor function. The symbol table in the object file prog (a.out by default) was read and correlated with a profile file (mon.out by default).

Regular Expression Compiler. By use of the regcmp command, compiled the regular expressions in file and placed the output in file.i.

Symbolic Debugger. By use of the sdb command, called a symbolic debugger that would be used with C and F77 programs. It could be used to examine their object files and core files and to provide a controlled environment for their execution.

Topological Sorter. By use of the tsort command, produced on the standard output a totally ordered list of items consistent with a partial ordering of items mentioned in the input file. If no file was specified, the standard input was understood.

Parser Generator. By use of the yacc command, converted a context-free grammar into set tables for a simple automation that executed an LR parsing algorithm. The grammar could be ambiguous; specified precedence rules were used to break ambiguities.

Object File Host Converter. Converted a WE32100 microprocessor object file with header information formatted for one host machine to an object file with header information for the specified host machine.

Object File Compressor. Attempted to reduce the size of an object file by removing duplicate structure and union descriptors from the symbol table.

Source Code Lister. Produced a C source listing under a C source file(s) with line number information supplied by its associated object file. Line numbers were printed for each breakpointable line generated by the compiler.

Object File Orderer. Used to order object file libraries for one-pass access by the link editor. The order was such that every symbol and function was referenced before it was defined. One or more object or library archive files could be passed to this command for ordering. This ordering was an optional feature for efficiency only. The link editor could process unordered libraries.

Object File (Format). A file format created by the assembler and the link editor and used by the other commands. The file contained a file header with general information about the file, one section header for each section contained in the object file, the raw section information (e.g., text, data), relocation information for each section, line number information for each section, the symbol table, and the string table for long user variable names. Depending on the assembler or link editor flag options used to produce the object file, the file could be devoid of relocation entries, line number entries, or the symbol table.

Archiver and Library Maintainer. Formatted one or more files into an archive file. Its main use was to create and update library files as used by the link editor.

Object and Archive File Converting. Converted a UNIX System V archive file to a UNIX System V portable archive, without altering the members of the archive file.

Environment Setup Script. Shell script provided for the Cross UNIX System and Cross SGP users. It modified its inherited environment according to its name-value pairs (of the form name-value) and then executed the command with the modified environment. If no command were given, /bin/sh was executed. The user could examine the script to determine which, if any, of the name-value pairs were meaningful to the particular use. In most cases, it served as an example but could be tailored to the user's need. The script also set the user's PATH variable to search for tools in the cross SGP directories first.

Make Command Script. A shell provided for Cross SGP and Cross UNIX System SGP users to allow them to use native makefiles containing default variables in the cross environment. The script executed the m32make command with the default parameters modified to point to the cross SGP products. The m32make command was used to run the user's makefiles. The user could examine the make script to determine if any of the macro definitions was useful.

Installation Script. A shell used by the system administrator to change the default floating-point mode MAU (Math Acceleration Unit) or FPE (Floating-Point Emulation). The SGP products were released with MAU default.

Compiler Command Generator. A shell used by the system administrator when directories other than the defaults were wanted. It created a cc command front-end that found tools in their designated directories.

Common Object File Access Library. Contained access routines that were a collection of functions for reading an object file. Although the calling program had to know the detailed structure of the parts of the object file it processed, the routines effectively isolated the calling program from the knowledge of the overall structure of the object file. There were 22 routines, which could be classified into four categories: open or close object files, read header or symbol table data, seek the data in a particular section, and return the index of a given entry in the symbol table. There was also a version of this library available to allow profiling of programs.

Floating-Point Emulation (FPE) Library (Cross SGP only). Contained routines that compute floating-point operations using WE32100 microprocessor integer instructions. In a UNIX System V Release 2.0 environment, a system call was executed to check the presence of a WE32106 Math Acceleration Unit in the system and utilize the increased performance of the math acceleration unit in a program-transparent way at execution.

Math Libraries for the FPE Mode

FPE Double-Precision Math Library. This math library contained the full set of transcendental and other functions in the UNIX System V standard math library. It conformed to the requirements set forth in the UNIX System V Release 2.1 manual.

FPE Single-Precision Math Library. This math library contained the sine, cos, tan, asine, acos, atan, exp, log, sqrt, and power functions. The polynomials and rational approximations gave a precision of 24-bit mantissa widths. Most internal computations used single-precision arithmetic, though subroutine linkage was double-precision.

Math Libraries for the Math Acceleration Unit Mode

A total of six math library alternatives were provided as follows:

Double-Precision Math Library. This math library contained the full set of functions in the UNIX System V standard math library. It conformed to the requirements described in the manual pages in the UNIX System V Release 2.0 *Programmer's Reference Manual.* Selected routines were hand optimized for performance. The optimized routines include sine, cos, tan, atan, exp, log10, pow, and sqrt. Most calculations were done using double-precision data types.

Single-Precision Math Library. This math library contained the sine, cos, tan, acos, atan, exp, log, log10, power, and sqrt functions. The polynomial and rational approximations had been modified from their UNIX System V math library counterparts to achieve precision of 24-bit mantissa widths. Most internal computations used single-precision arithmetic.

Fast Double-Precision Math Library. This math library contained the sine, cos, asine, acos, tan, atan, exp, log, log10, and sqrt functions. It conformed to the same requirements as the regular double-precision math library, but it was enhanced for specialized applications. These routines had no argument reduction or error checking.

Sine and cosine had to be in the range of $-\pi/2 \mu x \mu \pi/2$.

Tan had to be in the range of $-\pi/1 < x < \pi/2$.

Sqrt, log, and log10 had to be greater than 10.

Fast Single-Precision Math Library. This math library contained the sine, cos, tan, atan, exp, log, log10, and sqrt functions. It conformed to the same requirements as the regular single-precision math library, but it was enhanced for specialized applications. These routines had no argument reduction or error checking, and input restrictions were the same as for the fast double-precision math library.

Fast Double Precision Math Source Library. This math source library contained the assembly source of the functions in the fast double-precision math library. These source library routines were in-line expanded by the optimizer to provide fast execution by reducing the overhead of argument passing, function calling and returning, and return value passing (-J option must be invoked with this library pathname). The source library was designed for applications that sought an increase in speed at the potential cost of size.

Fast Single-Precision Math Source Library. This math source library contained the assembly source of the functions in the fast single-precision math library. The source library routines were in-line expanded by the optimizer to provide faster execution by reducing the overhead of argument passing, function calling and returning, and return value passing (-J option must be invoked with this library pathname). The source library was designed for applications that sought an increase in speed at the potential cost of size.

UNIX System Libraries. Two versions of each library were provided; one for the MAU mode and one for the FPE mode.

The Standard C Library (libc.a). This library contained the UNIX system calls, the standard I/O package, other C library functions, and floating-point emulation routines. The floating-point emulation routines implemented a useful set of floating-point operations using the WE32100 microprocessor instructions. In a UNIX System V Release 2 environment, a system call was executed at process startup to check for the presence of a WE32106 Math Acceleration Unit in the system and utilized the increased performance of the math acceleration unit, at execution, if the math acceleration unit was present. In addition to the routines mentioned above, this library included routines that completed compliance with the IEEE standard for binary floating-point arithmetic (ANSI/IEEE Standard 754-1985). The additional routines provided were single- and double-precision and binary-to-decimal and decimal-to-binary conversions. (There was also a version of libc.a available to allow profiling of programs.)

The PW Library. This library contained a set of routines that were used to support programs written for a prerelease version of the UNIX system.

The Debug Library. This library contained a startup routine for the symbolic debugger. This routine was loaded when specifying the -g option with the cc command.

The LEX library. This library contained routines for use with programs generated by the scanner generator, lex.

Startup Routines. Startup routines were responsible for setting up the stack, handling traps, and calling routines to determine the presence or absence of a WE32106 Math Acceleration Unit and for starting execution by calling main and gates to the exit routine.

Chapter 10

Built-in Development Support

The increasing complexity of both RISC and CISC microprocessors has resulted in greater complexity and cost of development support tools. In most such chip designs there are no special provisions to facilitate debugging or emulation. However, in a few designs steps have been taken to design into the microprocessor some special features to ease debugging and emulation operations. For instance, Intel's 80386 has eight registers set aside for debugging, and the follow-on 80486 has further enhancements of this feature. Intel's 80960KB microprocessor, which is aimed at embedded system applications, also has rather extensive debugging capability built into the chip.

Motorola's 68332 microcontroller chip carries the debugging feature a step further by implementing the debugger in CPU microcode. Wtih its internal debugger, the conventional emulator can be replaced with a bus state analyzer. This feature is described in more detail farther along in this chapter.

Built-in debugging features have not yet appeared in any RISC chips except Intel's 80860 (i860), where two control registers and a data access trap support debugging with both data and instruction breakpoints, and in Intel's 80960. It would appear that RISC chips would be likely candidates for built-in debugging features, if the most inexpensive development support is to be available for them.

Further details of the built-in development support available in the 80386, 80486, 80960KB, 68322, and 80860 follow in this chapter.

80386 Microprocessor Debug Registers

The debug registers provide the ability to set breakpoint addresses for either data or code segments. A breakpoint is a user-selected point in a computer program at which execution is temporarily suspended. In the early stages of program development, breakpoints provide a convenient means of determining whether specified instructions or memory addresses have been accessed. In the later stages of program development, the debug registers provide for the examination of register and memory contents when the breakpoint has been reached.

The Intel 80386 CPU has eight debug registers (see Figure 10-1). These consist of four address registers (DR0 through DR3), two reserved registers (DR4 and DR5), one status register (DR6), and one control register (DR7).

Debug Address Registers

The four debug address registers store the addresses of four memory locations used for breakpointing, where the contents of each location can be either an instruction or data. Each address is a 32-bit linear address. When paging is enabled, the linear address is automatically translated into

FIGURE 10-1
Debug registers of the 80386. Reprinted courtesy of Intel Corporation.

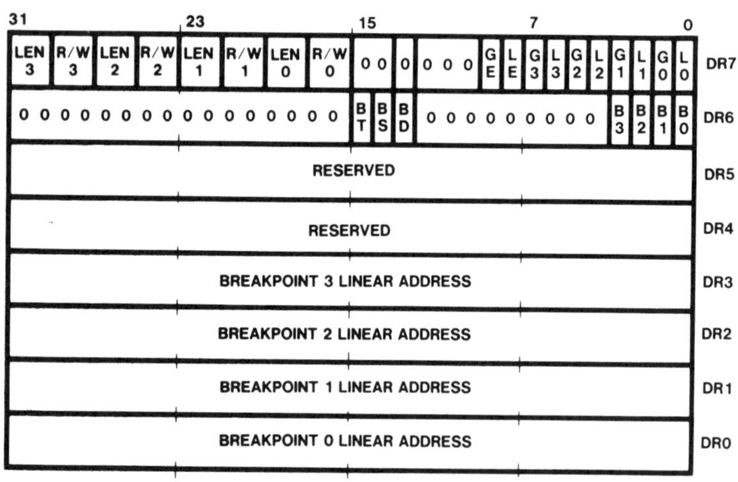

NOTE: 0 MEANS INTEL INTEL RESERVED. DO NOT DEFINE.

G30117

a physical address. If paging is not enabled, the linear address is treated as a physical address.

Debug Control Register

The debug control register, DR7, selectively enables specific debug conditions for the addresses contained in the debug address registers, DR0–DR3. Table 10-1 lists the functions of the bits contained in this register.

TABLE 10-1
Debug Control Register Fields

Symbol	*Bit(s)*	*Name*	*Description*
L0–L3	0,2, 4,6	Local active	If set (1), the breakpoint address in the associated debug address register is active for the current local task only. A task switch automatically clears these bits.
G0–G3	1,3, 5,7	Global active	If set (1), the breakpoint address in the associated debug address register is active for all tasks. These bits are unaffected by task switches.
LE	8	Local enable	When set (1), causes all local breakpoint interruptions to be reported in the debug status register. This bit is cleared on a task switch.
GE	9	Global enable	When set (1), causes all global breakpoint interruptions to be reported in the debug status register. This bit is unaffected by task switches.
R/W0 R/W1 R/W2 R/W3	16,17 20,21 24,25 28,29	Read/ write	Determines the type of action causing a breakpoint for the address in the associated debug address register. The codes are: 00 - Break only on instruction execution 01 - Break only on data writes 10 - Undefined 11 - Break only on data reads and writes
LEN0 LEN1 LEN2 LEN3	18,19 22,23 26,27 30,31	Length field	Determines the range of sequential bytes to be monitored starting from the address in the corresponding debug address register. The codes are: 00 - One-byte length 01 - Two-byte length 10 - Undefined 11 - Four-byte length

* Reprinted courtesy of AT&T.

Debug Status Register

The debug status register, DR6, defines 7 of its 32 bits as status flags. These are called B0–B3, BT, BS, and BD.

The B0, B1, B2, and B3 flags record the status of each breakpoint address set in the debug address registers. These bits can then be tested to cause execution of a debug exception (interrupt vector 1 in the Interrupt Descriptor Table). The BT flag is set prior to entering the debug exception handler if a task switch has been made and the T bit of the entering task state segment is set. The BS bit is set when the TF (trap flag) in the EFLAGS register indicates a single-step exception (TF = 1). The BD bit is set when the next instruction is to read or write one of the eight registers in the Intel Ice-386 In-Circuit Emulator debugging system.

By monitoring the state of the status and control register bits listed in Table 10-2, various breakpoint conditions can be recognized for further processing by an exception handler.

Test Registers

Two test registers are available for testing the on-chip translation look-aside buffer (TLB) cache. The CPU uses the TLB when translating linear addresses to physical addresses with paging enabled. Essentially, these two registers are used for writing test addresses and flag bits to the TLB and storing addresses and flag data that has been read from the TLB.

Typically, the test registers are used by the operating system only for a power-up confidence test of the 80386 CPU. As such, these two registers perform an extremely specialized function and are not normally considered part of the 80386's operating architecture.

TABLE 10-2
Debug Exception Conditions

Tested Bits	*Condition Indicated*
B0 = 1 AND (G0 = 1 OR L0 = 1)	Breakpoint occurred as defined by DR0, LEN0, and R/W0 (Table 10-1)
B1 = 1 AND (G1 = 1 OR L1 = 1)	Breakpoint occurred as defined by DR1, LEN1, and R/W1 (Table 10-1)
B2 = 1 AND (G2 = 1 OR L2 = 1)	Breakpoint occurred as defined by DR2, LEN2, and R/W2 (Table 10-1)
B3 = 1 AND (G3 = 1 OR L3 = 1)	Breakpoint occurred as defined by DR3, LEN3, and R/W3 (Table 10-1)
BS = 1	Single-step exception
BD = 1	Debug registers are not available due to use by ICE-386 debugger
BT = 1	Task switch has occurred

* Reprinted courtesy of AT&T.

80486(i486) Microprocessor Debugging Support

The 486 microprocessor provides several features that simplify the debugging process. The three categories of on-chip debugging aids are as folllows:

1. the code execution breakpoint opcode (0CCH)
2. the single-step capability provided by the TF bit in the flag register
3. the code and data breakpoint capability provided by the debug registers DR0–3, DR6, and DR7

Breakpoint Instruction

A single-byte-opcode breakpoint instruction is available for use by software debuggers. The breakpoint opcode is 0CCH, and it generates an exception 3 trap when executed. In typical use, a debugger program can "plant" the breakpoint instruction at all desired code execution breakpoints. The single-byte breakpoint opcode is an alias for the two-byte general software interrupt instruction, INT n, where n = 3. The only difference between INT 3 (0CCH) and INT n is that INT 3 is never IOPL-sensitive, but INT n is IOPL-sensitive in Protected Mode and Virtual 8086 Mode.

Single-Step Trap

If the single-step flag (TF, bit 8) in the EFLAG register is found to be set at the end of an instruction, a single-step exception occurs. The single-step exception is auto vectored to exception number 1. Precisely, exception 1 occurs as a trap after the instruction following the instruction that set TF. In typical practice, a debugger sets the TF bit of a flag register image on the debugger's stack. It then typically transfers control to the user program and loads the flag image with a signal instruction, the IRET instruction. The single-step trap occurs after executing one instruction of the user program.

Since the exception 1 occurs as a trap (that is, it occurs after the instruction has already executed), the CS:EIP pushed onto the debugger's stack points to the next unexecuted instruction of the program being debugged. An exception 1 handler, merely by ending with an IRET instruction, can therefore efficiently support single-stepping through a user program.

Debug Registers

The debug registers are an advanced debugging feature of the 486 microprocessor. They allow data access breakpoints as well as code execution breakpoints. Since the breakpoints are indicated by on-chip registers, an instruction execution breakpoint can be placed in ROM code or in code shared by several tasks, neither of which can be supported by the INT3 breakpoint opcode.

The 486 microprocessor contains six debug registers, providing the ability to specify up to four distinct breakpoint addresses, breakpoint control options, and read breakpoint status. Initially after reset, breakpoints are in the disabled state. Therefore, no breakpoints will occur unless the debug registers are programmed. Breakpoints set up in the debug registers are autovectored to exception number 1.

Linear Address Breakpoint Registers (DR0–DR3)

Up to four breakpoint addresses can be specified by writing into debug registers DR0–DR3, shown in Figure 10-2. The breakpoint addresses specified are 32-bit linear addresses. 486 microprocessor hardware continuously compares the linear breakpoint addresses in DR0–DR3 with the linear addresses generated by executing software (a linear address is the result of computing the effective address and adding the 32-bit segment base address). Note that if paging is not enabled, the linear address equals the physical address. If paging is enabled, the linear address is translated to a physical 32-bit address by the on-chip paging unit. Regardless of whether paging is enabled or not, however, the breakpoint registers hold linear addresses.

Debug Control Register (DR7)

A debug control register, DR7 shown in Figure 10-2, allows several debug control functions such as enabling the breakpoints and setting up other control options for the breakpoints. The fields within the debug control register, DR7, are as follows:

LENi (breakpoint length specification bits). A 2-bit LEN field exists for each of the four breakpoints. LEN specifies the length of the associated

FIGURE 10-2
Debug registers of the 80486. Reprinted courtesy of Intel Corporation.

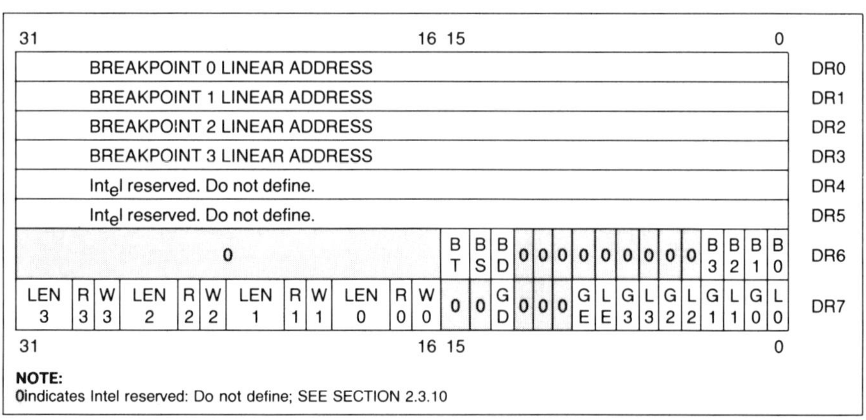

breakpoint field. The choices for data breakpoints are: 1 byte, 2 bytes, and 4 bytes. Instruction execution breakpoints must have a length of 1 (LENi = 00). Encoding of the LENi field is as follows:

LENi Encoding	*Breakpoint Field Width*	*Usage of Least Significant Bits in Breakpoint Address Register i, (i = 0 − 3)*
00	1 byte	All 32-bits used to specify a single-byte breakpoint field.
01	2 bytes	A1–A3 used to specify a two-byte word-aligned breakpoint field. A0 in Breakpoint Address Register is not used.
10	Undefined—do not use this encoding	
11	4 bytes	A2–A31 used to specify a four-byte, dword-aligned breakpoint field. A0 and A1 in Breakpoint Address Register are not used.

The LENi field controls the size of breakpoint field i by controlling whether all low-order linear address bits in the breakpoint address register are used to detect the breakpoint event. Therefore, all breakpoint fields are aligned; 2-byte breakpoint fields begin on word boundaries, and 4-byte breakpoint fields begin on dword boundaries.

The following is an example of various-sized breakpoint fields. Assume the breakpoint linear address in DR2 is 00000005H. In that situation, the following illustration indicates the region of the breakpoint field for lengths of 1, 2, or 4 bytes.

RWi (memory access qualifier bits). A 2-bit RW field exists for each of the four breakpoints. The 2-bit RW field specifies the type of usage that must occur in order to activate the associated breakpoint.

RW Encoding	*Usage Causing Breakpoint*
00	Instruction execution only
01	Data writes only
10	Undefined—do not use this encoding
11	Data reads and writes only

RW encoding 00 is used to set up an instruction execution breakpoint. RW encodings 01 and 11 are used to set up write-only or read/write data breakpoints.

Note that instruction execution breakpoints are taken as faults (i.e., before the instruction executes), but data breakpoints are taken as traps (i.e., after the data transfer takes place).

Using LENi and RWi to Set Data Breakpoint i. A data breakpoint can be set up by writing the linear address into DRi (i = 0–3). For data breakpoints, RWi can equal 01 (write only) or 11 (write/read). LEN can equal 00, 01, or 11.

If a data access entirely or partly falls within the data breakpoint field, the data breakpoint condition has occurred, and if the breakpoint is enabled, an exception 1 trap will occur.

Using LENi and RWi to Set Instruction Execution Breakpoint i. An instruction execution breakpoint can be set up by writing the address of the beginning of the instruction (including prefixes if any) into DRi (i = 0–3). RWi must equal 00 and LEN must equal 00 for instruction execution breakpoints.

If the instruction beginning at the breakpoint address is about to be executed, the instruction execution breakpoint condition has occurred, and if the breakpoint is enabled, an exception 1 fault will occur before the instruction is executed.

Note than an instruction execution breakpoint address must be equal to the beginning byte address of an instruction (including prefixes) in order for the instruction execution breakpoint to occur.

GD (Global Debug Register Access Detect). The debug registers can only be accessed in Real Mode or at privilege level 0 in Protected Mode. The GD bit, when set, provides extra protection against any debug register access even in Real Mode or at privilege level 0 in Protected Mode. This additional protection feature is provided to guarantee that a software debugger can have full control over the debug register resources when required. The GD bit, when set, causes an exception 1 fault if an instruction attempts to read or write any debug register. The GD bit is then automatically cleared when the exception 1 handler is invoked, allowing the exception 1 handler free access to the debug registers.

GE and LE (Exact Data Breakpoint Match, Global and Local). If either GE or LE is set, any data breakpoint trap will be reported exactly after completion of the instruction that caused the operand transfer. Exact reporting is provided by forcing the 486 microprocessor execution unit to wait for completion of data operand transfers before beginning execution of the next instruction.

If exact data breakpoint match is not selected, data breakpoints may not be reported until several instructions later or may not be reported at all. When enabling a data breakpoint, it is therefore recommended to enable the exact data breakpoint match.

When the 486 microprocessor performs a task switch, the LE bit is cleared. Thus the LE bit supports fast task switching out of tasks that have enabled the exact data breakpoint match for their task-local breakpoints. The LE bit is cleared by the processor during a task switch, to avoid having exact data breakpoint match enabled in the new task. Note that exact data breakpoint match must be reenabled under software control.

The 486 microprocessor GE bit is unaffected during a task switch. The GE bit supports exact data breakpoint match that is to remain enabled during all tasks executing in the system.

Note that instruction execution breakpoints are always reported exactly, whether or not exact data breakpoint match is selected.

Gi and Li (Breakpoint Enable, Global and Local). If either Gi or Li is set, then the associated breakpoint (as defined by the linear address in DRi, the length in LENi, and the usage criteria in RWi) is enabled. If either Gi or Li is set, and the 486 microprocessor detects the ith breakpoint condition, then the exception 1 handler is invoked.

When the 486 microprocessor performs a task switch to a new Task State Segment (TSS), all Li bits are cleared. Thus the Li bits support fast task switching out of tasks that use some task-local breakpoint registers. The Li bits are cleared by the processor during a task switch, to avoid spurious exceptions in the new task. Note that the breakpoints must be reenabled under software control.

All 486 microprocessor Gi bits are unaffected during a task switch. The Gi bits support breakpoints that are active in all tasks executing in the system.

Debut Status Register (DR6)

A debug status register, DR6 shown in Figure 10-2, allows the exception 1 handler to easily determine why it was invoked. Note the exception 1 handler can be invoked as a result of one of several events:

1. DR0 breakpoint fault/trap
2. DR1 breakpoint fault/trap
3. DR2 breakpoint fault/trap
4. DR3 breakpoint fault/trap

5. Single-step (TF) trap
6. Task switch trap
7. Fault due to attempted debug register access when GD = 1.

The debug status register contains single-bit flags for each of the possible events invoking exception 1. Note below that some of these events are faults (exception taken before the instruction is executed), while other events are traps (exception taken after the debug events occurred).

The flags in DR6 are set by the hardware but never cleared by hardware. Exception 1 handler software should clear DR6 before returning to the user program to avoid future confusion in identifying the source of exception 1.

The fields within the debug status register, DR6, are: Bi (debug fault/trap due to breakpoint 0–3). Four breakpoint indicator flags, B0–B3, correspond one-to-one with the breakpoint registers in DR0–DR3; a flag Bi, which is set when the condition described by DRi, LENi, and RWi occurs.

If Gi or Li is set, and if the ith breakpoint is detected, the processor will invoke the exception 1 handler. The exception is handled as a fault if an instruction execution breakpoint occurred or as a trap if a data breakpoint occurred.

Important Note: A flag Bi is set whenever the hardware detects a match condition on enabled breakpoint i. Whenever a match is detected on at least one enabled breakpoint i, the hardware immediately sets all Bi bits corresponding to breakpoint conditions matching at that instant, whether enabled or not. Therefore, the exception 1 handler may see that multiple Bi bits are set, but only set Bi bits corresponding to enabled breakpoints (Li or Gi set) are true indications of why the exception 1 handler was invoked.

BD (Debug Fault Due to Attempted Register Access When GD Bit Set). The BD bit is set if the exception 1 handler was invoked due to an instruction attempting to read or write to the debug registers when the GD bit was set. If such an event occurs, then the GD bit is automatically cleared when the exception 1 handler is invoked, allowing handler access to the debug registers.

BS (Debug Trap Due to Single-step). The BS bit is set if the exception 1 handler was invoked due to the TF bit in the flag register being set (for single-stepping).

BT (Debug Trap Due to Task Switch). The BT bit is set if the exception 1 handler was invoked due to a task switch occurring to a task having a 486 microprocessor TSS with the T bit set. Note the task switch into the new task occurs normally, but before the first instruction of the task is executed, the exception 1 handler is invoked. With respect to the task switch operation, the operation is considered to be a trap.

Use of the Resume Flag (RF) in the Flag Register

The Resume Flag (RF) in the flag word can suppress an instruction execution breakpoint when the exception 1 handler returns to a user program at a user address that is also an instruction execution breakpoint.

80960KB Debugging Features

Debug Capabilities

The 80960KB microprocessor provides extensive system debug capabilities, an important feature for embedded computing where the ability to instrument an application may be limited. The 80960KB microprocessor allows breakpoint instructions that stop program execution on various events, such as procedure calls or certain instructions. Another debug facility traces the activity of the processor while it is executing a program. Tracing is done by recording the addresses of instructions that cause trace events to occur. For example, a trace event can occur on the execution of a specific instruction, branch, or procedure call. To ensure that the 80960KB is operating properly, the processor performs a self-test when it is reset. If the self-test is successful, the 80960KB begins operation; otherwise, it enters the stopped state.

The tracing facilities of the 80960KB processors, which allow the monitoring of instruction execution, are described below.

Overview of the Trace-Control Facilities

The 80960KB processor provides facilities for monitoring the activity of the processor by means of trace events. A trace event in the 80960KB is a condition where the processor has just completed executing a particular instruction or type of instruction, or where the processor is about to execute a particular instruction.

By monitoring trace events, debugging software is able to display or analyze the activity of the processor or of a program. This analysis can be used to locate software or hardware bugs or for general system monitoring during the development of system or applications programs.

The typical way to use this tracing capability is to set the processor to detect certain trace events either by means of the trace-controls word or a set of breakpoint registers. An alternate method of creating a trace event is with the mark and force-mark (fmark) instructions. These instructions cause an explicit trace event to be generated when the processor detects them in the instruction stream.

If tracing is enabled, the processor signals a trace fault when it detects a trace event. The fault handler for trace faults can then call the debugging monitor software to display or analyze the state of the processor when the trace event occurred.

Required Software Support for Tracing

To use the processor's tracing facilities, software must provide trace-fault handling procedures, perhaps interfaced with a debugging monitor. Software must also manipulate several control flags to enable the various tracing modes and to enable or disable tracing in general. These control flags are located in the system-data structures described in the next section.

Trace Controls

The following flags or fields control tracing:

Trace controls word

Trace-enable flag in the process controls

Trace-fault-pending flag in the process controls

Trace flag (bit 0) in the return-status field of register r0

Trace-control flag in the supervisor-stack-pointer field of the system table or a procedure table

Trace-Controls Word

The trace-controls word is cached internally in the processor. The trace controls allow software to define the conditions under which trace events are generated. Figure 10-3a shows the structure of the trace-controls word.

This word contains two sets of bits: the mode flags and the event flags. The mode flags define a set of trace modes that the processor can use to generate trace events. A mode represents a subset of instructions that will cause trace events to be generated. For example, when the call-trace mode is enabled, the processor generates a trace event whenever a call or branch-and-link operation is executed. To enable a trace mode, the kernel sets the mode flag for the selected trace mode in the trace controls. The trace modes are described later in this chapter.

The processor uses the event flags to keep track of which trace events (for those trace modes that have been enabled) have been detected.

A special instruction, the modify-trace controls (modtc) instruction, allows software to set or clear flags in the trace controls. On initialization,

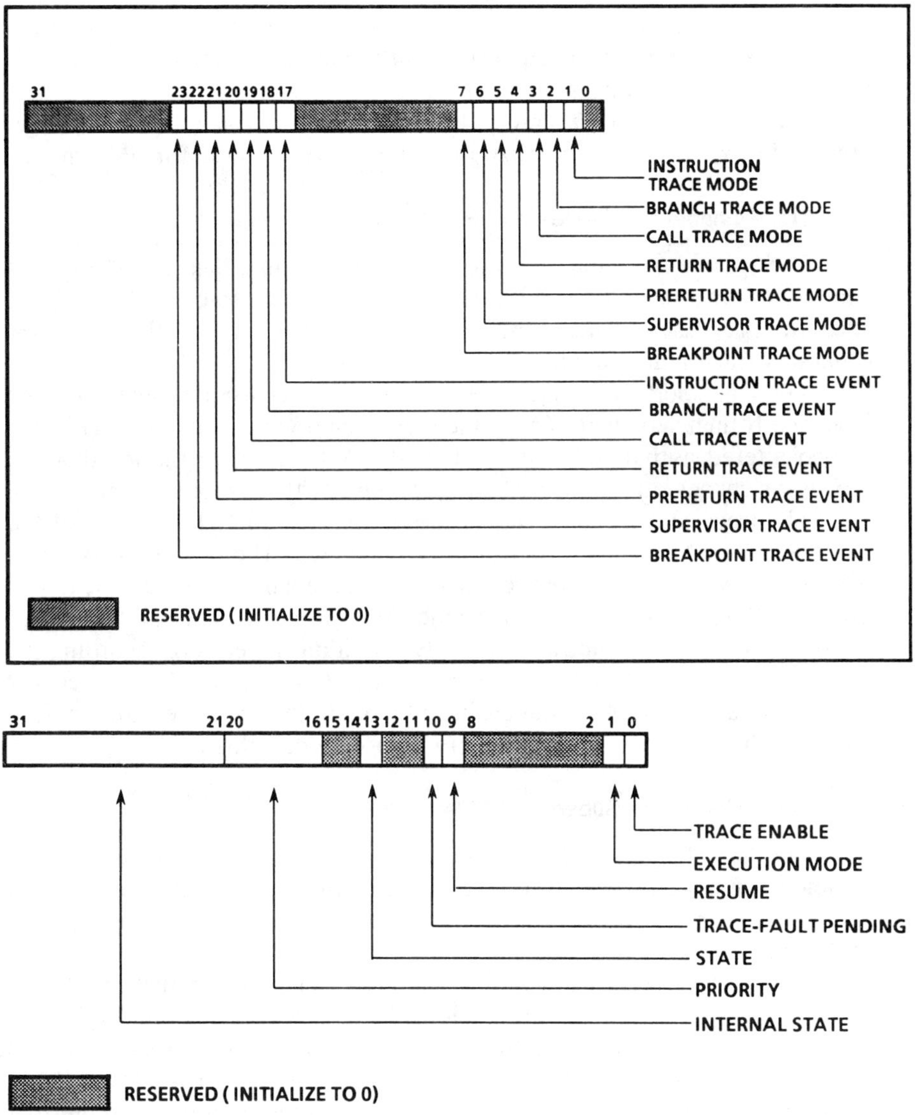

FIGURE 10-3
Trace-controls word of the 80960. Reprinted courtesy of Intel Corporation.

all the flags in the processor's internal trace controls are cleared. The modtc instruction can then be used to set or clear trace mode flags as required.

Software can access the event flags using the modtc instruction, however, there is no reason to. The processor modifies these flags as part of its trace-handling mechanism.

Bits 0, 8 through 16, and 24 through 31 of the trace controls are reserved. Software should initialize these bits to zero and not modify them.

Trace-Enable and Trace-Fault-Pending Flags

The trace-enable flag and the trace-fault-pending flag, in the process controls (shown in Figure 10-3b), control tracing. The trace-enable flag enables the processor's tracing facilities. When this flag is set, the processor generates trace faults on all trace events.

Typically, software selects the trace modes to be used through the trace controls. It then sets the trace-enable flag when tracing is to begin. This flag is also altered as part of some of the call and return operations that the processor carries out, as described at the end of this chapter.

The trace-fault-pending flag allows the processor to keep track of the fact that an enabled trace event has been detected. The processor uses this flag as follows. When the processor detects an enabled trace event, it sets this flag. Before executing an instruction, the processor checks this flag. If the flag is set, it signals a trace fault. By providing a means of recording the occurrence of a trace event, the trace-fault-pending flag allows the processor to service an interrupt or handle a fault other than a trace fault, before handling the trace fault. Software should not modify this flag.

Trace Control on Supervisor Calls

The trace flag and the trace-control flag allow tracing to be enabled or disabled when a call-system instruction (calls) is executed that results in a switch to supervisor mode. This action occurs independent of whether tracing is enabled prior to the call.

When a supervisor call is executed (calls an instruction that references an entry in the system procedure table with an entry type 11_2), the processor saves the current state of the trace-enable flag (from the process controls) in the trace flag (bit 0) of the return-status field of register r0. Then, when the processor selects the supervisor procedure from the procedure table, it sets the trace-enable flag in the process controls according to the setting in the trace-control flag in the procedure table (bit 0 of the word that contains the supervisor-stack pointer).

On a return from the supervisor procedure, the trace-enable flag in the process controls is restored to the value saved in the return-status field of register r0.

Trace Modes

The following trace modes can be enabled through the trace controls:

Instruction trace
Branch trace
Call trace
Return trace
Prereturn trace
Supervisor trace
Breakpoint trace

These modes can be enabled individually, or several modes can be enabled at once. Some of these modes overlap, such as the call-trace mode and the supervisor-trace mode. The section later in this chapter titled ''Handling Multiple Trace Events'' describes what the processor does when multiple trace events occur.

The following sections describe each of the trace modes.

Instruction Trace

When the instruction-trace mode is enabled, the processor generates an instruction-trace event each time an instruction is executed. This mode can be used within a debugging monitor to single-step the processor.

Branch Trace

When the branch-trace mode is enabled, the processor generates a branch-trace event any time a branch instruction that branches is executed. A branch-trace event is not generated for conditional-branch instructions that do not branch. Also, branch-and-link, call, and return instructions do not cause branch-trace events to be generated.

Call Trace

When the call-trace mode is enabled, the processor generates a call-trace event any time a call instruction (call, callx, or calls) or a branch-and-link instruction (bal or balx) is executed. An implicit call, such as the action used to invoke a fault handler or an interrupt handler, also causes a call-trace event to be generated.

When the processor detects a call-by-trace event, it also sets the prereturn-trace flag (bit 3 of register r0) in the new frame created by the call operation or in the current frame if a branch-and-link operation was performed. The processor uses this flag to determine whether or not to signal a prereturn-trace event on a return instruction.

Return Trace

When the return-trace mode is enabled, the processor generates a return-trace event any time a ret instruction is executed.

Prereturn Trace

The prereturn-trace mode causes the processor to generate a prereturn-trace event prior to the execution of any ret instruction, providing the prereturn-trace flag in r0 is set. (Prereturn tracing cannot be used without enabling call tracing.)

The processor sets the prereturn-trace flag whenever it detects a call-trace event (as described above for the call-trace mode). This flag performs a prereturn-trace-pending function. If another trace event occurs at the same time as the prereturn-trace event, the prereturn-trace flag allows the processor to fault on the non-prereturn-trace event first, then come back and fault again on the prereturn-trace event. The prereturn trace is the only trace event that can cause two successive trace faults to be generated between instruction boundaries.

Supervisor Trace

When the supervisor-trace mode is enabled, the processor generates a supervisor-trace event any time (1) a call-system instruction (calls) is executed, where the procedure table entry is a supervisor procedure, or (2) when a ret instruction is executed and the return-status field is set to 010_2 or 011_2 (i.e., return from supervisor mode).

This trace mode allows a debugging program to determine the boundaries of kernel procedure calls within the instruction stream.

Breakpoint Trace

The breakpoint-trace mode allows trace events to be generated at places other than those specified with the other trace modes. This mode is used in conjunction with the mark and force-mark (fmark) instruction and the breakpoint registers.

The mark and fmark instructions allow breakpoint-trace events to be generated at specific points in the instruction stream. When the breakpoint-trace mode is enabled, the processor generates a breakpoint-trace event any time it encounters a mark instruction. The fmark causes the processor to generate a breakpoint-trace event regardless of whether the breakpoint-trace mode is enabled or not.

The processor has two one-word breakpoint registers, designated as breakpoint 0 and breakpoint 1. Using the set-breakpoint-register IAC, one instruction pointer can be loaded into each register. The processor then generates a breakpoint trace any time it executes an instruction referenced in a breakpoint register.

Trace-Fault Handler

A fault handler is a procedure that the processor calls to handle faults that occur. A trace-fault handler has one additional restriction. It must be called with an implicit supervisor call, and the trace-control flag in the system-procedure-table entry must be clear. This restriction ensures that tracing is turned off when a trace fault is being handled, which is necessary to prevent an endless loop.

Signaling a Trace Event

To summarize the information presented in the previous sections, the processor signals a trace event when it detects any of the following conditions:

- An instruction included in a trace-mode group is executed or about to be executed (in the case of a prereturn trace event) and the trace mode for that instruction is enabled.
- An implicit call operation has been executed and the call-trace mode is enabled.
- A mark instruction has been executed and the breakpoint-trace model is enabled.
- An fmark instruction has been executed.
- An instruction specified in a breakpoint register is executed and the breakpoint-trace mode is enabled.

When the processor detects a trace event and the trace-enable flag in the process controls is set, the processor performs the following action:

1. The processor sets the appropriate trace-event flag in the trace controls. If a trace event meets the conditions of more than one of the enabled trace modes, a trace-event flag is set for each trace mode condition met.
2. The processor sets the trace-fault-pending flag in the process controls.

Note: The processor may set a trace-event flag and the trace-fault-pending flag before it has completed execution of the instruction that caused the event. However, the processor only handles trace events in between the execution of instructions.

If, when the processor detects a trace event, the trace-enable flag in the processor controls is clear, the processor sets the appropriate event flags but does not set the trace-fault-pending flag.

Handling Multiple Trace Events

If the processor detects multiple trace events, it records one or more of them based on the following precedence, where 1 is the highest precedence:

1. Supervisor-trace event
2. Breakpoint- (from mark or fmark instruction, or from a breakpoint register), branch-, call-, or return-trace event
3. Instruction-trace event

When multiple trace events are detected, the processor may not signal each event; however, it will signal at least the one with the highest precedence.

Trace Handling Action

Once a trace event has been signaled, the processor determines how to handle the trace event, according to the setting of the trace-enable and trace-fault-pending flags in the process controls and to other events that might occur simultaneously with the trace event such as an interrupt or a nontrace fault.

The following sections describe how the processor handles trace events for various situations.

Normal Handling of Trace Events

Prior to executing an instruction, the processor performs the following action regarding trace events:

1. The processor checks the state of the trace-fault-pending flag. If this flag is clear, the processor begins execution of the next instruction. If the flag is set, the processor performs the following actions.
2. The processor checks the state of the trace-enable flag. If the trace-enable flag is clear, the processor clears any trace event flags that have been set, prior to starting execution of the next instruction. If the trace-enable flag is set, the processor performs the following action.
3. The processor signals a trace fault and begins the fault handling action.

Prereturn Trace Handling

The processor handles a prereturn-trace event the same as described above except when it occurs at the same time as a nontrace fault. Here, the nontrace fault is handled first.

On returning from the fault handler for the nontrace fault, the processor checks the prereturn-trace flag in register r0. If this flag is set, the processor generates a prereturn-trace event, then handles it as described above.

Tracing and Interrupt Handlers

When the processor invokes an interrupt handler to service an interrupt, it disables tracing. It does this by saving the current state of the process controls, then clearing the trace-enable and trace-fault-pending flags in the current process controls.

On returning from the interrupt handler, the processor restores the process controls to the state they were in prior to handling the interrupt, which restores the state of the trace-enable and trace-fault-pending flags. If these two flags were set prior to calling the interrupt handler, a trace fault will be signaled on the return from the interrupt handler.

Tracing and Fault Handlers

The processor can invoke a fault handler with either an implicit local call or an implicit supervisor call. On a local call, the trace-enable and trace-fault-pending flags are neither saved on the call nor restored on the return. The state of these flags on the return is thus dependent on the action of the faulthandler.

On a supervisor call, the trace-enable and trace-fault-pending flags are saved, as part of the saved process controls, and restored on the return. So, if these two flags were set prior to calling the fault handler, a trace fault will be signaled on the return from the fault handler.

Note: On a return from an interrupt handler or a fault handler (other than the trace-fault handler), the trace-fault-pending flag is restored. If this flag is set as a result of the handler's ret instruction, the detected trace event is lost.

68332 Microcontroller Development Support

The following features have been implemented on the CPU32 to enhance the instrumentation and development environment:

M68000 Family Development Support

Background Debug Mode

Deterministic Opcode Tracking
Hardware Breakpoints

M68000 Family Development Support

All M68000 family members include features to facilitate applications development. These features include:

Trace on Instruction Execution. M68000 processors include an instruction-by-instruction tracing facility as an aid to program development. The MC68020, 030, 040, and CPU32 also allow tracing only those instructions causing a change in program flow. In the trace mode, a trace exception is generated after an instruction is executed, allowing a debugger program to monitor the execution of a program under test.

Breakpoint Instruction. An emulator may insert software breakpoints into the target code to indicate when a breakpoint has occurred. On the MC68010, MC68020, MC68030, and CPU32, this function is provided via illegal instructions, $4848–$484F, to serve as breakpoint instructions.

Unimplemented Instruction Emulation. During instruction execution, when an attempt is made to execute an illegal instruction, an illegal instruction exception occurs. Unimplemented instructions (F-line, A-line, etc.) utilize separate exception vectors to permit efficient emulation of unimplemented instructions in software.

Background Debug Mode

Microcomputer systems generally provide a debugger, implemented in software, for system analysis at the lowest level. The background debug mode on the CPU32 is unique in that the debugger has been implemented in CPU microcode. Registers can be viewed and/or altered, memory can be read or written to, and test features can be invoked. Incorporating these capabilities on-chip simplifies the environment in which the in-circuit emulator operates. With an integrated debugger, the traditional emulator configuration, as shown in Figure 10-4, may be replaced by a bus state analyzer (BSA), shown in Figure 10-5. The BSA simply monitors the traffic on the internal bus, whereas the in-circuit emulator replaces the processor in the target system with hardware in the emulator, and all external bus traffic must traverse the cable between the emulator and the target system.

As each command is accumulated in the serial shifter, a microaddress is generated that points to the microcode routine corresponding to that command. If the command can complete without additional serial traffic, it does; however, if addresses or operands are required, the microcode waits as each

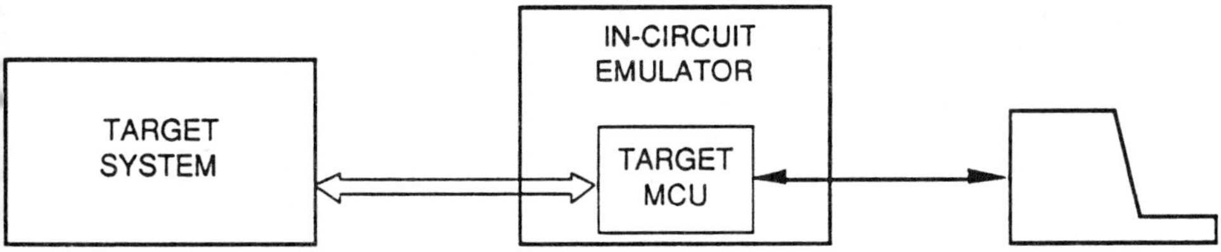

FIGURE 10-4
Traditional in-circuit emulator diagram. Reprinted with permission of Motorola, Inc.

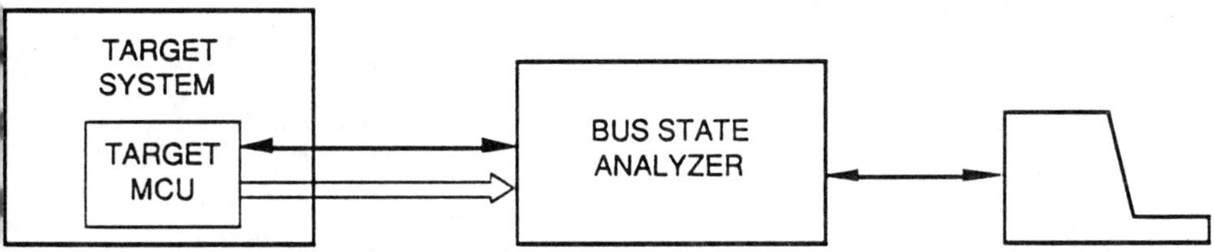

FIGURE 10-5
Bus state analyzer configuration. Reprinted with permission of Motorola, Inc.

word is assembled. Result operands are loaded into the output shift register to be shifted out as the next command is read. Table 10-3 summarizes the command set available in the Background Debug Mode.

The Debug Mode feature on the MC68332 MCU significantly reduces the external logic necessary to provide emulation. Additionally, the use of the M68000 architecture for the CPU and external bus (based on the 68020) permits the use of many existing or slightly modified existing development tools. Motorola and third-party development support companies take advantage of these features to provide new emulation and debugging options. The user may utilize these new features, at minimum system cost, by providing an interface connector on their target system.

Deterministic Opcode Tracking

CPU32 function code outputs are augmented by two supplementary signals to monitor the instruction pipeline. The instruction pipe (IPIPE) output indicates the start of each new instruction and each mid-instruction pipeline advance. The instruction fetch (IFETCH) output identifies the bus cycles in which the operand is loaded into the instruction pipeline. Pipeline flushes are also signalled with IFETCH. Monitoring these two signals allows a bus analyzer to synchronize itself to the instruction stream and monitor its activity.

TABLE 10-3
Background Debug Mode Command Summary

Command	*Mnemonic*	*Description*
Read D/A Register	RDREG/RAREG	Read the selected address or data register and return the results via the serial interface.
Write D/A Register	WDREG/WAREG	The data operand is written to the specified address or data register.
Read System Register	RSREG	The specified system control register is read. All registers that can be read in supervisor mode can be read in background mode.
Write System Register	WSREG	The operand data is written into the specified system control register.
Read Memory Location	READ	Read the sized data at the memory location specified by the longword address. The source function code register (SFC) determines the address space accessed.
Write Memory Location	WRITE	Write the operand data to the memory location specified by the longword address. The destination function code (DFC) register determines the address space accessed.
Dump Memory Block	DUMP	Used in conjunction with the READ command to dump large blocks of memory. An initial READ is executed to set up the starting address of the block and retrieve the first result. Subsequent operands are retrieved with the DUMP command.
Fill Memory Block	FILL	Used in conjunction with the WRITE command to fill large blocks of memory. An initial WRITE is executed to set up the starting address of the block and supply the first operand. Subsequent operands are written with the FILL command.
Resume Execution-Flush Pipe	GO	The pipe is flushed and refilled before instruction execution may continue.
Patch User Code	CALL	Current program counter is stacked at the location of the current stack pointer.
Reset Peripherals	RST	Asserts RESET for 512 clock cycles. The CPU is NOT reset by this command. Synonymous with the CPU RESET instruction.
No Operation	NOP	NOP performs no operation and may be used as a null command.

On-Chip Breakpoint Hardware

An external breakpoint input and on-chip breakpoint hardware allow a breakpoint trap on any memory access. Off-chip address comparators preclude breakpoints unless show cycles are enabled. Breakpoints on instruction prefetches that are ultimately flushed from the instruction pipeline are not acknowledged; operand breakpoints are always acknowledged. Acknowledged breakpoints initiate exception processing at the address in exception vector number 12.

80860(i860) Microprocessor Debugging

The i860 microprocessor supports debugging with both data and instruction breakpoints. The features of the i860 architecture that support debugging include the following:

db (data breakpoint register), which permits specification of data addresses that the i860 microprocessor will monitor.

BR (break read) and BW (break write) bits of the psr, which enable trapping of either reads or writes (respectively) to the address in db.

DAT (data access trap) bit of the psr, which allows the trap handler to determine when a data breakpoint was the cause of the trap.

Trap instruction that can be used to set breakpoints in code. Any number of code breakpoints can be set. The values of the src1 and src2 fields help identify which breakpoint has occurred.

IT (instruction trap) bit of the psr, which allows the trap handler to determine when a trap instruction was the cause of the trap.

Chapter 11

Boundary Scan

Boundary scan is a design for testability (DFT) technique proposed by the Joint Test Action Group (JTAG). It is the application of a partitioning scan ring at the boundary of IC designs to provide controllability and observability access via scan operations. It had the support of about two hundred companies, which form JTAG, and has been refined and balloted by the IEEE to become Standard P1149.1. Boundary scan is applied at the board level, but some built-in self-test (BIST) elements are required in the IC to make it functional. The addition of these elements results in some cost increase for the chip and can also affect its performance. So while there is a growing use of BIST techniques in chip design, they do not necessarily accommodate the boundary scan technique because of cost and performance considerations. More conformance could be expected if the IEEE standard is adopted.

The boundary-scan technique requires the inclusion of a shift-register latch (contained in a boundary-scan cell) adjacent to each component pin so that signals at the component boundaries can be controlled and observed using certain scan testing principles.

One possible implementation of a boundary-scan cell, which could be used for an input connection to an integrated circuit, is shown in Figure 11-1. Depending on the control signal applied to the multiplexer, data can

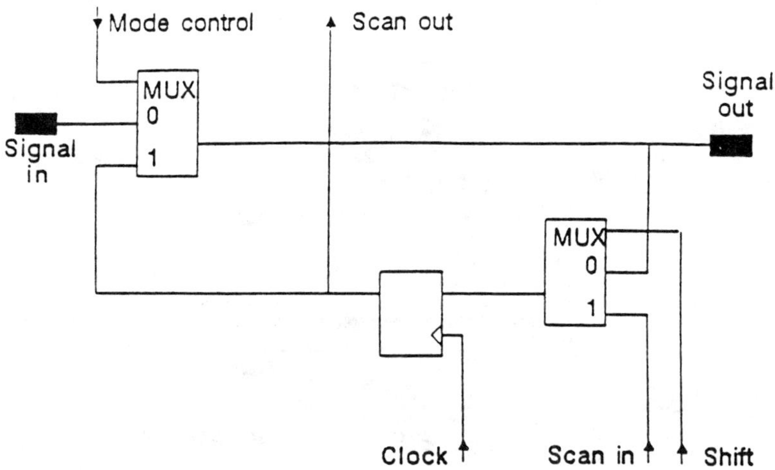

FIGURE 11-1
A simple boundary-scan cell design

either be loaded into the scan register from the input port or driven from the register through the output port of the cell into the core of the component design. An analogous design could be used for boundary-scan cells associated with output or bidirectional connections to the component.

The boundary-scan cells for the pins of a component are interconnected so as to form a shift-register chain around the border of the design. This path is provided with serial input and output connections and appropriate clock and control signals. Within a product assembled from several integrated circuits the boundary-scan registers for the individual components could be connected in series to form a single path through the complete design, as shown in Figure 11-2. As an alternate, a board design could contain several independent boundary-scan paths.

If all of the components in a circuit have the boundary-scan capability, the resulting serial path through the complete design can be used in two ways—an external boundary-scan test and an internal boundary-scan test.

With external boundary scan, test data can be shifted into all of the boundary-scan cells associated with component output pins and loaded in parallel through the component interconnections into these cells associated with input pins. By this procedure the interconnections between the various components can be tested.

With internal boundary-scan, the boundary-scan register can be used as a means of isolating system logic from stimuli received from surrounding components while an internal self-test is being performed. Thus components

* Figures in this chapter are reprinted courtesy of AT&T.

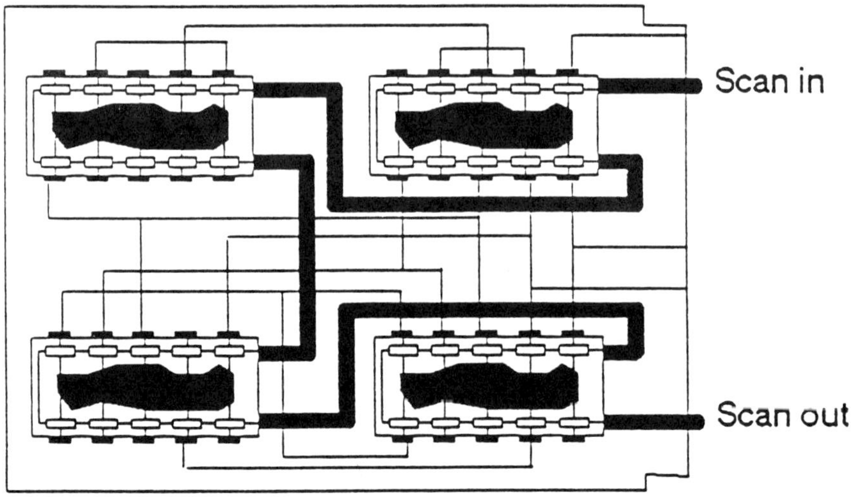

FIGURE 11-2
A boundary-scan board

on the board are tested. Alternatively, the register can permit a slow-speed static test of the system logic, since it allows the delivery of test data to the component and examination of the test results.

By parallel loading the cells at both the inputs and outputs of a component and shifting out the results, the boundary-scan register provides a means of sampling the data flowing through a component without interfering with its behavior. Such a sample boundary-scan test is valuable for design debugging and fault analysis.

Through the use of the boundary-scan register, internal and external boundary-scan tests allow confirmation that each module of a circuit performs its required function and that the modules are interconnected correctly. In effect, tests applied using the register can detect many of the faults that in-circuit testers currently address, but without the need for extensive bed-of-nails access and without the need for expensive test equipment. The sample test can be used to examine the states of connections not normally accessible to other test systems.

A technical description of JTAG Boundary-Scan Architecture Standard Proposal Version 2 follows.

Architecture: Block Structure

The JTAG architecture allows access to and control of test-support features within integrated circuits through a test access port (TAP). The following elements are required components of the design, (Figure 11-3): test access

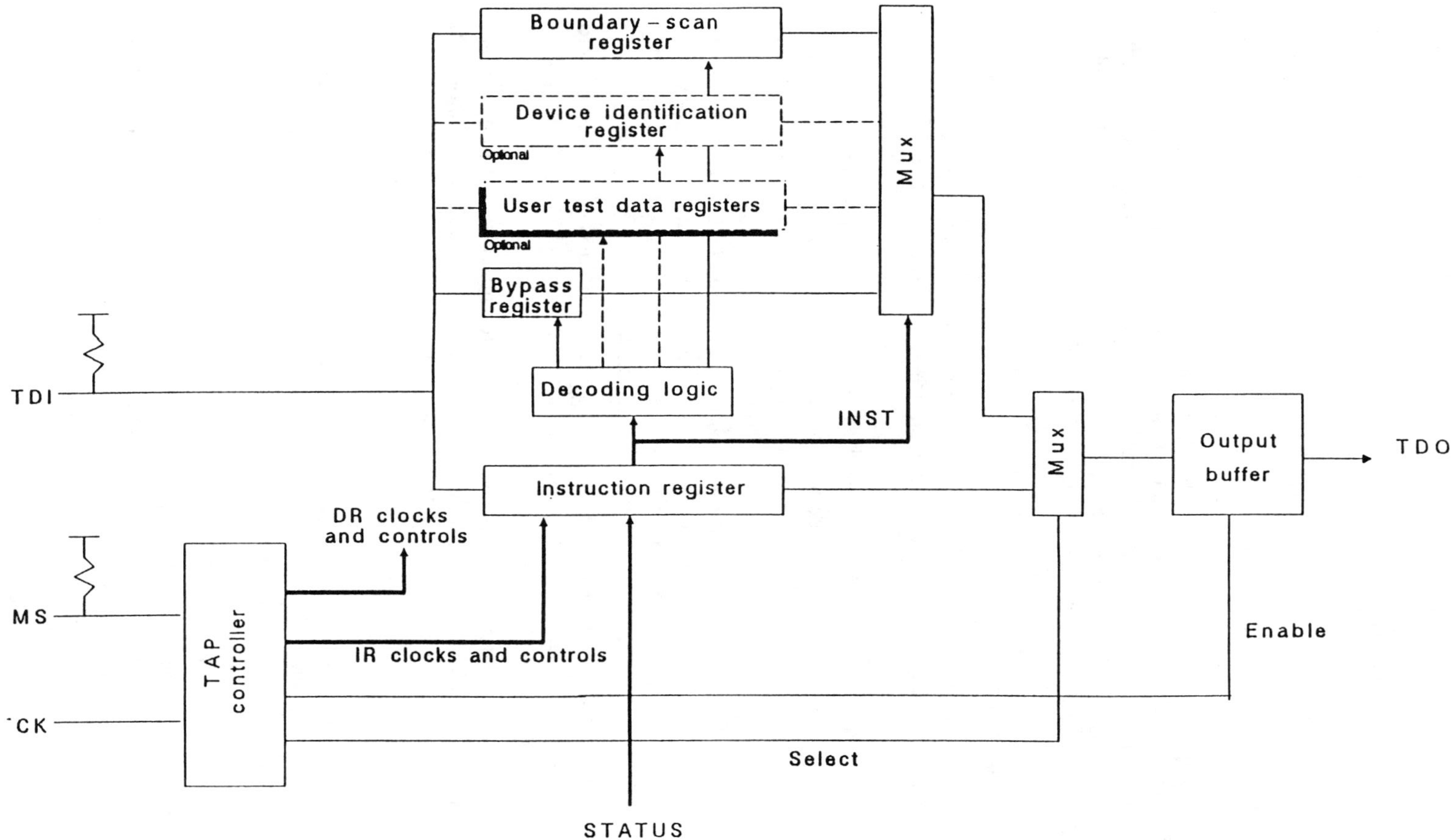

FIGURE 11-3 JTAG boundary-scan architecture

port (TAP), a comprising four pins, a TAP controller, and a scannable instruction register. Each circuit must also include at least two scannable test data registers: a bypass register and a boundary-scan register. Optionally, a third test data register—the device identification register—may be included. Other user-designed test data registers may be provided optionally to give access to design-specific features.

The instruction and test data registers are parallel shift-register-based paths, having a common serial data input and a common serial data output, both of which are connected to the TAP. The selection between these parallel shift-register-based paths is made under control of the TAP controller, when necessary taking into account the contents of the instruction register.

Description

The top-level design of the JTAG test support circuitry is shown in Figure 11-3. Key features of the design are as follows:

- The test access port (TAP). The TAP consists of the three input connections and one output connection for the test logic shown in Figure 11-3. The TAP is a general-purpose port that can provide access to many test support functions built into the circuit design.
- The TAP controller, which receives the dedicated test clock (TCK) and interprets the signals on the test mode select (TMS) line. The TAP controller generates clock and control signals for the instruction and test data registers and other parts of the architecture.
- The instruction register, which allows the instruction to be shifted into the design. The instruction is used to select the test to be performed and/or the test data register to be accessed.
- The boundary-scan register, which allows testing of board interconnections—detecting typical production defects such as opens and shorts. It also allows access to the inputs and outputs of devices to test their system logic, or to allow sampling of signals flowing through the systems inputs and outputs.
- The bypass register, which provides a single-bit serial connection through the circuit when none of the other test data registers is selected. This register can, for example, be used to allow test data to flow through a particular device to other devices on a board without affecting the normal operation of the particular device.
- The device identification register, which is an optional test data register that allows the manufacture, part number, and variant of a component to be determined. If this register is included, then it should conform to the design requirements listed on page 309.

Additional test data registers can be provided to allow access to design-specific test-support facilities in the integrated circuit, such as self-test facilities.

Prime and Second Sources

The test logic in prime and second source devices must share a common architecture and a common data sheet. The only exception to this requirement is the device identification register.

Description

It is essential that both the system and the test logic of prime and second source devices conform to a common data sheet. This ensures that test programs created for a printed circuit board containing multiple-sourced devices produce consistent results regardless of the source of individual components.

This rule requires that the following design features are shared by devices from all sources:

a. instruction register length and instruction coding
b. the number of test data registers and their function
c. the structure of each test data register and its operation

The only exception to this requirement is in the design of the optional device identification register. Here the identification code must vary to identify the source of the particular device, its part number, and its revision.

Test Access Port (TAP)

The test access port (TAP) is composed of four signals. These support serial communication of test data and instructions and allow control of test execution.

The signals in the TAP form a bus that interconnects a slave device with other slave devices and with a bus master.

The minimum bus configuration contains two broadcast signals (TMS and TCK) fed from the bus master to all slaves in parallel and a serial path formed by a "daisy-chain" connection of the serial test data pins of the master and slave devices (TDI and TDO).

Description

Two example configurations are shown in Figure 11-4a. The bus master may be either an ATE system or a component that interfaces the TAP bus to a higher-level test bus.

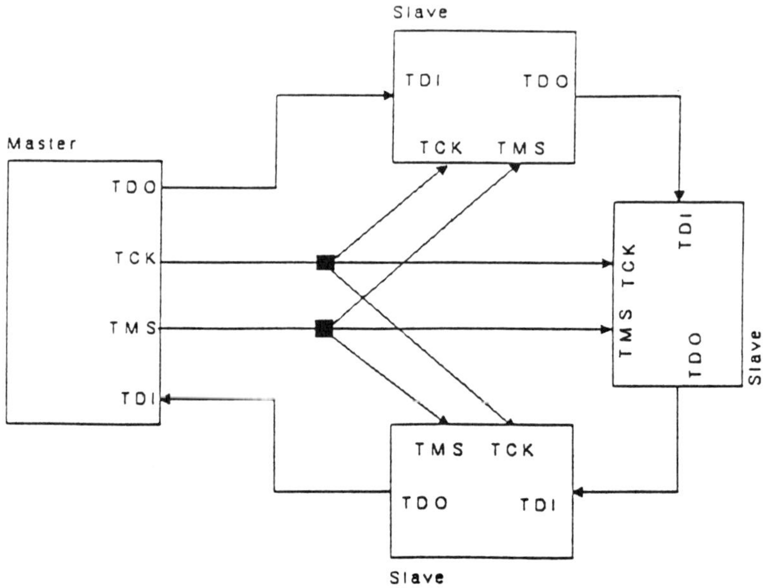

FIGURE 11-4(a)
Ring configuration

TAP Signal Definition

The signals in the TAP for a slave component are defined as follows:

TCK *T*est *C*lock. This is nominally a free-running clock signal with a 50 percent duty cycle. The changes on TAP input signals (TMS and TDI) are clocked into the TAP controller, instruction register, or selected test data register on the rising edge of TCK. Changes at the TAP output signal (TDO) occur on the falling edge of TCK. The design of the test logic in an integrated circuit must not require or prohibit stopping of the clock for any purpose.

TDI *T*est *D*ata *I*nput. TDI is clocked into the selected register (instruction or data) on a rising edge of TCK. The TDI input must have a built-in pull-up resistor of a value which ensures that an unterminated input is seen by the test logic as a high signal level.

TDO *T*est *D*ata *O*utput. The contents of the selected register (instruction or data) are shifted out of TDO on the falling edge of TCK. TDO drivers must be sent to high impedance except when scanning of data is in progress.

TMS *T*est *M*ode *S*elect. This serial control input is clocked into the TAP controller on the rising edge of TCK. The TMS input must have a built-in pull-up resistor of a value which ensures that an unterminated input is seen by the test logic as a high signal level.

Test data received at TDI must appear without inversion at TDO following an appropriate number of rising and falling edges of TCK, regardless of the instruction or test data register selected. For merchant devices (that is, devices where the complete application environment is not defined at design time) all connections must be made through dedicated package pins. The dedicated pin requirement can be reduced—*with loss of functionality*—for applications where the application environment can be completely defined.

Description

The signal applied at the test mode select pin (TMS) is decoded by the TAP controller and distinguishes the principal operations of the test-support circuitry. The built-in pull-up is included to force the controller to the Test-Logic-Reset controller state, which ensures normal operation of the complete design in the case of an unterminated TMS pin.

The dedicated test clock input (TCK) is included so that the serial test data path between devices can be used independently of chip-specific clocks, which may vary significantly in frequency from one chip to the next. It also permits shifting of test data concurrently with normal operation of the design. The latter facility is required to support the use of the TAP and test data registers in a design for on-line monitoring (a key requirement of several JTAG member companies). Having an independent clock ensures that test data can be moved to or from a chip without changing the state of the system logic. The independent clock is also essential if boundary-scan facilities are to be usable for board interconnect testing in all circumstances—including cases where clock signals are derived in one chip for use in others.

While TCK is nominally driven by a free-running clock, there may be situations where the clock needs to stop in a controlled manner. One example is when an ATE needs to fetch test data from backup memory (e.g., disk), since not all test systems are able to maintain the clock during such an operation.

The data pins (TDI and TDO) provide for serial movement of test data through the circuit. The requirement for data to be propagated from TDI to TDO without inversion is included to simplify the operation of components linked on a TAP bus.

The built-in pull-up on the TDI pin is included so that open-circuit faults in the board-level serial test data path cause a defined logic value to be shifted into the TAP. Note that when this constant value is shifted into the instruction register the bypass register will be selected.

The high-impedance capability of the TDO pin is required to allow parallel, rather than serial, connection of board-level test data paths in cases where this is required (see Figure 11-4b). In the illustrations given in this document, edge-operated circuit designs are generally used. For an edge-operated implementation, note that the TDO output changes must be delayed until the falling edge of TCK, which can be achieved by including a flip-flop clocked by the falling edge of TCK in the TDO output buffer in Figure 11-3.

TAP Controller

The TAP controller is a synchronous finite state machine whereby sequencing through the various operations of the JTAG circuitry occurs under control of the TMS signal.

Main Controller State Diagram

The main state diagram for the TAP controller must be as shown in Figure 11-5. In Figure 11-5, the signal values shown adjacent to the state transition arcs represent the value of TMS at the time of a rising edge on

FIGURE 11-4(b)
Hybrid star/ring configuration

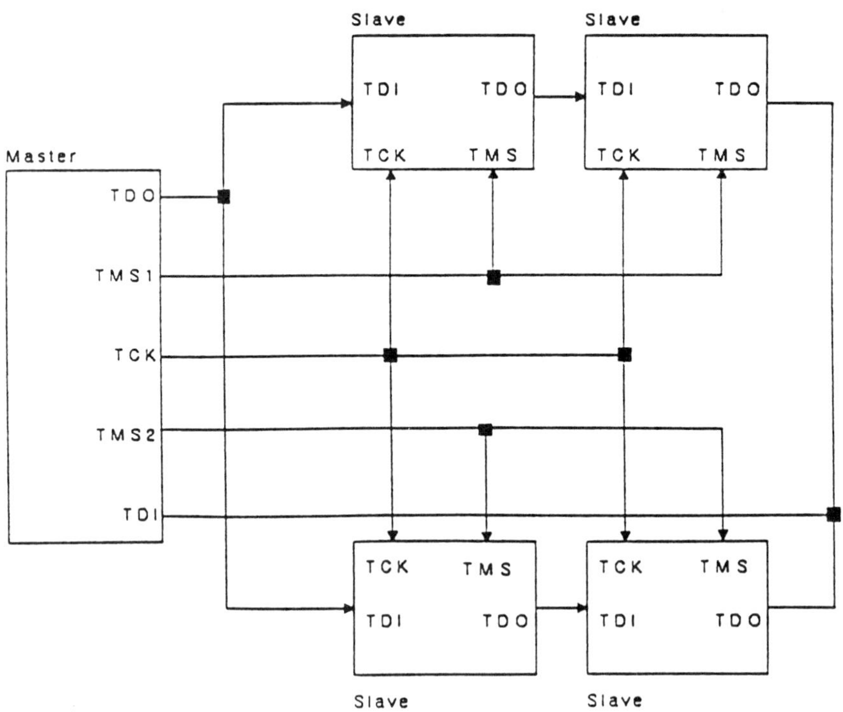

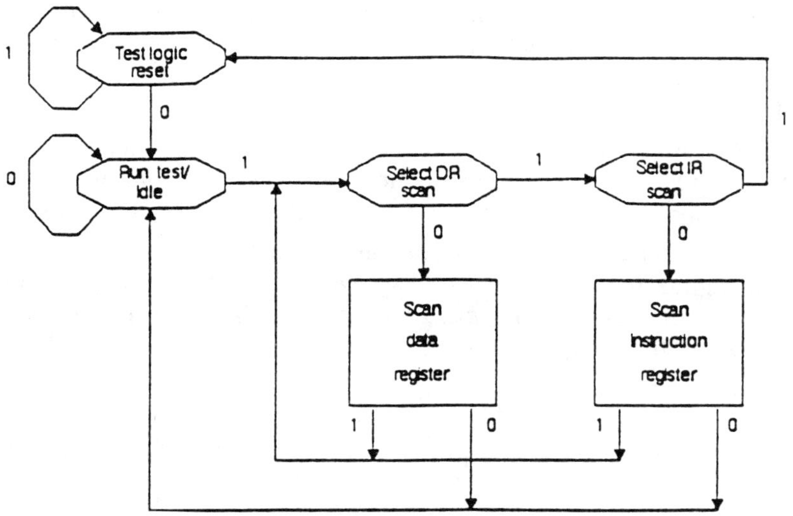

FIGURE 11-5
Main state diagram

TCK. The figure references two similar subsidiary state diagrams, which describe the scanning sequences for the instruction register and for a previously selected test data register.

All state transitions occur based on the value of TMS at the time of a rising edge of TCK. All operations of the test logic occur on the rising edge of TCK following entry into a controller state. Logic level changes at TDO occur on the falling edge of TCK following entry into a controller state.

The states in the main controller state diagram are as follows:

- Test-Logic-Reset—A state wherein the test logic is disabled so that normal operation of the system logic in the component can continue unhindered. No matter what the original state of the controller may be, the controller will enter Test-Logic-Reset when the TMS input is held high for at least five rising edges of TCK. The controller may also be forced into this state at power-up. The controller remains in this state while TMS is high.
- Run-Test-Idle—A controller state between scan operations where an internal test (such as a built-in self-test) previously selected by setting the instruction register may be executed (Run-Test). Registers not involved in the application of the test should retain their previous state. If the data in the instruction register does not indicate that a test should be executed, then all test logic (the TAP controller and instruction and test data registers) must retain its previous state (idle). Once entered, the controller will remain in the Run-Test/Idle state as long as TMS is held low.

- Select-DR-Scan—This is a temporary controller state in which all test logic retains its previous state. If TMS is held low when the controller is in this state, then a scan sequence for the selected test data register is initiated.
- Select-IR-Scan—This is a temporary controller state in which all test logic retains its previous state. If TMS is held low when the controller is in this state, then a scan sequence for the instruction register is initiated.
- Scan-Data-Register—The scanning process defined in the controller scan state diagram is applied to the test data register identified by the current instruction. The instruction register and all test data registers not addressed retain their previous state.
- Scan-Instruction-Register—The scanning process defined in the controller scan state diagram (see Figure 11-6) is applied to the instruction register. All test data registers retain their previous state.

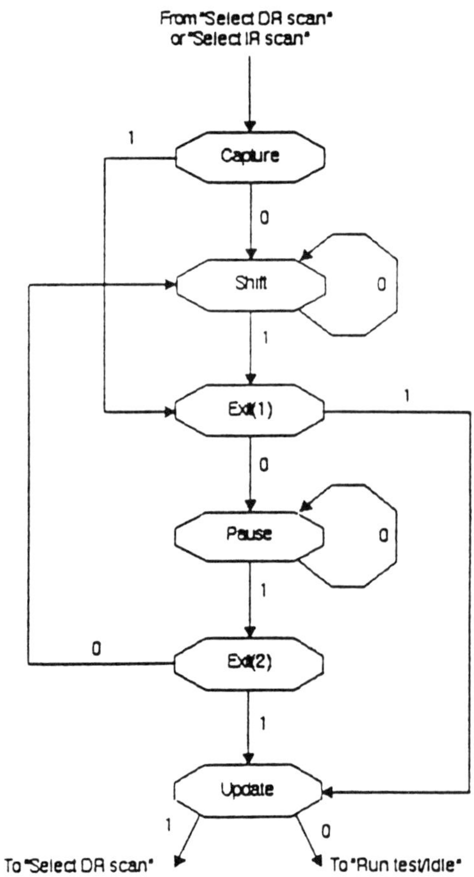

FIGURE 11-6
Scan state diagram

Description

The Test-Logic-Reset controller state is entered following a maximum of five rising edges on TCK if TMS is held high. An erroneous low signal on the TMS line at the time of a rising edge on TCK (for example, a glitch due to external interference) will cause the controller to move on to the Run-Test/Idle controller state; however, no action will occur in this state, since the instruction register is initialized to select scanning of the bypass register in Test-Logic-Reset. Following the error, the controller will return to the Test-Logic-Rest controller state once three rising edges of TCK have been applied.

Background

The version 1.0 JTAG proposal provided the equivalent of the Test-Logic-Reset controller state through an asynchronous enable pin. Reduction of the basic pin requirement to four (from five in version 1.0) has necessitated a change to a synchronous reset for the test logic. Note that the version 2.0 design is tolerant of errors during normal operation, since a single bit error on TMS only permits scanning of the bypass register, which has no connection to the system logic.

Scan State Diagram

The scan state diagrams for the TAP controller must be as shown in Figure 11-6. In that figure, the signal values shown adjacent to the state transition arcs represent values on TMS at the time of a rising edge on TCK.

All controller state transitions occur based on the value of TMS at the time of a rising edge of TCK. All operations of the test logic occur on the rising edge of TCK following entry into a state. TDO changes state on the falling edge of TCK following entry into a state.

The scanning process for the selected test data (instruction) register contains six states, which are defined as follows:

- Capture—In this controller state data is parallel loaded into the selected test data (instruction) register. If the register does not have a parallel input, or if capturing is not required for the selected test, then the register retains its previous state unchanged. All registers other than the selected test data (instruction) register retain their previous state unchanged.
- Shift—In this controller state, the selected test data (instruction) register shifts data one state towards its serial output on each rising edge of TCK. All registers other than the selected test data (instruction) register retain their previous state unchanged.
- Exit (1)—This is a temporary controller state. All test data registers and the instruction register retain their previous state unchanged. A high signal on the TMS line while in this state causes termination of the scanning process; a low causes entry into the Pause state.

- Pause—This controller state allows shifting of the selected test data (instruction) register to be temporarily halted. All test data registers and the instruction retain their previous state unchanged. The controller remains in this state while TMS is low. When TMS goes high, the controller moves on to the Exit(2) state.
- Exit(2)—This is a temporary controller state. All test data registers and the instruction register retain their previous state unchanged. A high signal on the TMS line while in this state causes termination of the scanning process; a low causes a return to the Shift state.
- Update—During this controller state, data is transferred from each shift-register stage into the corresponding parallel output latch (if the selected register includes a parallel output latch). All shift-register stages in the selected register retain their previous state unchanged. All registers other than the selected test data (instruction) register retain their previous state unchanged.

TAP Controller Operation

The TAP controller must generate signals to control the operation of the test data and instruction registers and the associated multiplexers as defined in Table 11-1. In addition, the table shows how the TDO output buffer and the preceding multiplexers that select the register output fed to TDO are controlled.

All operations of the instruction and test data registers occur on the rising edge of TCK following entry into a state.

Changes at the TDO output occur on the falling edge of TCK following entry into the state.

Description

An example of a circuit that meets the above requirements is shown in Figures 11-7a and 11-7b. Note that while this figure and the other examples given in this section assume that the instruction and test data registers are controlled by gated clock signals, this is only one of many possible implementations and should not be regarded as a preferred implementation style. The assignment of controller states in the example implementation is given in Table 11-2.

The Boolean equations for the next state logic in Figure 11-7b are as follows (lowercase characters indicate inverse variables):

$$ND := Dc + tCb + dCba$$

$$NC := Cb + CA + Tb$$

$$NB := tBa + tc + tdB + tda + TCb + TDCA$$

$$NA := tcA + Tb + Ta + TDC$$

TABLE 11-1
Test Logic Operation in Each Controller State

State	*TDO*	*Instruction Register*	*Selected Data Register*	*Other Data Registers*
Test-Logic-Reset	High-Z	Set to all-is	.. Reset to achieve normal .. circuit function	
Run-Test/Idle	High-Z	Retain last state	*Execute test	Retain last state
Select-DR-Scan	High-Z	Retain last state	..... Retain last state	
Select-IR Scan	High-Z	Retain last state	..... Retain last state	
Capture-IR	MSB of IR	Load Status	..... Retain last state	
Shift-IR	MSB of IR	Shift	..... Retain last state	
Exit(1)-IR	MSB of IR	Retain last state	..... Retain last state	
Pause-IR	MSB of IR	Retain last state	..... Retain last state	
Exit(2)-IR	MSB of IR	Retain last state	..... Retain last state	
Update-IR	High-Z	Load parallel output latch	..... Retain last state	
Capture-DR	MSB of selected DR	Retain last state	*Load from parallel input	Retain last state
Shift-DR	MSB of selected DR	Retain last state	Shift	Retain last state
Exit(1)-DR	MSB of selected DR	Retain last state	..... Retain last state	
Pause-DR	MSB of selected DR	Retain last state	..... Retain last state	
Exit(2)-DR	MSB of selected DR	Retain last state	..... Retain last state	
Update-DR	High-Z selected DR	Retain last state	*Load parallel output latch	Retain last state

Operations marked by an asterisk (*) occur only if required by the test data register or by the current instruction.

Figure 11-8 shows the operation of this controller implementation through instruction and test data register scan cycles.

The TAP controller states include the three basic actions required for testing: stimulus application (Update-DR), execution (Run-Test/Idle), and response capture (Capture-DR). However, not all these actions are required for every type of test, as shown in Table 11-3.

For scan testing, the stimulus is made available for use at the end of shifting or, if a parallel output latch is included, by updating the parallel output in the Update-DR state. The results of the test are captured into the test data register during the Capture-DR state.

For BILBO (built-in logic block observation)-based self-test, the starting values of the BILBOs are available at the end of shifting: there is no parallel output latch to update. The BILBOs should operate in their linear feedback shift-register modes during Run-Test/Idle. Since the result is already contained in a test data register, no action is required during Capture-DR.

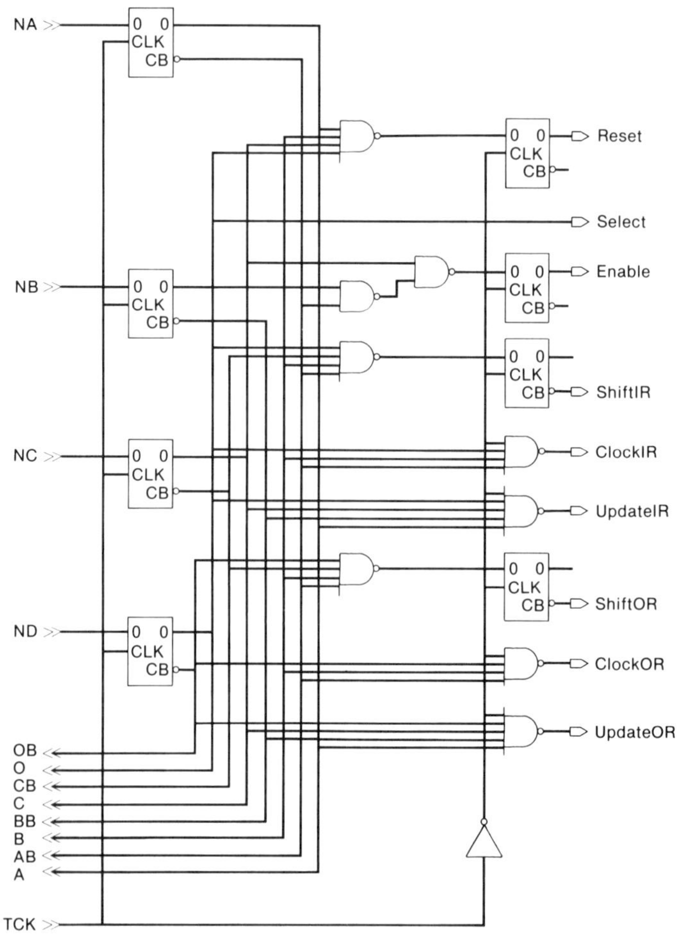

FIGURE 11-7(a)
Example of TAP controller implementation

TAP Controller Initialization

Rules

When a device has tristate or bidirectional system pins, the TAP controller must be forced into the Test-Logic-Reset controller state at power-up independently of the signals applied at the TAP inputs.

The TAP controller must not be initialized by operation of any system input, such as a system reset.

Where a dedicated reset pin is provided to allow initialization of the TAP controller, this must be active in the low state.

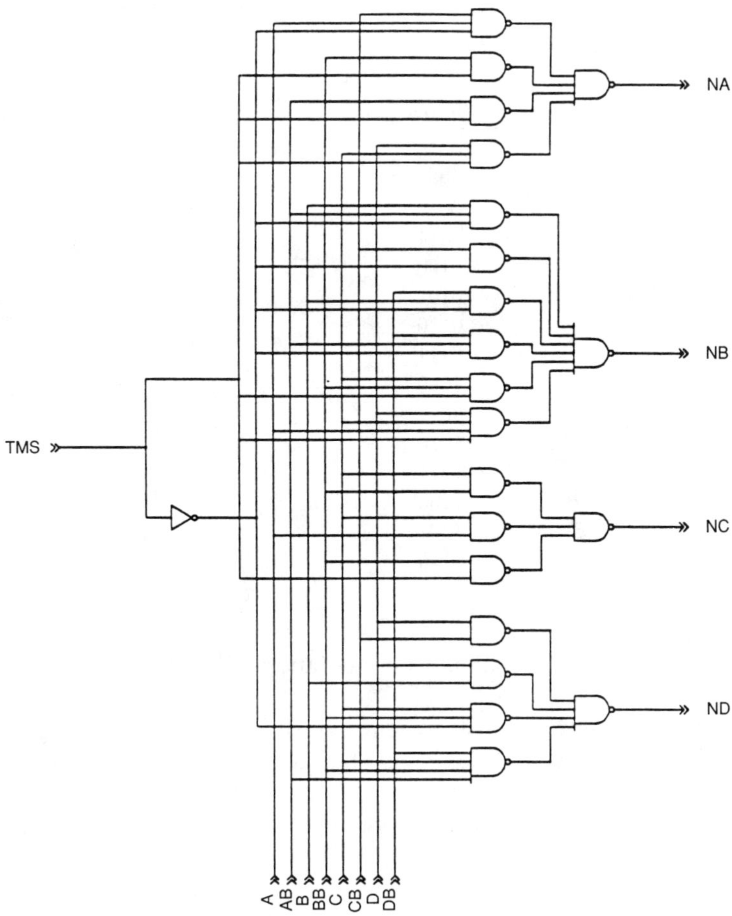

FIGURE 11-7(b)
Next state logic for example TAP controller

Description

In a board design that contains wired junctions or buses, provision must be made to ensure that at power-up any period of contention between drivers on the bus is kept within limits that ensure that no damage occurs to the components on the board.

When boundary-scan circuitry is inserted between the system logic and package pins, it becomes essential to ensure that shortly following power-up this circuitry enters a state where buses and wired junctions are controlled by the system circuitry—the Test-Logic-Reset controller state. Rules relating to the boundary scan register ensure that boundary-scan circuitry at

TABLE 11-2
State Assignments for Example TAP Controller

Controller State	*Number*	*DCBA*
Exit(2)-DR	0	0000
Exit(1)-DR	1	0001
Shift-DR	2	0010
Pause-DR	3	0011
Select-IR-Scan	4	0100
Update-DR	5	0101
Capture-DR	6	0110
Select-DR-Scan	7	0111
Exit(2)-IR	8	1000
Exit(1)-IR	9	1001
Shift-IR	10	1010
Pause-IR	11	1011
Run-Test/Idle	12	1100
Update-IR	13	1101
Capture-IR	14	1110
Test-Logic-Reset	15	1111

system pins does not interfere with system operation when the Test-Logic-Reset controller state is selected.

While the TAP controller will synchronously enter the Test-Logic-Reset controller state following five rising edges at TCK (provided TMS is held high), the worst-case time taken to reach this state will exceed that at which damage is likely to occur. Therefore, the "reset at power-up" requirement is included.

The requirement can be met in a variety of ways, for example by inclusion of a power-up reset within the integrated circuit or by asymmetric design of the latches or registers used to construct the TAP controller. It could also be met by inclusion of a dedicated Reset pin for the TAP controller. Note, however, that a system reset cannot also be used to initialize the TAP controller, since this would compromise the ability to test system interconnections at the board level using the boundary-scan circuitry.

Note that, while it is not a requirement for devices without tristate or bidirectional system pins to initialize at power-up, this facility can be included in all device designs.

Instruction Register

The instruction register allows an instruction to be shifted into the design. The instruction is used to select the test to be performed and/or the test data register to be accessed. Optionally, the instruction register also allows examination of status information generated within the design.

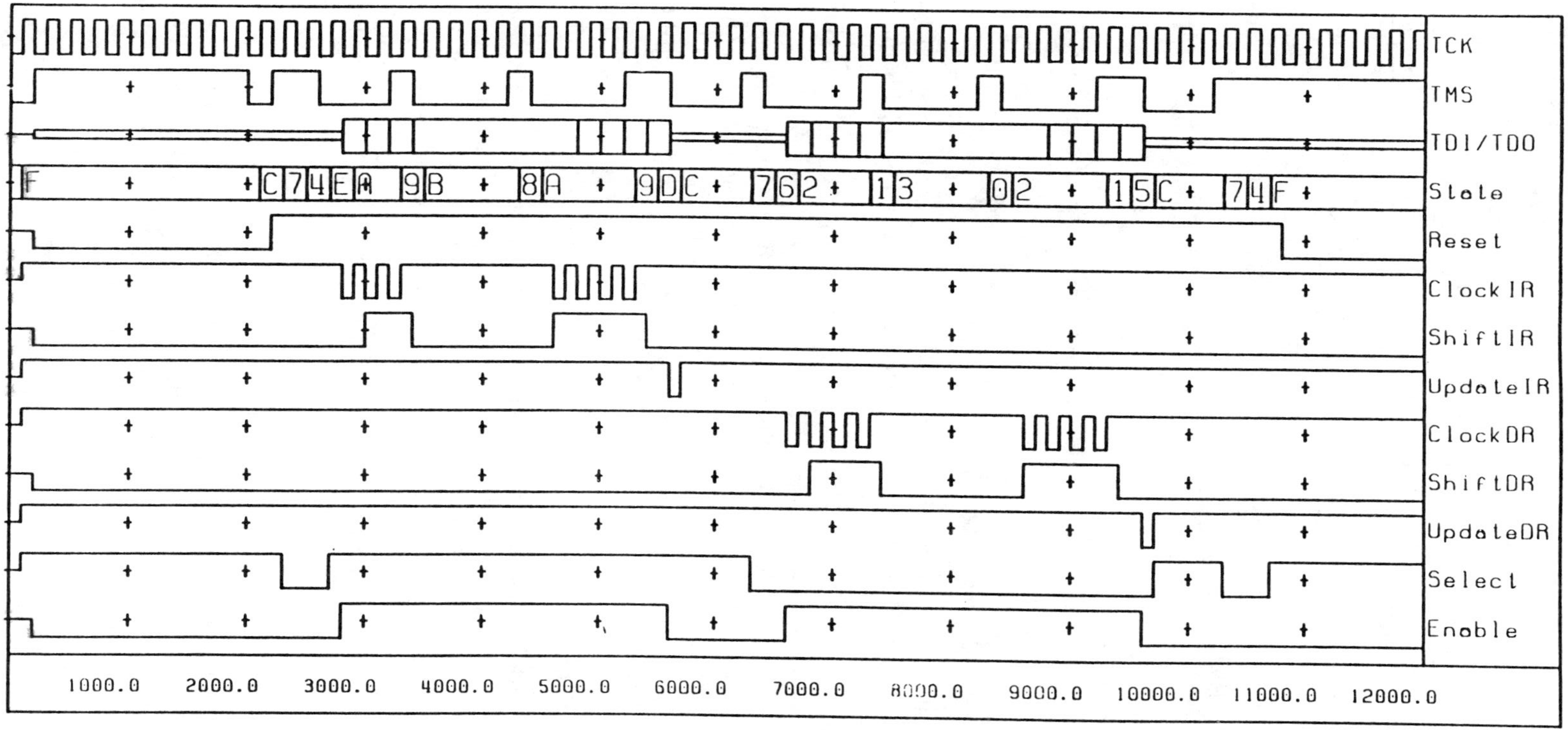

FIGURE 11-8 TAP controller operation

TABLE 11-3
Use of Controller States for Different Test Types

Test Type	*Update-DR*	*Run-Test/Idle*	*Capture-DR*
Boundary-scan external test	Yes	No	Yes
Internal scan test	Maybe	No	Yes
Internal self-test (e.g., BILBO)	No	Yes	No
Boundary-scan sample	No	No	Yes

This section contains the design requirements for the instruction register.

Design and Construction

The instruction register must include at least three shift-register-based cells. Each cell must contain a shift-register stage and a parallel output register/latch.

There must be no inversion of data between the serial input and the serial output of an instruction register cell.

The two most significant cells must load a fixed binary "10" pattern (the 1 into the most significant bit location) in the Capture-IR controller state. Other cells may load design-specific status information.

The instruction registers in prime and second source devices must have the same length.

Description

The parallel output register/latch is included to ensure that the instruction only changes at the end of the instruction register scanning sequence. The values set onto the instruction register outputs at the end of a scan sequence define the test to be applied and/or the test data register that must be accessed.

The minimum size (three instruction register cells) is set because of the following requirements:

- The instruction register must allow selection of the bypass register.
- The instruction register must allow selection of three or more boundary-scan register tests.
- The instruction register must allow selection of the optional device identification register.

- Fault isolation of the board-level serial test data path must be supported. This requires a constant binary "10" pattern to be loaded into the most significant bits of the instruction register at the start of the instruction-scan cycle.

The inclusion of the optional status input to the instruction register allows key data signals within the device to be examined at the start of testing, with future test actions potentially depending on the status information gathered.

Operation

The behavior of the instruction register in each TAP controller state must be as defined in Table 11-4. All operations occur on a rising edge of TCK.

Description

Figure 11-9 shows an implementation of an instruction register cell that satisfied the following requirements:

- The parallel output register/latch is updated at the end of the instruction-scan cycle during the Update-IR controller state.
- The clock input to the register in the serial path is only applied during the Capture-IR and Shift-IR controller states.
- The parallel output register/latch is set to all-ones during the Test-Logic-Reset controller state.

Note that the STATUS inputs to the two most significant stages must be tied to fixed logic levels (1 for the most significant bit, 0 for the next most significant bit).

TABLE 11-4
Instruction Register Operation in Each Controller State

Controller State	*Shift-Register Stage*	*Parallel Output Register/Latch*
Test-Logic-Reset	Retain last state	All outputs set high
Capture-IR	Load data from STATUS input	Retain last state
Shift-IR	Shift towards TDO	Retain last state
Update-IR	Retain last state	Load from shift-register stage
All other controller states	Retain last state	Retain last state

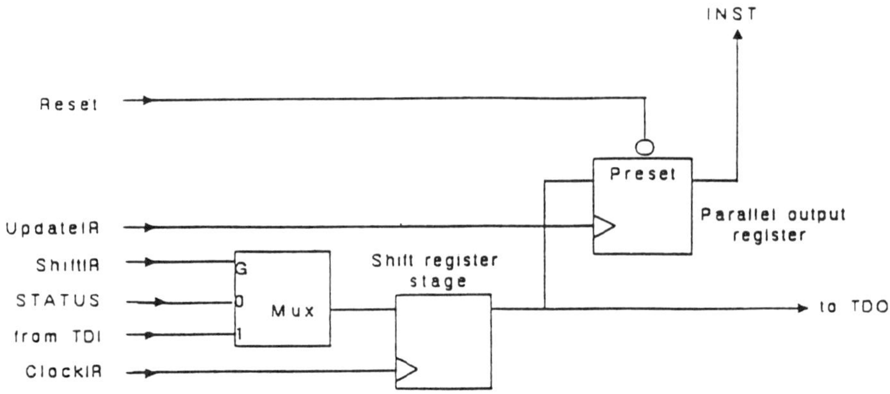

FIGURE 11-9
Instruction register cell design

Instructions

Every instruction must select one of the test data registers included in a design. The bypass register should be selected when a particular instruction does not select the boundary-scan register, the device identification register, or a user-defined test data register.

At a minimum, instructions must be provided to achieve the following:

- selection of the bypass register
- selection of the boundary-scan register, with the boundary-scan mode controls set for External test
- selection of the boundary-scan register, with the boundary-scan mode controls set for Sample test
- selection of the boundary-scan register, with the boundary-scan mode controls set for Internal test, together with any other test data registers required to perform a test of the system logic—for example, using self-test facilities
- selection of the device identification register (if included)

Three instructions are reserved, as defined in Table 11-5.

Instructions that are not reserved can be assigned to give access to the boundary-scan register with its mode controls set to the remaining required states (i.e., Sample and Internal), or to built-in self-test facilities, or to design-specific scannable test data registers.

Prime and second source devices must have the same instruction definitions.

TABLE 11-5
Reserved Instructions

Instruction MSB.....LSB	*Register Selected*	*Setting of Boundary-Scan Mode Controls*
All-ones {111...1}	Bypass	Sample
All-zeros {000...0}	Boundary scan	External
{10X...X}	Device identification (Bypass if no device identification register in the design)	Sample

Description

It is essential that all possible instructions should cause selection of one or the other of the test data registers in a design. For this reason, the first rule requires that the bypass register is selected when an instruction does not select any other test data register. There must be no "illegal" instructions.

The reserved instructions are defined to allow easy selection of three key operations within requiring complex instruction data patterns to be shifted into the design. The key operations are: bypass register scanning, boundary-scan register scanning (External setting of the boundary-scan mode controls), and device identification register scanning.

In the first two cases, the appropriate instruction can be entered by holding TDI at a constant value (1 or 0, respectively) and completing an instruction-scan cycle. The demands on the host test system are consequently reduced. Note also that since a built-in pull-up resistor is included at the TDI input, an open circuit fault in the serial board-level test data path will cause the bypass register to be selected following an instruction-scan cycle.

The third reserved instruction (more specifically, the range of instructions {10X. . .X}) selects the device identification register, or the bypass register if there is no device identification register in the design. This allows the device identification register to be selected by manipulation of the broadcast register scan operation. The means of initiating device identification register scanning using the TMS and TCK signals is shown in Table 11-6, starting at the Test-Logic-Reset controller state.

The reserved instruction definition permits designers to segment the instruction register into a test data register selection field (the most significant bits) and a test operation selection field (the least significant bits). This may allow the design of the test logic to be simplified in some circuits,

TABLE 11-6
Method of Accessing Device Identification Registers for Unknown Devices

TMS Value	*State Following Rising Edge on TCK*	*Action*
1	Test-Logic-Reset	
0	Run-Test/Idle	None
1	Select-DR-Scan	
1	Select-DR-Scan	
0	Capture-IR	Load {10X...X} into IR shift-register
1	Exit(1)-IR	
1	Update-IR	Transfer {10X...X} into IR parallel output
1	Select-DR-Scan	
0	Capture-DR	Load device ID code
1	Shift-DR	Begin shift of device ID code
1	Shift-DR	
...	...	

potentially at the expense of inefficiency in the use of the instruction register outputs.

Bypass Register

The bypass register provides a short-circuit route for test data in the data register scanning cycle that can be used when access to other test data registers in an integrated circuit is not required. Use of the bypass register speeds access to test data registers in other integrated circuits on a board-level test data path.

Design and Construction

The bypass register must consist of a single shift-register state.

There must be no inversion of data between the serial input and the serial output of the bypass register.

When the bypass register is selected by the current instruction, it must load a logic zero into the shift-register state on a rising edge of TCK in the Capture-DR controller state.

Description

The bypass register may be implemented as shown in Figure 11-10. The provision of this register allows bypassing of segments of the board-level serial test data register that are not required for a specific test. Test access times to the segments of interest are reduced.

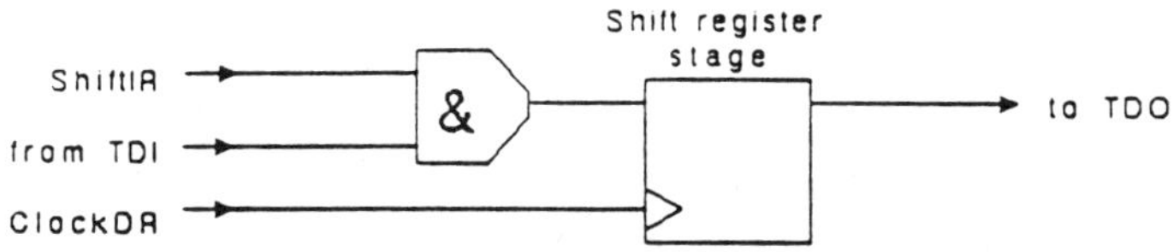

FIGURE 11-10
Bypass path design

Consider as an example a circuit board containing 100 integrated circuits, each of which has 100 bits in its boundary-scan register. The boundary-scan path on the assembled board would include 10,000 shift-register stages if all the segments were connected in series simultaneously. This would give protracted test times, for example when accessing just one of the integrated circuits on the path.

The ability to bypass segments of the shift-register path under control of the appropriate instruction register allows considerable shortening of the overall path in such circumstances. Continuing the example, 99 of the chips could be set to shift only through their bypass register, with the integrated circuit under test having its full boundary-scan register in circuit. This would give a total serial path length of 199 stages—a considerable reduction compared to 10,000.

There must be no inversion of data between the serial input and the serial output of the bypass register, so that the presence or absence of a device identification register in the test logic can be determined by examination of the serial output data. The bypass register (which is selected in the absence of a device identification register) loads a logic 0 at the start of a scan cycle, whereas a device identification register loads a constant logic 1 into its most significant bit. When the "scan device identification register" instruction is loaded into the instruction register, a subsequent data register scan cycle will allow the first bit of data shifted out of each device to be examined—a logic 1 showing that a device identification register is present. This allows "blind" interrogation of device identification registers by setting the {10X. . .X} instruction as outlined on page 288.

Operation

As a minimum, the bypass register must be selected by the all-ones instruction. It must also be selected by instructions that do not select any other test data register in the design, or by the "select device identification register" instruction when no device identification register is provided.

When the current instruction selects the bypass register, its behavior must be as defined in Table 11-7.

TABLE 11-7
Operation of the Bypass Register in Each Controller State

Controller State	*Action*
Test-Logic-Reset	Retain last state
Capture-DR	Load a logic Ø
Shift-DR	Shift data from serial input to serial output
All other controller states	Retain last state

When the current instruction does not select the bypass register, it should retain its last state unchanged.

All bypass register operations occur on the rising edge of TCK.

Boundary Scan Register

The design must include a boundary-scan register as a test data register. The boundary-scan register allows testing of circuitry external to the integrated circuit package (primarily the board interconnect) and provides for defined conditions to be established at the periphery of the system logic during internal testing. It also permits the signals flowing through the system pins to be sampled and examined without impacting the operation of the system logic.

Design and Construction

The boundary-scan register is a single shift-register-based path containing cells connected to all system inputs and outputs of the circuit and the system logic, including clock inputs to the system logic of the design, and to output enable or direction control signals associated with system pins.

There must be no inversion of data between the serial input and the serial output of a boundary-scan register cell.

The boundary-scan cells for the various input/output pins of the system logic may be assembled in the register in any order. However, the ordering must be clearly defined in the integrated circuit's data sheet and must be the same for prime and second source versions of a given component type.

For merchant designs, the logic required to implement the boundary-scan register must *not* also be used in the system logic of the design.

Description

The last requirement is included to ensure that the boundary-scan register supports on-line monitoring when its mode controls are set to the state that allows execution of a Sample test. Reuse of boundary-scan register logic to implement part of the intended function of the design would prevent execution of an on-line Sample test and is therefore acceptable *only* for in-company implementations—that is, where this capability is guaranteed not to be required.

Tests Using the Boundary-Scan Register

Boundary-scan cells must allow execution of at least three types of test as determined by the state of mode controls supplied to each cell. These mode controls are generated by decoding of the instruction register's parallel output. The required test types are as follows:

a. External test. The mode controls must set cells at system logic outputs so that the connected system pins are driven from the shift-register path. Cells at system logic inputs must load the data being conveyed between the connected system pins and the system logic.
b. Sample test. The mode controls must set cells at system logic inputs and outputs so that they load the data being conveyed between the connected system output pins and the system logic or between the system logic and the connected system input pins.
c. Internal test. The mode controls must set cells at system logic inputs so that the system logic is driven from the shift-register path. Cells at system logic outputs must load the data being conveyed between the system logic and the connected system output pins.

Description

The key types of test that can be executed by appropriate setting of the boundary-scan mode controls are illustrated by Figure 11-11. In Figure 11-11, the signal paths used within the boundary-scan cells for each test type are shown as bold lines. The three required test types are as follows:

a. External test: (Figure 11-11a) The boundary-scan register is used to test circuitry external to the integrated circuit package—typically the board interconnect. Cells at output pins are used to apply test stimuli, while those at input pins capture tests results. This test also allows testing of blocks of components that do not themselves incorporate the complete boundary-scan facility.
b. Sample test: (Figure 11-11b) The boundary-scan mode controls are set to allow normal operation of the system logic. Signals entering and leaving the circuit can be sampled without affecting circuit operation. The boundary-scan mode controls are forced into this state in the Test-Logic-Reset controller state and when the boundary-scan register is not selected by current instruction. This ensures that the presence of the boundary-scan

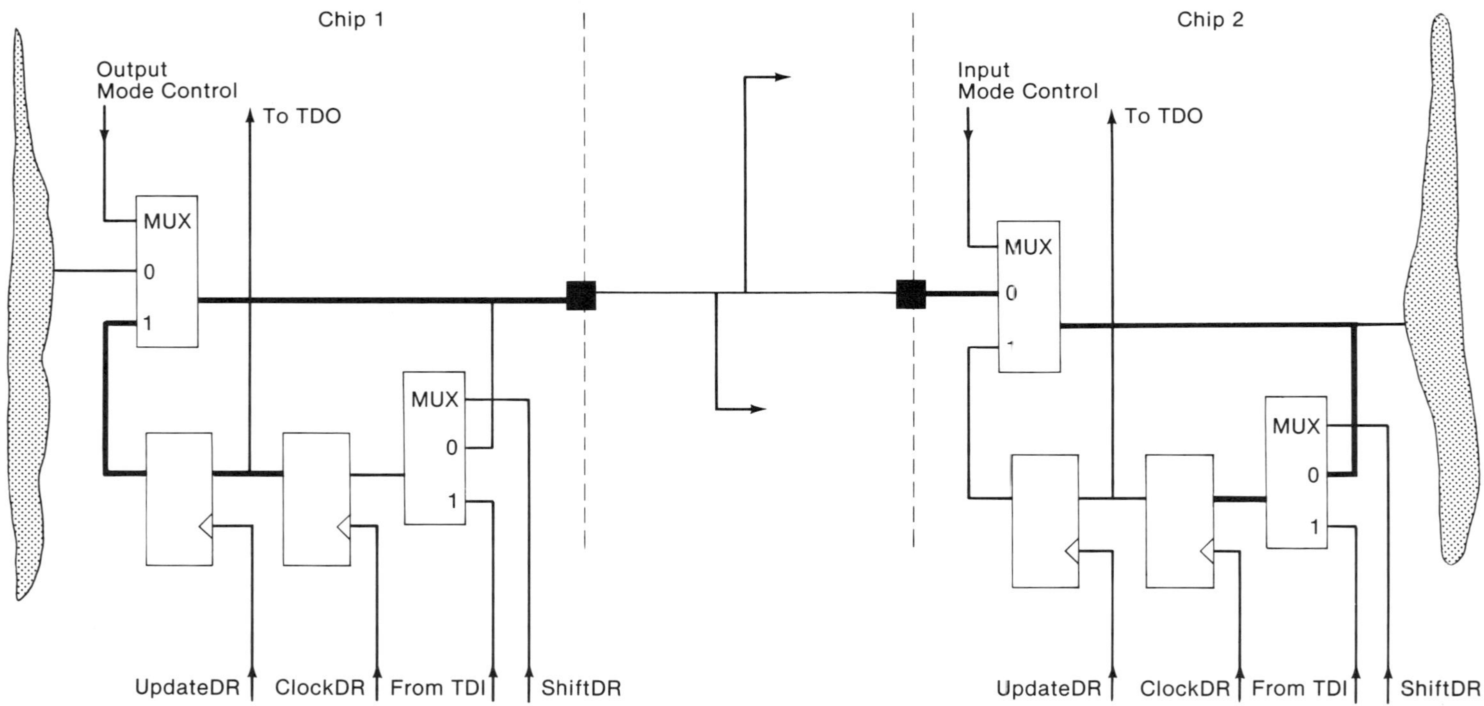

FIGURE 11-11(a) External test

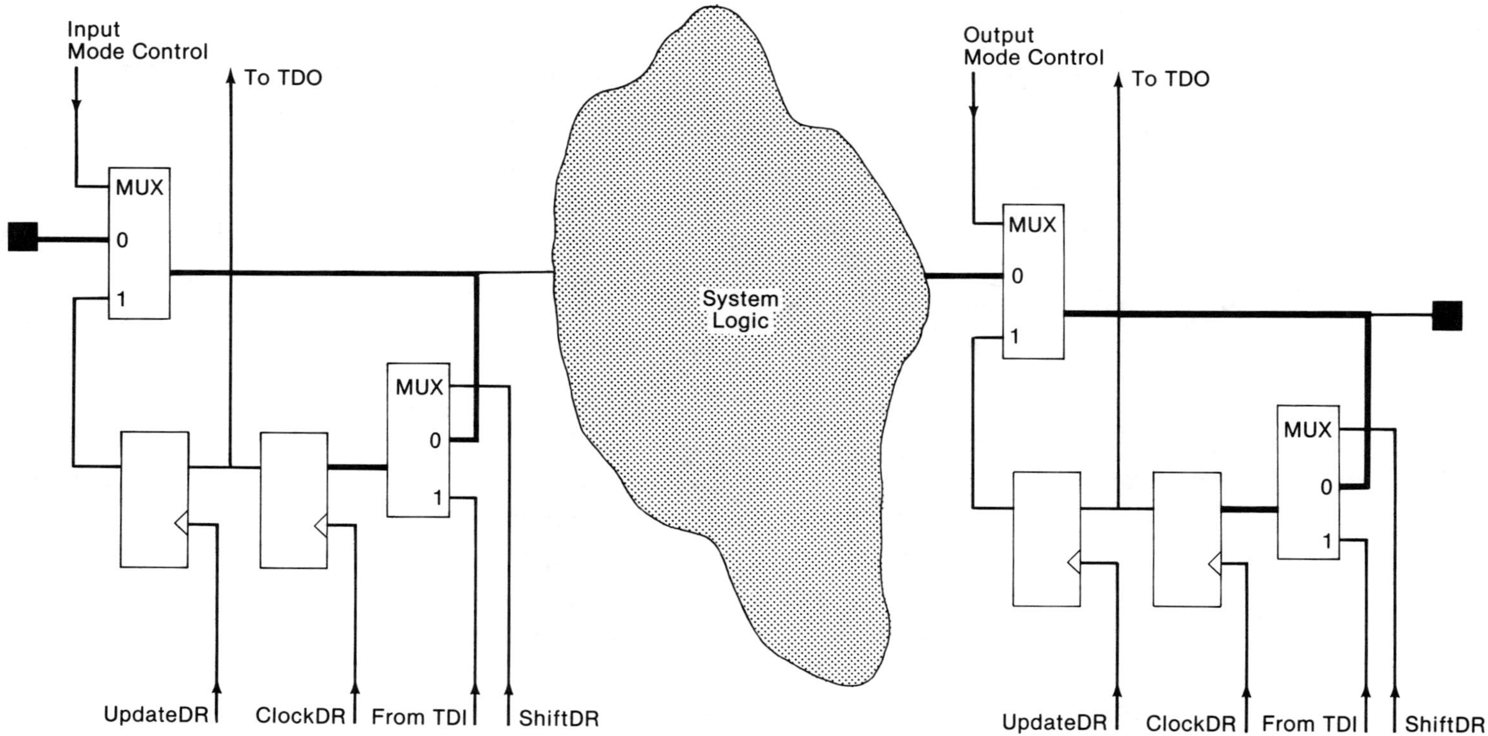

FIGURE 11-11(b) Test example

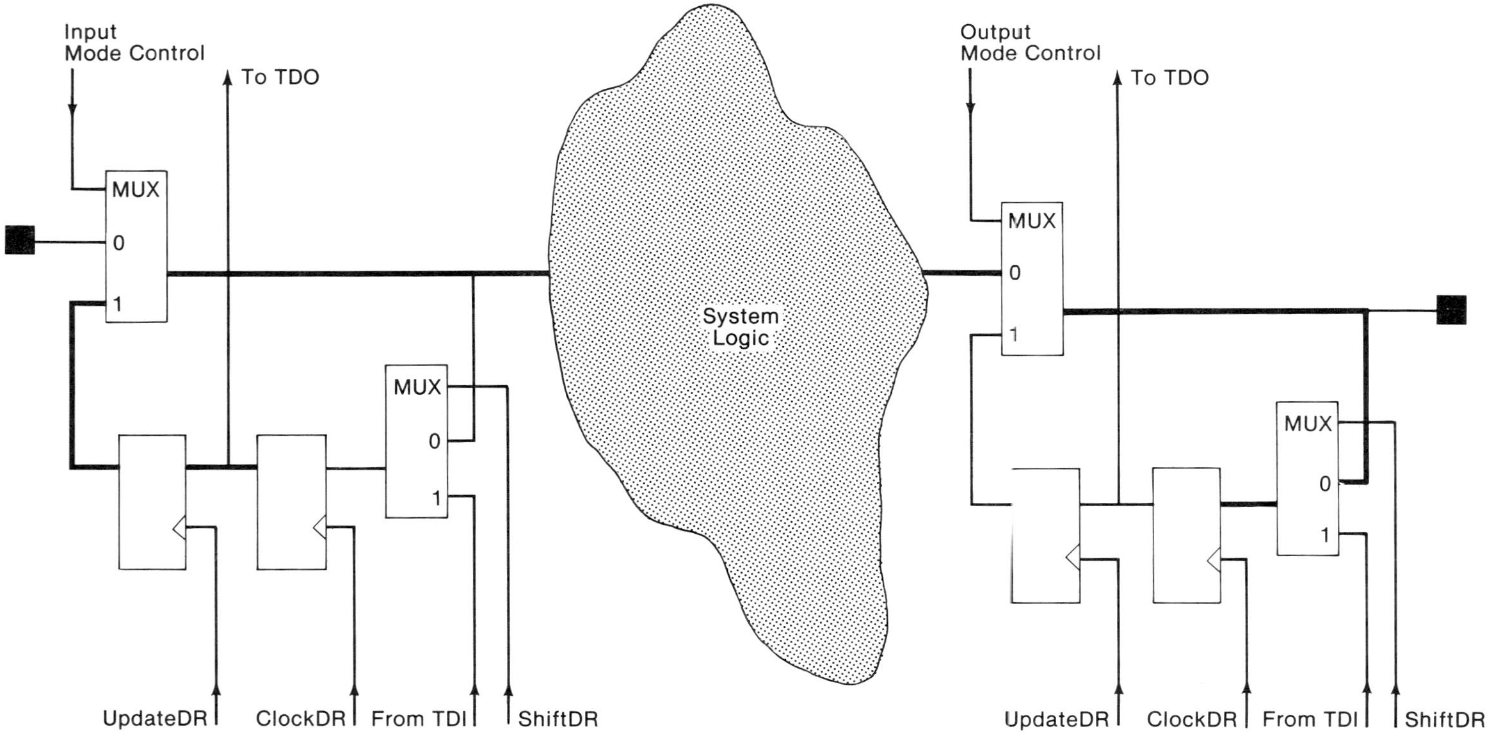

FIGURE 11-11(c) Internal test

cells in the signal path between the system pins and the system logic does not interfere with the operation of the system logic.

c. Internal test: (Figure 11-11c) The design of the complete circuit should provide the capability to test the system logic once the component has been assembled onto the board. This test should be performed under control of the instruction register and make its results available at the STATUS port of the instruction register or through the serial test data register(s). The internal test may be accomplished using scan or self-test techniques, in which case the boundary-scan register must be able to be set to shield the system logic from unwanted external interference. Alternatively, a limited static test of the system logic could be provided at slow speed in a manner analogous to the External test already described. In this case, cells at input pins apply test stimuli, and those at output pins capture test results.

Note that in the figures of this chapter, two boundary-scan mode controls are used to set the various cells as required for the three required test types. The states of these mode controls are coded as shown in Table 11-8.

The ability to execute a Sample test using the boundary-scan register allows two important test functions to be achieved. First, it allows concurrent monitoring of signals flowing into or out of the circuit without affecting its operation. This feature allows the development of system-level maintenance and support functions based around the TAP and boundary-scan register, or for contactless "guided probing" of the circuit during functional test diagnosis.

Second, execution of a Sample test in components surrounding functional blocks within a circuit allows the interaction of components in the functional block to be checked.

Operation

When the current instruction selects the boundary-scan register, its behavior in each TAP controller state must be as defined in Table 11-9.

When the current instruction does not select the boundary-scan register, the boundary-scan mode controls should be set to the Sample state and all boundary-scan register cells should retain their last state.

Description

In order to ensure that there is no unwanted interference between the boundary-scan register contents and the normal operation of the design, the boundary-scan mode controls should be set to the state required for the Sample test when:

a. the TAP controller is in the Test-Logic Reset state
b. the boundary-scan register is not selected by the current instruction

Input Pins

Each system input pin must have a boundary-scan cell.
When the boundary-scan register is selected by the current instruction,

a. If the mode controls are set for External or Sample test, the signal applied to each system input pin must be loaded into the corresponding shift-register stage during the Capture-DR controller state on the rising edge of TCK.
b. If the mode controls are set for Internal test, the signal applied to the system logic must change only during the Update-DR controller state.

TABLE 11-8
Input and Output Mode Controls Used in Figures 11-11 to 11-15

Boundary-Scan Test	*Input Mode Control*	*Output Mode Control*
External	0	1
Sample	0	0
Internal	1	0
⟨undefined⟩	1	1

TABLE 11-9
Operation of the Boundary-Scan Register in Each Controller State

Controller State	*State of Mode Controls*	*Action*
Test-Logic-Reset	Sample	Retain last state
Capture-DR	As defined by instruction	Load data at the cell output into the shift-register stage
Shift-DR	As defined by instruction	Shift data between boundary-scan register cells
Update-DR	As defined by instruction	Transfer data from shift-register stage to the parallel output register/latch (if provided)
All other controller states	As defined by instruction	Retain last state

Description

The designs illustrated in Figure 11-12 all satisfy the second rule. The circuits in Figures 11-12a and 11-12b allow the system logic to be driven from the boundary scan register when set for the internal test, while the circuit of Figure 11-12c cannot drive signals into the system logic. It is recommended that the latter configuration be used only in circumstances where the delay introduced in the signal path by the multiplexer would cause a design target to be exceeded. (An example would be a high-performance clock pin.)

The design in Figure 11-12a includes an output register/latch that is updated from the shift-register stage in the Update-DR controller state. This register is included to prevent changes at the output of the shift-register stage during shifting from being applied to the system logic of the chip (which could cause unwanted operation).

Output Pins (Two-State and Open-Collector)

Each unidirectional two-stage or open-collector system output pin must have a boundary-scan cell.

When the boundary-scan register is selected by the current instruction:

a. If Capture-DR occurs and the mode controls are set for Internal or Sample test, the cell must load the signal received from the system logic into the shift-register stage on the rising edge of TCK.
b. If the mode controls are set for the External test, the signal fed to the system pin must change only during the Update-DR controller state.

Description

Figure 11-13 shows a design that meets the above rules.

When the mode controls are set to allow execution of the Internal or Sample test, the multiplexer is set to allow data from the system logic to be captured. When the mode controls are set for External test, the feedback around the registers via the multiplexer ensures that their state does not change.

The last rule is met by the addition of a parallel output register/latch, which is updated from the shift-register path only when shifting is complete (i.e., during the Update-DR controller state).

Achieving raceless operation when testing interpackage wiring is essential. Difficulty may arise due to skew in the TCK signals supplied to various chips on a board but can be avoided if the registers at output pins hold their state when those at input pins load the signals from the interconnect.

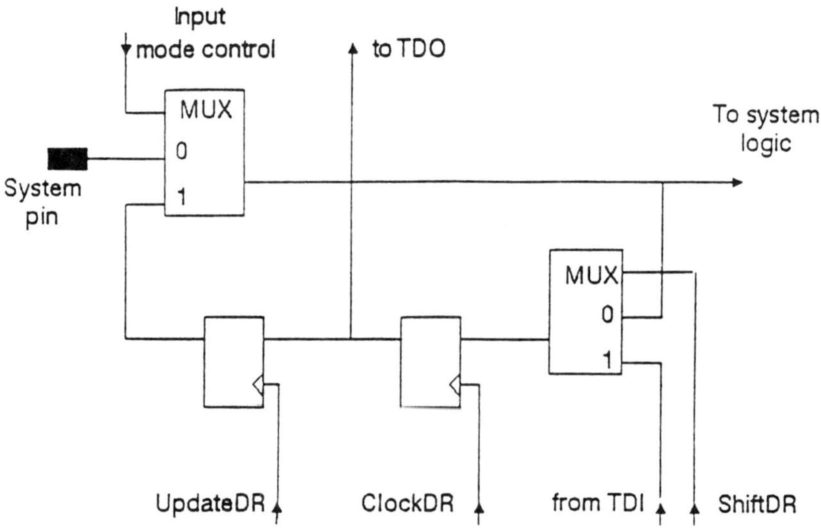

FIGURE 11-12(a)
Input pin cell with output holding register

FIGURE 11-12(b)
Input pin cell without output holding register

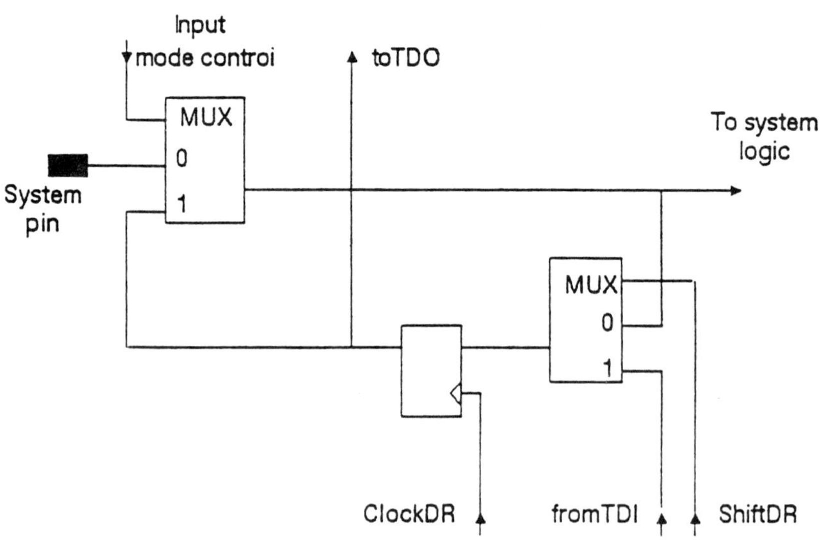

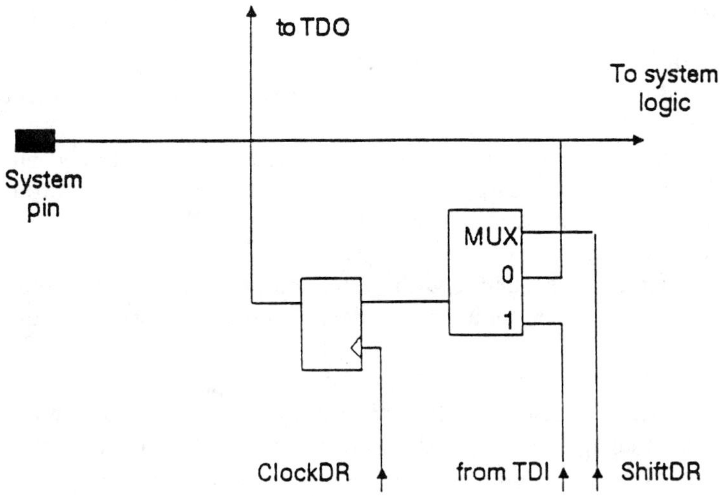

FIGURE 11-12(c)
Input pin cell allowing capture only

FIGURE 11-13
Output pin cell

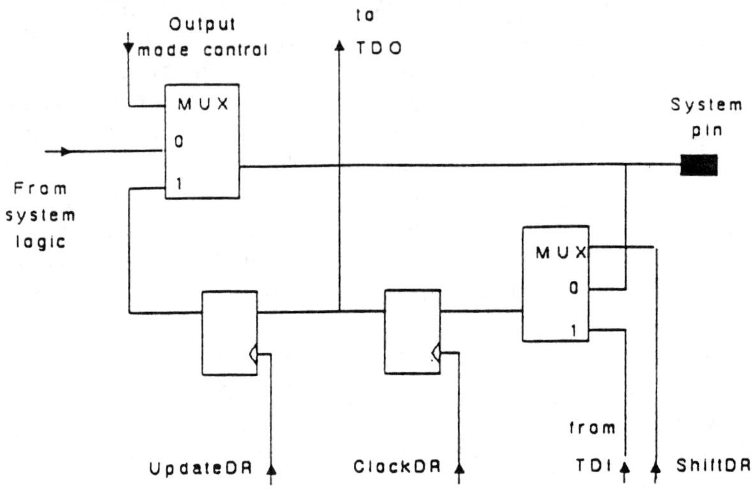

Tristate Pins

Each tristate system pin must have a boundary-scan cell prior to the tri-state buffer fed from the system data output. Any such cell should meet the requirements listed on page 301.

Additionally, each system output enable must feed its associated output buffers through a boundary-scan cell. In cases where the system output enable is fed to output buffers at several system pins, only one boundary-scan cell is required.

Boundary-scan cells associated with system output enable signals must behave as follows when the boundary-scan register is selected by the current instruction:

a. If the mode controls are set for Internal or Sample test, the state of the system output enable must be loaded into the shift-register stage on a rising edge of TCK in the Capture-DR controller state.
b. If the mode controls are set for External test, the signal fed to connected output buffer(s) must change only during the Update-DR controller state.

In the Test-Logic-Reset controller state or when the boundary-scan mode controls are set for Internal test, boundary-scan cells associated with system output enable signals must be initialized to the state which, when fed to the connected output buffer(s), will force the system pin(s) to the high-impedance state.

Description

Where a component has tristate output pins, these may feed onto a wired junction at the board level. In order to test the interconnections forming the wired junction, it must be possible to independently drive onto the junction from each of the possible driving pins using the boundary-scan register. To achieve this, it is necessary to be able to control the tristate enable signal for each tristate system output pin. This must be done from the boundary-scan register when the boundary-scan mode controls are set for External test. Figure 11-14 shows how this might be achieved.

Although it would appear that the additional data bits might significantly increase the overhead needed to implement boundary scan, it is only necessary to provide one additional bit for each tristate enable signal generated in the circuit. Thus where many tristate output pins are controlled from a single source, as for example in a microprocessor address bus, only one additional bit is required to give the necessary control. Since, given the basic design of the circuit, it would be a design error if such tristate pins were wired together, there is no need for the boundary-scan architecture to take account of this possibility.

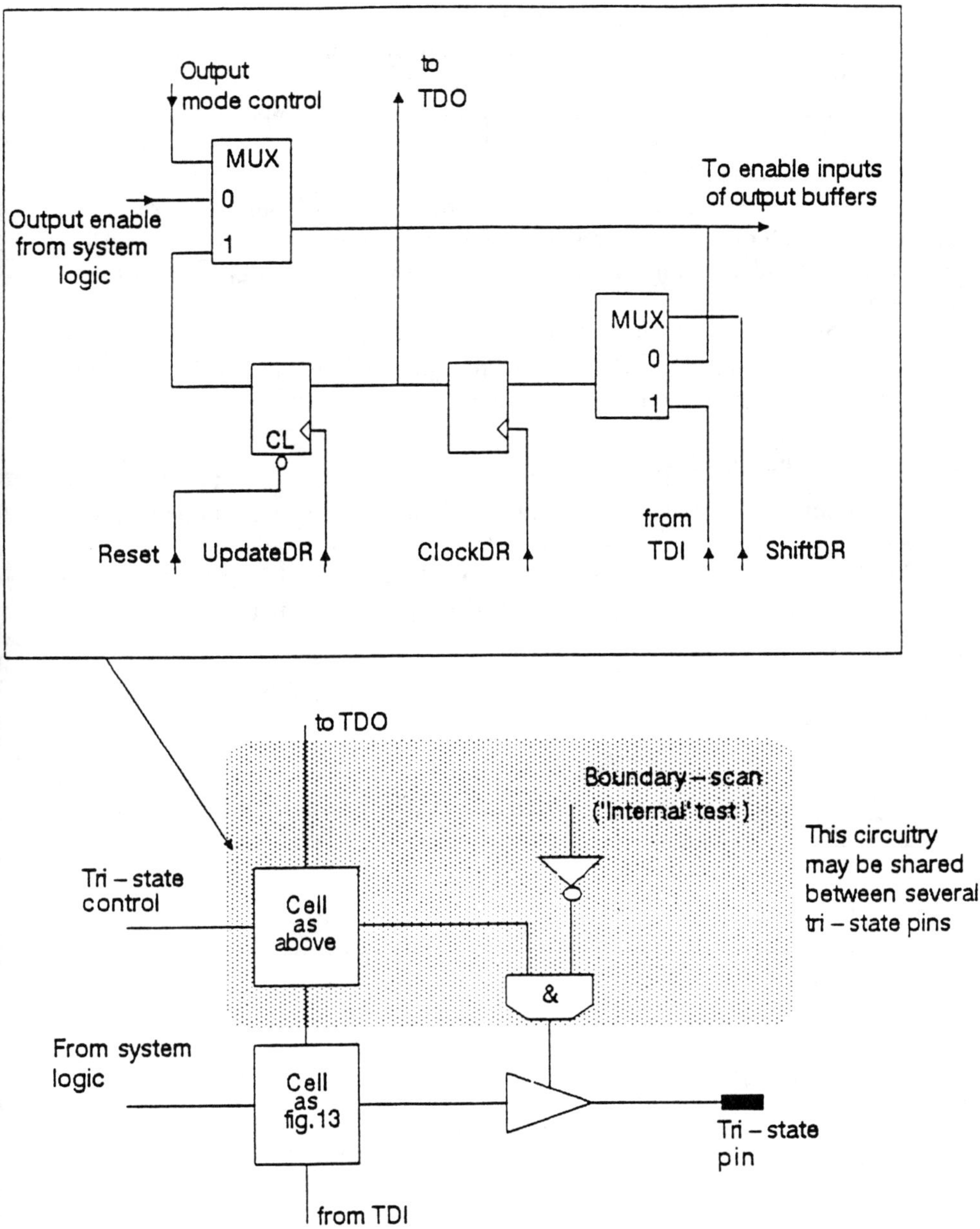

FIGURE 11-14
Tristate outputs

Figure 11-15 shows a wired junction between tristate outputs from a number of chips. To test this junction, a series of tests must be performed, each of which checks that one of the outputs can drive either a 0 or 1 to the receiving devices. During each test the other outputs must be set to the opposite data value (1 or 0, respectively) with a high-impedance drive. The figure shows the pair of tests needed to check the operation of one of the outputs connected to the junction.

To apply the test, it is necessary to be able to control both the data value at each output and whether the output is enabled. This must be done via the boundary-scan register independently of the system logic of each chip.

The last rule is included to ensure that there is no contention between devices on a wired bus while the system logic within chips is being tested, or on entry into boundary-scan testing.

Bidirectional Pins

Each bidirectional system pin must have one or more boundary-scan cells placed between the system pin and the system logic such that data flowing in either direction can be controlled or observed from the boundary-scan cell(s). Any such cell(s) should meet the requirements listed on pages 300 and 301.

FIGURE 11-15
Testing board-level bus lines

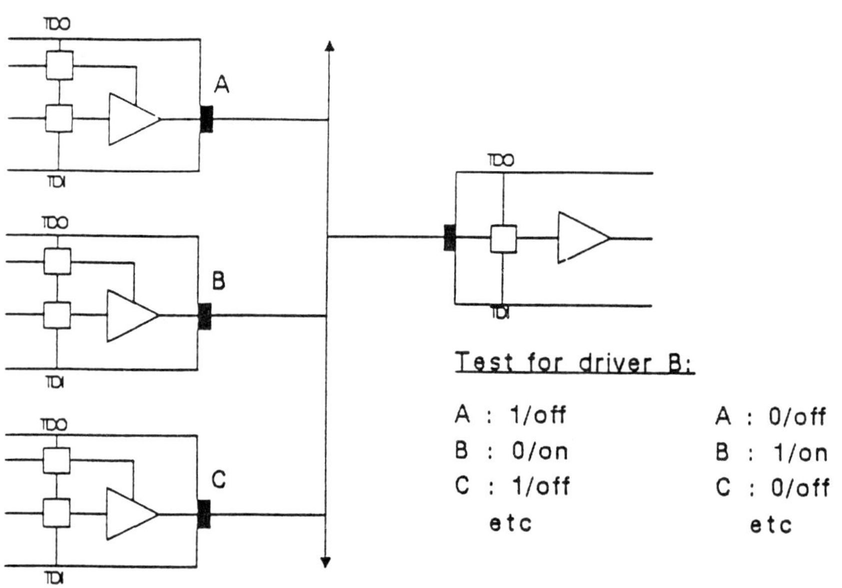

Additionally, each system direction control must feed its associated output buffers through a boundary-scan cell. In cases where the system output enable is fed to output buffers at several system pins, only one boundary-scan cell is required.

Boundary-scan cells associated with system direction control signals must behave as follows when the boundary-scan register is selected by the current instruction:

a. If the mode controls are set for Internal or Sample test, the state of the system direction control must be loaded into the shift-register stage on a rising edge of TCK in the Capture-DR controller state.
b. If the mode controls are set for External test, the signal fed to connected output buffer(s) must change only during the Update-DR controller state.

In the Test-Logic-Reset controller state or when the boundary-scan mode controls are set for Internal test, boundary-scan cells associated with system direction control signals must be initialized to the state which, when fed to the connected output buffer(s), will force the system pin(s) to the high-impedance (input) state.

Description

For reasons similar to those stated on page 304, there must be an additional boundary-scan cell associated with each bidirectional system pin that controls whether it operates as input or output. As in the case of tristate system pins, this cell may be shared across a bus or between any group of bidirectional system pins that obtain their direction control signal from a single source.

An example configuration is shown in Figure 11-16.

Device Identification Register

This section defines the design and operation of the optional device identification register. If provided, this register allows the manufacturer, part number, and version of a component to be determined through the JTAG TAP. One application of the device identification register is to distinguish the manufacturer(s) of devices on a board when multiple sourcing is used. As more merchant devices emerge with JTAG ports, it may become desirable to allow for a system diagnostic controller unit to "blindly" interrogate a board design in order to determine the physical location and type of each JTAG device. The need to do this becomes more than apparent if one considers systems that are configurable by the addition of option boards and the like. This information is also available for factory process monitoring and for failure mode analysis of assembled boards.

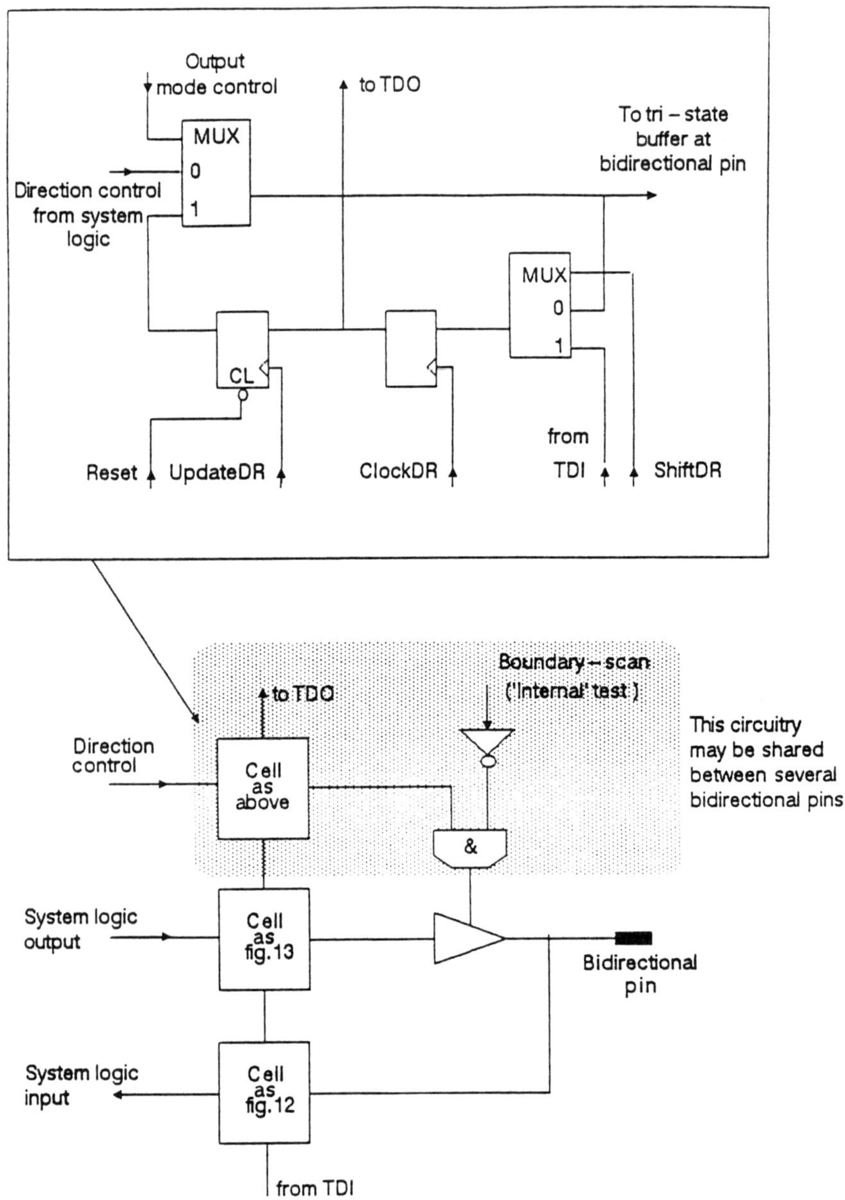

FIGURE 11-16
Bidirectional pin

Design and Construction

The device identification register is a single shift-register-based path that has a parallel input but no parallel output.

There must be no inversion of data between the serial input and the serial output of a device identification register cell.

The device identification register allows an identification code to be loaded from its parallel input, which consists of a minimum of four fields, as illustrated in Figure 11-17. User-programmable devices may optionally contain a fifth user-programmable field. Table 11-10 shows the function and size of these fields.

The binary pattern loaded into each field of the device identification register must be clearly defined in the integrated circuit's data sheet. The overall length of the device identification register must be fixed for any given manufacturer and clearly defined in the integrated circuit's data sheet.

In prime and second source devices the device identification registers may have a different overall size or load different device identification codes.

Operation

If the design contains a device identification register, then this must be selected by any instruction of the form {10X. . .X}. If no device identification register is included, then instructions of the form {10X. . .X} must select the bypass register.

When the current instruction selects the device identification register, its behavior in each controller state must be as defined in Table 11-11.

When the current instruction does not select the device identification register, the register should retain its last state.

Description

Figure 11-18 shows a design for a device identification register cell that satisfies these requirements.

Since the bypass register (which is selected in the absence of a device identification register) loads a logic 0 at the start of a scan cycle, whereas

FIGURE 11-17
Structure of device indentification register

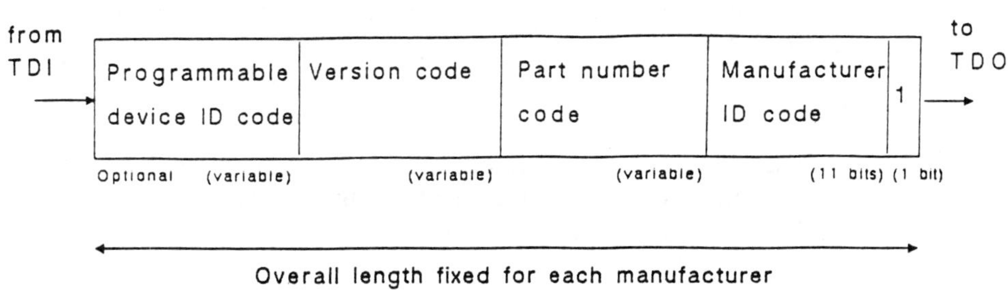

TABLE 11-10
Sizes of Fields Within the Device Identification Register

Field Name	*Size (bits)*
Fixed '1'	1
Manufacturer ID	11
Part number	> −1
Version	> −1
Programmable device ID	> −0

TABLE 11-11
Operation of the Device Identification Register in Each Controller State

Controller State	*Action*
Test-Logic-Reset	Retain last state
Capture-DR	Load device identification code
Shift-DR	Shift data between device identification register cells from TDI towards TDO
All other controller states	Retain last state

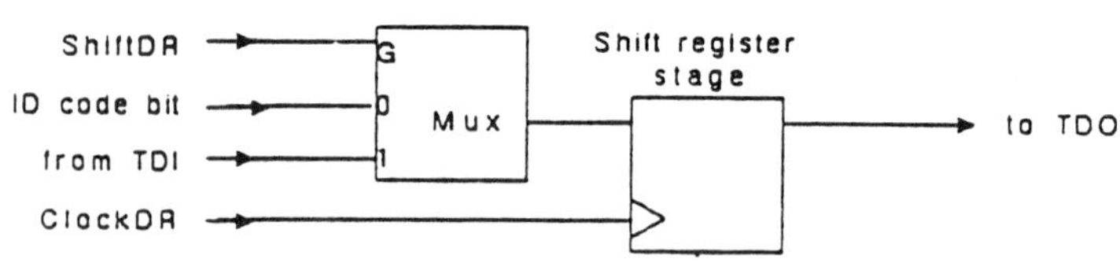

FIGURE 11-18
Device identification register cell design

a device identification register will load a constant logic 1 into its most significant bit, examination of the first bit of data shifted out of a device when the device identification register is selected will show whether such a register is included in the design. This allows "blind" interrogation of device identification registers by setting the {10X. . .X} instruction as outlined on page 290.

Manufacturer ID Code

The manufacturer identification code is a compressed form of the JEDEC (the Joint Electron Device Engineering Council) Standard Manufacturer's Identification Code. It is generated from the JEDEC code as follows:

a. Bits 6–0. The seven least significant bits are derived from the last byte of the JEDEC code by discarding the parity bit.
b. Bits 10–7. The four most significant bits provide a binary count of the number of bytes in the JEDEC code that contain the continuation character (hex 7F).

Bit 10 of the manufacturer ID code is the most significant bit.
The device data sheet must clearly identify the manufacturer ID code.

Description

This scheme utilizes the manufacturer coding scheme administered by JEDEC. The JEDEC code is formed from a variable number of 8-bit bytes. Each byte contains 7 data bits and an odd parity bit (the most significant bit). Bytes other than the last contain continuation characters (hex 7F), while the last contains one of 127 different codes which, together with a knowledge of the number of preceding continuation code bytes, allows the manufacturer's identity to be determined.

The compressed form of the JEDEC code used within the device identification register limits the number of bits needed in the device identification register to contain the manufacturer's code and allows the length of the code to be standardized. The length of the compressed JEDEC code is fixed at 11 bits. Sixteen codes are unused, since these correspond to the hex 7F code in the 7 least significant bits—the JEDEC continuation character.

Part Number, Revision, and Programmable Device ID Codes: Rules

The lengths of the part number and revision fields not defined should be fixed for a given device manufacturer.

The device data sheet must clearly identify the lengths of the part number and revision fields and the contents of the part number field.

User-programmable devices may specify a number of user-programmable device ID code bits within the overall device identification code. Where this is done, the number of bits in the programmable device ID code field and the method of programming must be clearly identified in the device data sheet.

User-Defined Test Data Registers

The overall architecture is intended to be extensible to allow access to any test-support features embedded in the system logic of the design. These features might include scan-test, self-test registers, or access to key data registers in the design (for example via scannable shadow registers).

This section defines the design requirements for additional test data registers and how they can be incorporated in the overall architecture.

Design and Construction

There must be no inversion of data between the serial input and the serial output of a user-defined test data register.

Inversions may occur between shift-register cells within a user-defined test data register provided the preceding rule is satisfied. Where a test data register contains inversions in the shift-register path, this must be clearly documented in the device data sheet.

Operation

When the current instruction selects a particular test data register, the register's behavior in each TAP controller state must be as defined in Table 11-12. All test operations are performed on the rising edge of TCK.

When the current instruction does not select a given user-defined test data register, it should perform its system function (if one exists); otherwise it should retain its last state.

Description

These requirements ensure that test data registers operate correctly in conjunction with the TAP and TAP controller.

Interconnection of Test Data Registers

Where several test data registers are required to support a single type of test, then these registers may be connected in series when that test is selected.

TABLE 11-12
Test Data Register Operation in Each Controller State

Controller State	*System or Test Operation*	*Action*
Test-Logic-Reset	System	Test operation inhibited
Capture-DR	Test	Load data from parallel data input
Shift-DR	Test	Shift data between register cells
Update-DR	Test	Transfer data from shift-register latch to parallel output latch (if provided)
Run-Test/Idle	Test	Run in test mode (e.g., self-test)
All other controller states	Test	Retain last state

Description

To perform some types of test, two or more test data registers may need to be used. For example, to perform an internal self-test when the component is mounted on a board with other devices, access may be required to both the boundary-scan register and user-defined test data registers. The purpose of the boundary-scan register is to shield the system logic from signal changes arriving at the component inputs from other devices.

To speed access in these situations, test data registers may be serially connected when the appropriate test is selected by the instruction register. See, for example, Figure 11-19.

Test Execution

Where a test can be executed during the Run-Test/Idle controller state, a minimum number of rising edges of TCK can be defined to ensure test completion. Following that number of rising edges of TCK, the test result should remain constant.

No maximum number of rising edges on TCK may be defined for a test that is to be executed in the Run-Test/Idle controller state.

Description

These rules ensure that sufficient clock edges can be applied to allow completion of all self-tests and the like that are being executed concurrently in different components on an assembled board.

FIGURE 11-19
Serial interconnection of test data registers

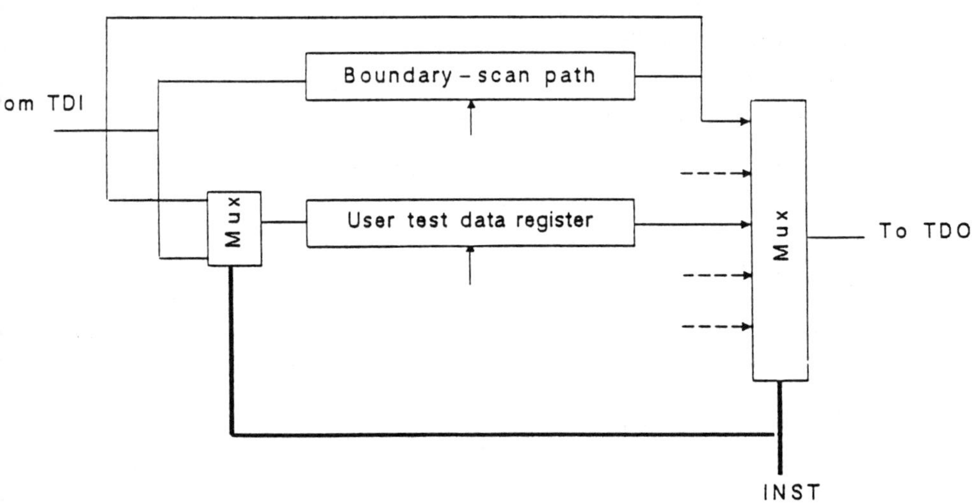

Thus in a circuit containing devices with self-test lengths of 1,000, 5,000, 10,000, and 50,000 rising clock edges, the complete board must be left in the Run-Test/Idle controller state for at least 50,000 rising clock edges to ensure that all tests complete satisfactorily.

Performance

The required performance of the test logic is detailed below.

Clock Performance

The test logic must be able to operate with TCK in the range shown in Table 11-13.

The maximum frequency acceptable at TCK must be clearly identified in the integrated circuit's data sheet.

The nominal duty cycle of the supplied TCK signal will be 50 percent.

Description

The clock performance requirement is included to ensure an acceptable minimum level of performance for the test logic, since the overall performance of a board-level test data path will be dictated by the lowest performance device in it. The time taken to shift data along the board-level test data path increases as the maximum achievable clock frequency is reduced.

Setup and Hold Times

Any setup and hold times required by components at the TMS and TDI pins must be less than 15 ns.

Output Drive Capability

The TDO pin must be capable of driving at least two TAP data input pins (TDI) in the same technology.

Logic levels used at the TAP pin should conform to those used at the other package pins of the device.

TABLE 11-13
Clock Performance Requirement

Minimum	*Maximum*
0 Hz	at least 6.25 MHz

Description

The additional drive capability of the TDO pin is included to allow for bed-of-nails or hand-held probes to be connected to the board-level serial test data paths.

Implementation Costs

Any change made to a design to improve its testability carries with it one or most costs. Typically, factors such as increased circuit size and reduced performance must be considered. For boundary-scan, the main cost areas are circuit size, pin requirements, and circuit performance.

Circuit Size

The effect on circuit size of adding boundary-scan capability to a design is, as for other design-for-test changes, difficult to predict, since much depends on the detail of the implementation. The following discussion provides some estimates for the overall size of the JTAG circuitry related to a 6 mm $\times$ 6 mm integrated circuit built in a 2.0-micron single-layer metal CMOS process.

TAP Controller

The TAP controller implementation shown in Figure 11-7 requires on the order of 80 NAND gates. A more efficient implementation could, however, be achieved using a transistor-level state machine design. An initial implementation in the state technology requires a silicon area of approximately 0.3 mm^2.

Instruction and Bypass Registers

An instruction register containing three bits would occupy on the order of 0.03 mm^2.

The bypass register is approximately one-third the size of the minimum instruction register, or 0.02 mm^2.

Boundary-Scan Register

An estimate of the total size of the boundary-scan register can be obtained by looking at the size of the data bit cells illustrated in Figures 11-12 and 11-13. A circuit that implements the design of Figure 11-12a requires around 0.015 mm^2.

For an integrated circuit with 40 system pins (input or output), the boundary-scan register would therefore require some 0.6 mm^2 silicon area.

Overall Silicon Requirement

The combined silicon area for a minimal implementation comprising a TAP controller, a three-bit instruction register, a bypass register, and a 40-bit boundary-scan register would be approximately 1 mm^2 from the above figures (including a small allowance for the multiplexers and other items in Figure 11-3). This represents an increase in size of 3 percent for the 36 mm^2 device. Clearly, this figure is significantly affected by changes in circuit size, component geometries, and other changes (such as the use of two metal layers). It also does not take into account any increase in the size of the system logic, for example due to increased separation of cells caused by increased pitch between input/output pads.

Reducing the Silicon Overhead

The overhead can, however, be reduced in several ways dependent on the circuit design, for example:

a. The data bit cells of the boundary-scan register can be integrated with the input or output buffer stages in the circuit design.
b. The TAP and the boundary-scan register cells can be implemented in "dead" area around the periphery of the circuit. For example, the cells are located beneath power distribution buses.

Pin Requirements

The implementation described requires a minimum of four dedicated pins to provide the TAP on merchant devices. As illustrated by the following application notes, the TAP on merchant devices. As illustrated by the following application notes, the TAP can allow access to all testability features within a design—which might otherwise require package pins for additional data or control access. The four pins required by the TAP may therefore frequently provide for all test purposes.

Circuit Performance

The greatest impact on circuit performance is likely to be due to the inclusion of additional circuitry immediately before the output buffers in the design. The multiplexer in Figure 11-12a, for example, could add two gate delays which, together with the additional delay due to the input loading of the boundary-scan register, would increase the propagation delay of signals leaving the circuit. Similarly, the delays experienced by signals entering the circuit would be increased.

The importance of these additional delays clearly depends on the application for which the component is intended. However, the impact of the additional circuitry can be minimized by careful design or by combining input buffers with the boundary-scan register cells.

When Only Some Components Have Boundary Scan

In cases where not all components have the boundary-scan facility, the presence of a partial boundary-scan register allows improved control and observation of the connections around non-boundary-scan components and can assist in testing such areas of logic. This fact is extremely important, since it is unlikely that the situation where all components in a product have boundary scan will exist for many years. Clearly there will always be some components in use that do not have the facility, either because the effort in redesigning them is not justified or because there are sound technical reasons for excluding the facility.

Advantages gained through having a partial boundary scan at the board level include a reduction in the need for physical test access (e.g., through a bed of nails), partitioning of the design, and the ability to electrically remove a complex boundary-scan device from the circuit. For example, the presence of the facility in a microprocessor support chip would allow a block of 74-series TTL components connected it to be tested as a small functional unit without having to consider the operation of the support chip itself.

Figure 11-20 illustrates the use of the boundary-scan register with its mode controls set for External test to allow tests to be applied to external logic. The figure also highlights problems that might be encountered in applying tests to logic blocks in this way. For example:

a. The logic block being tested may contain asynchronous logic that will be sent into undesirable states if shifting patterns appear at its inputs.
b. The signals applied from the boundary-scan register may feed into clock inputs on the logic block being tested, which again will produce undesirable effects if the logic is not shielded from shifting patterns.

Since, in a generally applicable architecture, it cannot be guaranteed that such features do not exist in the circuitry under test, the boundary-scan design must be such that these problems are guaranteed to be avoided. The JTAG design ensures this by requiring a parallel output register/latch in each boundary-scan cell that can drive out of a component (e.g., at unidirectional, tristate, or bidirectional pins).

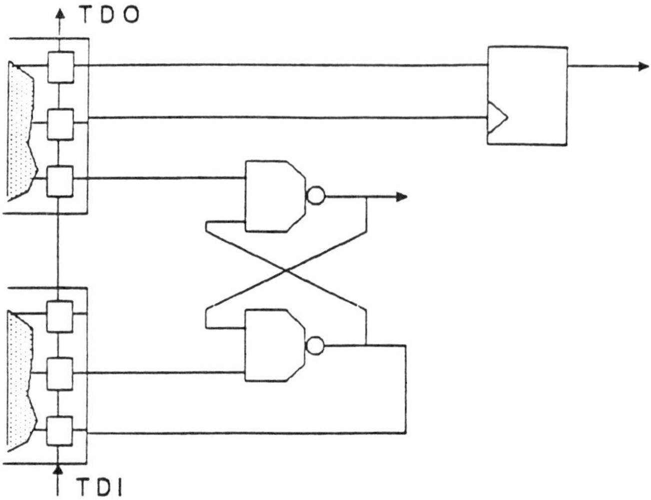

FIGURE 11-20
Testing external logic via the boundary-scan path

Diagnosing Faults in the Serial Test Data Path

Objective

Before the serial test data path can be used to test the components on a board and, through the boundary-scan registers, their normal functional interconnections, it must itself be tested for common production defects—for example solder shorts and opens. The design of the instruction register within the TAP includes facilities to assist in this task.

A Basic Path Test

The first step is to initialize the TAP and instruction register. This can be achieved by holding the TMS signal (which is broadcast to all devices) high and applying five rising edges to TCK. At the end of this process the TAP controller in each component will be in the Test-Logic-Reset controller state, which will cause some other circuitry (e.g., the instruction register outputs) to be initialized.

The second step is to move through the Run-Test/Idle, Select-DR-Scan, and Select-IR-Scan controller states and enter the instruction register scanning sequence. The initial Capture-IR controller state will cause the instruction register to be loaded with the {10X. . .X} pattern.

Note that so far the only connections that have been involved are TMS and TCK. No shifting is needed during initialization.

By entering and remaining in the Shift-IR controller state, the constant patterns loaded into the most significant bits of the instruction registers in each component will be observed at the board's serial output. For example, for a board containing four components each having a four-bit instruction register, the pattern at the serial output for a fault-free board would be as follows:

10XX10XX10XX10XX

$\longleftarrow$

Shift direction

In the event of faults, the patterns from several components will be observed followed by erroneous data—allowing the nearest fault to the board's serial output to be localized. For example, the following would indicate an open-circuit fault between components 2 and 3, since the pull-up on the TDI input of component 2 would cause a constant high to be shifted in instead of the expected pattern:

10XX10XX11111111

$\longleftarrow$

Shift direction

In this way, faults can be diagnosed and removed from the board one at a time.

Use of the Device Identification Register

If a device identification register is present in the design, it is possible to combine a check of the assembly process with a test of the integrity of the board-level serial test data path by shifting all device identification codes out of the board in one pass.

More Complex Methods

The above diagnostic method adequately deals with most of the faults in the serial test data path (e.g., internal faults in the instruction registers and external faults resulting from solder opens and shorts). However, it does not guarantee the diagnosability of internal faults in the scannable test data registers. In the following sections, we describe two diagnostic methods that can alleviate this problem.

Method 1

This method requires the test data registers to be designed in such a way that, for each of the registers except the single-bit bypass register, the two bits nearest to the serial output (TDO) can be initialized to a binary

"10" pattern upon Test-Logic-Reset. With this added hardware provision, the method of testing the integrity of the serial test data path consisting of test data registers can be easily derived from the preceding paragraphs. However, since the bypass register consists of only a single bit, we cannot locate a fault in the bypass bit of an arbitrary component within "one scan pass". (Method 2 can be adapted here for locating a fault in the bypass register within two scan passes.)

Method 2

This method saves the hardware overhead required by method 1 at the expense of decreased diagnostic throughput. In this method, a board is set up so that alternating ICs on the serial test data path between bus master TDO and bus master TDI receive TMS signals from two different, but coordinated, sources. This allows one set of ICs (say those in the even-numbered spots along the serial path) to be set up for instruction register scanning while the other set (those in the odd-numbered spots) are set up for test data register scanning. Then the fixed bits of the instruction registers in the even-numbered ICs can be used to diagnose the integrity of the selected test data registers in the odd-numbered ICs.

As an example, we will apply this method to the diagnosis of a fault in the bypass register of an IC. Any fault located in an IC's bypass register can be located in no more than two scan passes. Since the bypass registers and the instruction registers are scanned in two different controller states—Shift-DR and Shift-IR, respectively—this approach requires that two separate TMS wires (call them TMS1 and TMS2) be distributed. TMS1 should control the odd-numbered ICs, while TMS2 controls the even-numbered ICs. Note that TMS1 and TMS2 can come directly from the TAP bus master (see Figure 11-4b), or they can come from a board-level controller. To illustrate this approach, let us assume that all components on a board's test data path have been initialized to the Test-Logic-Reset controller state. Then we can apply the control sequences on TMS1 and TMS2 shown in Table 11-14 so that, on each of two separate scan passes, half of the components at a time are bypassed and the other half have their instruction registers scanned.

The instruction register is initialized in the Test-Logic-Reset controller state to all-ones, which selects the bypass registers for a following data register scan sequence automatically.

Note that, in the first scan pass, TMS1 holds the odd-numbered components in the Test-Logic-Reset state for one more cycle than the even-numbered components, so that shifting of the data/instruction registers of all components is synchronized. In the second scan pass, TMS2 holds the even-numbered components in the Test-Logic-Reset state for one more cycle than the even-numbered components, so that shifting of the data/instruction

TABLE 11-14
TMS1 and TMS2 Values for Method 2

Clock Cycles	*TMS1 Value*	*Odd Component State*	*TMS2 Value*	*Even Component State*
1	1	Test-Logic-Reset	0	Run-Test/Idle
2	0	Run-Test/Idle	1	Select-DR-Scan
3	1	Select-DR-Scan	0	Select-IR-Scan
4	0	Capture-DR	0	Capture-IR
5	0	Shift-DR	0	Shift-IR
6	0	Shift-DR	0	Shift-IR
.	.	...	.	...
.	.	...	.	...
N + 1	0	Run-Test/Idle	1	Test-Logic-Reset
N + 2	1	Select-DR-Scan	0	Run-Test/Idle
N + 3	1	Select-IR-Scan	1	Select-DR-Scan
N + 4	0	Capture-IR	0	Capture-DR
N + 5	0	Shift-IR	0	Shift-DR
N + 6	0	Shift-IR	0	Shift-DR
.	.	...	.	...
.	.	...	.	...

registers of all components is synchronized. In the second scan pass, TMS2 holds the even-numbered components in the Test-Logic-Reset state for one more cycle than the odd-numbered components. Also, note that a fault in the bypass register of an odd-numbered component will be located in the first scan pass, and a similar fault in an even-numbered component will be located in the second scan pass.

In the following example, control sequences similar to those just described are applied to diagnose a stuck-at-1 fault in IC4's bypass register. All devices are depicted as having instruction registers that are 5 bits long. *Note*: Brackets are used to delimit the single-bit serial outputs from those components that are bypassed.

Component ID	IC1	IC2	IC3	IC3	IC5	IC6	. . .
Expected output 1:	[X]	10XXX	[X]	10XXX	[X]	10XXX	. . .
Observed output 1:	[X]	10XXX	[X]	10XXX	[X]	10XXX	. . .
Expected output 2:	10XXX	[X]	10XXX	[X]	10XXX	[X]	. . .
Observed output 2:	10XXX	[X]	10XXX	[1]	10XXX	[1]	. . .

Achieving Other Test Goals

The basic JTAG architecture can be extended to achieve other test goals for particular circuit designs.

Simplified Parametric Testing

The boundary-scan register can be used to simplify the creation and application of parametric tests for the integrated circuits themselves. For example, a test of the input switching thresholds of an IC would normally require paths to be set up through the circuit that allowed the state of each input to be observed by monitoring chip outputs. The resulting test sequence could be extremely long, say up to 50,000 vectors.

To perform a parametric test using the boundary-scan register, a loop is entered where first the boundary-scan mode controls are set for External test by shifting the appropriate instruction into the instruction register. A data register scan cycle is then entered, which causes the data applied at the system input pins to be loaded into the boundary-scan register. The logic signal value perceived at each input pin can be examined by shifting the latched values through the TDO pin. This load-shift cycle is repeated for different input voltage.

Similarly, the boundary-scan register can be used to facilitate measurements on output drive capability, slew rates, and the like. Further, the inclusion of cells in the boundary-scan register that allow each tristate output pin or bidirectional pin to be forced to the high-impedance output state allows this aspect of device performance to be tested easily.

Circuit Partitioning

Partitioning is a valuable tool in reducing test costs for integrated circuits as well as other products. For this reason, VLSI circuits are often designed such that they can be divided into several smaller blocks that can be tested independently before the functionality of the design as a whole is checked. The use of multiplexers and scanlike techniques to achieve this partitioning is well documented.

One way of supplying the signals necessary to set up the various partitions within a circuit is to extend the instruction register in the design. To allow the circuit to be tested on an ATE, the instruction register is set to disable the boundary-scan register while allowing access to other test features, for example, to achieve the desired circuit partitioning. To allow the circuit to be tested via the boundary-scan register once it has been assembled onto a board, the instruction register is set to select the boundary-scan register Internal test with partitioning being achieved in the same way as for an ATE-based test.

Other design-for-test improvements can be achieved through the boundary-scan register in a similar way. The key point is that they can be achieved without requiring the use of any further package pins, which in many applications are at a premium. The cost of extending the instruction register to provide the additional capability is relatively low, although the size of the instruction decoding logic may increase.

Mixed Analog/Digital Circuits

This proposal is directed primarily at digital circuits. However, boundary scan can assist in testing circuits that contain some element of analog circuitry—which may be on- or off-chip. Examples of analog features that may occur on-chip include crystal oscillators, analog data inputs and/or outputs, balanced line receivers, and logic signals encoded for transmission (e.g., RS-232, Ethernet).

Figure 11-21 shows how a boundary-scan register might be arranged in a case where analog circuitry is included on-chip. Here the boundary-scan register is distorted so that it intersects the digital signals emerging from the analog/digital interface. This allows separation of the analog tests from those for the digital part of the circuit as follows:

a. The analog circuitry is considered to be external to the device. When the boundary-scan cells at the interface to the digital circuitry are set for External test, test data can be applied to the analog circuitry without having to propagate signals through the (potentially complex) digital block.

FIGURE 11-21
Mixed analog/digital circuits

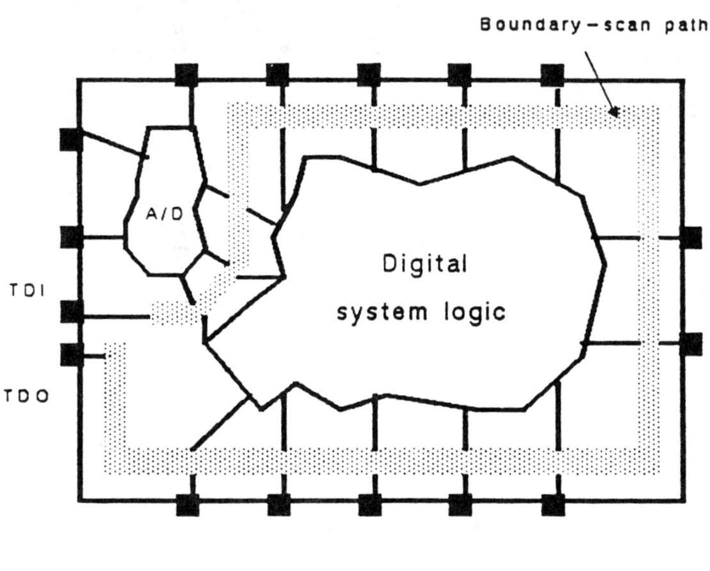

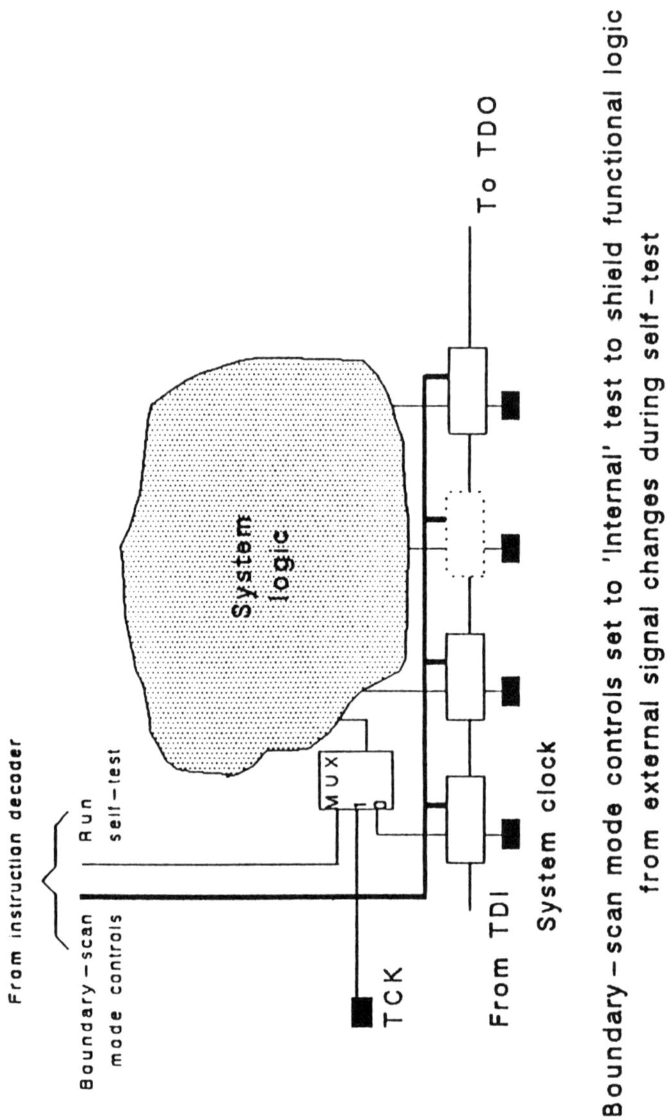

FIGURE 11-22 Access to self-test features

b. When the boundary-scan cells at the interface are set for Internal test, tests can be applied to the digital block without having to propagate signals through the analog circuitry. This removes the complication introduced by the analog circuitry.

A similar test approach can be used in cases where the analog circuitry is contained in separate devices. In this case, the boundary-scan facilities in the digital components provide the necessary access.

Boundary-Scan as a Basis for Self-Test

The boundary-scan framework provides a basis for the use of self-test features in integrated circuits during board testing. As in the control of partitioning features, self-test capability in a circuit can be triggered by setting an appropriate instruction, and the results of such tests can be made observable through the boundary-scan register or at package pins. The self-test should execute in the Run-Test/Idle controller state.

Figure 11-22 illustrates the decoding use of an instruction to initiate self-test of the enclosed circuit. This instruction is also decoded to control a multiplexer, which ensures that the test clock (TCK) is used during self-test in place of the normal clock signal. At the conclusion of the test, the results would remain in the test data register(s) and be shifted out of the circuit for examination during a scan cycle. Note that for self-test applications, it is likely that there would be no activity in the Update-DR and Capture-DR controller states.

FIGURE 11-23
Problem in sharing TDI with a system configuration

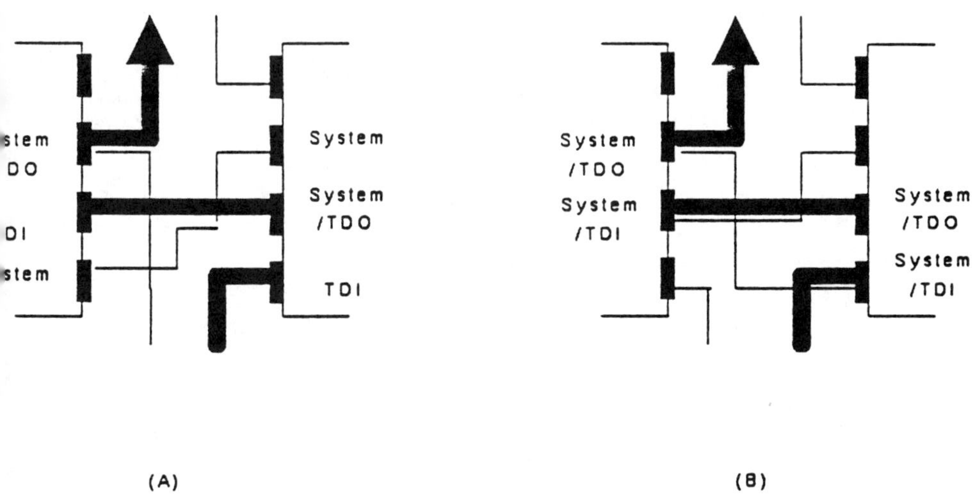

Other means of enabling the circuit during self-test are possible, including the use of the normal circuit clock signal. Where this is done, direct control of the clock signal at board level must be possible, and a data bit cell such as that shown in Figure 11-12c may be used for the clock input pin.

Note that where self-test features can be controlled through the JTAG TAP the requirements must be satisfied. This is to ensure that, on a board containing many self-testable components, all self-test routines can be executed concurrently.

Chapter 12

BIST and Boundary Scan

In this chapter the built-in self-test (BIST) features of Intel's 80486 and Motorola's 68332 are described in detail. However, neither of these processors incorporates BIST circuitry that is compatible with boundary scan. The BIST features of Intel's RISC μP, the 80860, which are described in less detail, include a boundary-scan modc. It is thought that the 80486 and the 68332 may not have opted for boundary-scan compatibility because that might have affected their performance. However, the 80486 design manager stated that it could accommodate boundary scan if it is standardized by the IEEE. It is said that less than 2 percent of the chip's gates are devoted to BIST.

Texas Instruments has been the first to offer a range of products for ASIC and board-level design compatible with boundary-scan testing. TI incorporates test cells, which are called SCOPE, an abbreviation for System Compatibility/Observability Partitioning Environment. SCOPE cells offer ASIC designers board partitioning without adding any devices. Using a series of bit slice test elements, designers can build structured hierarchical test systems that use the JTAG four-wire serial architecture. An application of the SCOPE test cells in a multiprocessor-based controller card is shown in Figure 12-1. Within SCOPE a crucial part is called ASSET, an abbreviation

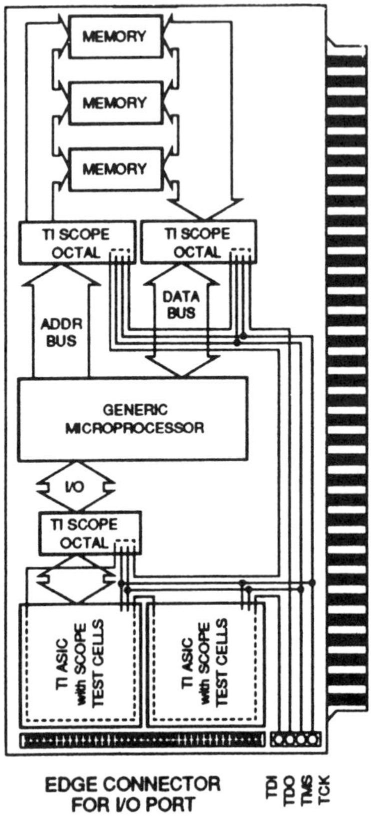

FIGURE 12-1
SCOPE cells provide boundary-scan test capabilities. M.H. Mickle, New Products Dept., "Cell Promotes Board Level Testability," *MICRO*, August 1989 © 1989 IEEE.

of Advanced Support System for Emulation and Test, which helps by isolating the details of the boundary-scan serial scan protocol and the target system. ASSET software runs on an IBM PC AT. Other SCOPE cells deal with pattern generation, signature analysis, and internal scan. A TAP and a scan bus master, which interfaces its DSP processors to the four-wire test bus, are important parts of TI's boundary-scan support.

Motorola has announced embedded JTAG cells for forthcoming CMOS and ECL arrays. Its MCA-4 family will support boundary-scan testing with custom circuitry in its I/O cells. BIST continues to grow throughout the industry, but wide acceptance of JTAG boundary scan depends in a large measure on its adoption as an IEEE standard.

Detailed descriptions of the BIST features of Intel's 80486, Motorola's 68332, and Intel's 80860 RISC processor follow.

(80486) i486 Microprocessor

Testing in the 486 microprocessor can be divided into two categories: BIST and external testing. The BIST tests the nonrandom logic, control ROM (CROM), translation lookaside buffer (TLB), and on-chip cache memory. External tests can be run on the TLB and the on-chip cache. The 486 microprocessor also has a test mode in which all outputs are tristated.

Built-In Self-Test (BIST)

The BIST is initiated by holding the AHOLD (address hold) pin high in the clock prior to reset going from high to low, as shown in Figure 12-2. The BIST takes approximately 42 milliseconds with a 25-MHz 486 microprocessor. No bus cycles will be run by the 486 microprocessor until the BIST is concluded.

The results of BIST are stored in the EAX register. The 486 microprocessor has successfully passed the BIST if the contents of the EAX register are zero. If the results in EAX are not zero then the BIST has detected a flaw in the microprocessor.

The microprocessor performs reset and begins normal operation at the completion of the BIST. The DX register will contain a component identifier at the conclusion of reset, with or without the BIST. The upper byte of DX will contain 04, and the lower byte will contain a stepping identifier (see Figure 12-3).

The nonrandom logic, control ROM, on-chip cache, and TLB are tested during the BIST.

The cache portion of the BIST verifies that the cache is functional and that it is possible to read and write to the cache. The BIST manipulates test registers TR3, TR4, and TR5 while testing the cache. These test registers are described on page 332.

The cache testing algorithm writes a value to each cache entry, reads the value back, and checks that the correct value was read back. The algorithm may be repeated more than once for each of the 512 cache entries using different constants.

The TLB portion of the BIST verifies that the TLB is functional and that it is possible to read and write to the TLB. The BIST manipulates test registers TR6 and TR7 while testing the TLB. TR6 and TR7 are described on pages 336 and 337.

FIGURE 12-2
BIST initiation. Reprinted courtesy of Intel Corporation.

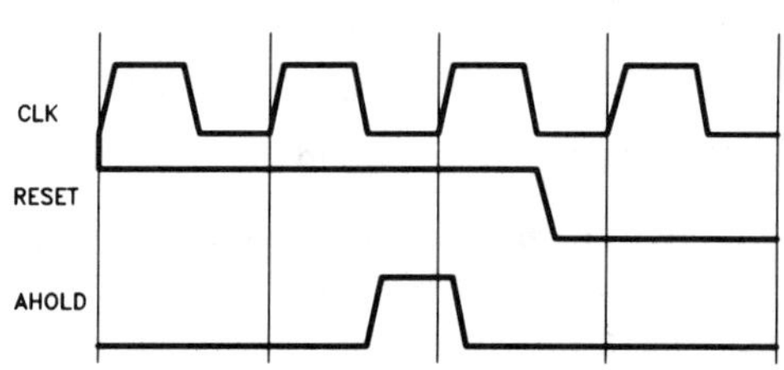

240440–41

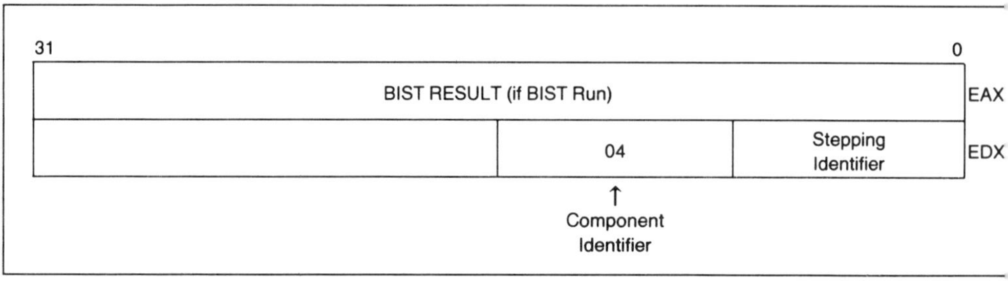

FIGURE 12-3
BIST results. Reprinted courtesy of Intel Corporation.

On-Chip Cache Testing

The on-chip cache testability hooks are designed to be accessible during the BIST and for assembly-language testing of the cache. The following subsections describe the 486 microprocessor cache organization, its relation to the testability hardware, and the algorithms used for accessing the cache.

Cache Organization

The cache is logically organized into three blocks, as shown in Figure 12-4. The cache holds 8K of data and is four-way set associative. Each line in the cache is 16 bytes wide. The cache contains 128 sets.

The largest block is the data block. The data block is logically organized as 128 sets, each containing four 16-byte lines or entries. Addressing into the data block consists of two components, the set select and the entry select. The set select is 7 bits wide, since there are 128 sets. The entry select is 2 bits wide to select one of the four entries.

The second block is the tag array. The tag array is physically organized as 128 sets, each containing four 21-bit-wide data words.

The third block consists of 128 seven-bit quantities, which contain the valid bits and the bits used in the pseudo least recently used (LRU) replacement algorithm called the LRU bits. There are 4 valid bits, one for each line in the set. There are 3 LRU bits for each of the 128 sets.

The 486 microprocessor contains a cache fill buffer and a cache read buffer. For testability writes, data must be written to the cache fill buffer before it can be written to a location in the cache. Data must be read from a cache location into the cache read buffer before the microprocessor can access the data. The cache fill and cache read buffer are both 128 bits wide.

Cache Testing Registers TR3, TR4, and TR5

Figure 12-5 shows the three cache testing registers: the Cache Data Test Register (TR3), the Cache Status Test Register (TR4), and the Cache Control Test Register (TR5). External access to these registers is provided through MOV reg,TREG and MOV TREG,reg instructions.

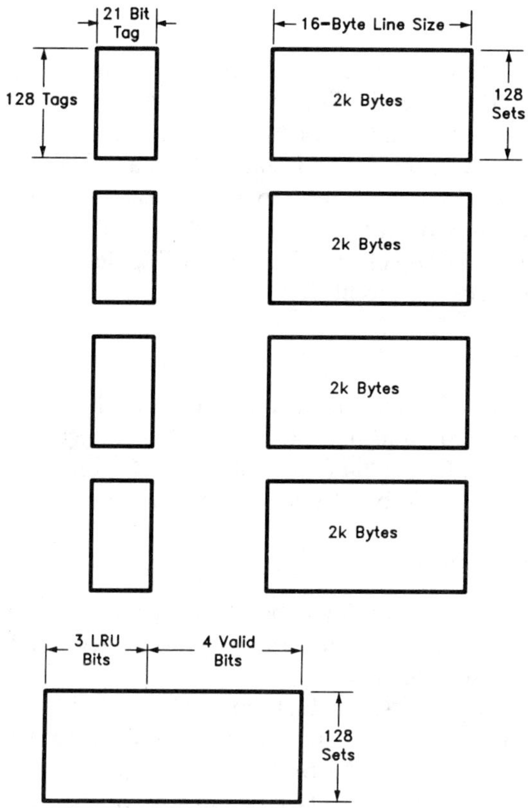

FIGURE 12-4
On-chip cache physical organization. Reprinted courtesy of Intel Corporation.

FIGURE 12-5
Cache test registers. Reprinted courtesy of Intel Corporation.

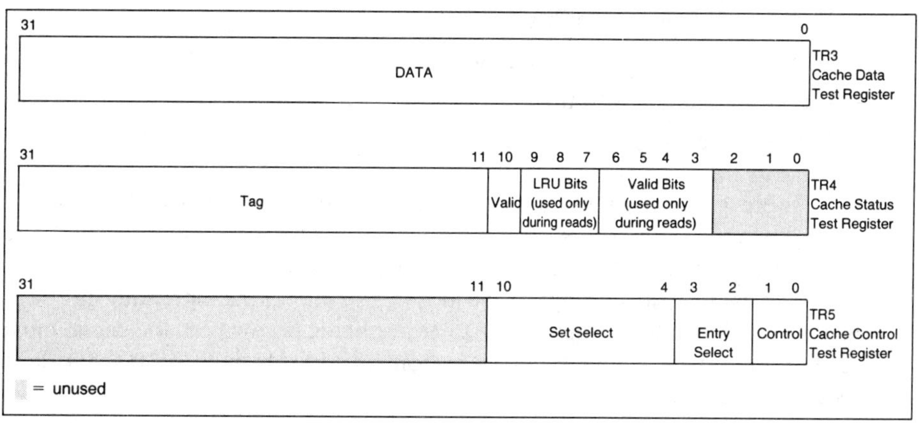

Cache Data Test Register: TR3

The cache fill buffer and the cache read buffer can only be accessed through TR3. Data to be written to the cache fill buffer must first be written to TR3. Data read from the cache read buffer must be loaded into TR3.

TR3 is 32 bits wide, while the cache fill and read buffers are 128 bits wide, and therefore 32 bits of data must be written to TR3 four times to fill the cache fill buffer. Thirty-two bits of data must be read from TR3 four times to empty the cache read buffer. The entry select bits in TR5 determine which 32 bits of data TR3 will access in the buffers.

Cache Status Test Register: TR4

TR4 handles tag, LRU, and valid bit information during cache tests. TR4 must be loaded with a tag and a valid bit before a write to the cache. After a read from a cache entry, TR4 contains the tag and valid bit from that entry, and the LRU bits and 4 valid bits from the accessed set.

Cache Control Test Register: TR5

TR5 specifies which testability operation will be performed and the set and entry within the set that will be accessed. The 7-bit set select field determines which of the 128 sets will be accessed.

The functionality of the two entry select bits depend on the state of the control bits. When the fill or read buffers are being accessed, the entry select bits point to the 32-bit location in the buffer being accessed. When a cache location is specified, the entry select bits point to one of the four entries in a set. Refer to Table 12-1.

Five testability functions can be performed on the cache. The two control bits in TR5 specify the operation to be executed. The five operations are as follows:

1. Write cache fill buffer.
2. Perform a cache testability write.
3. Perform a cache testability read.
4. Read the cache read buffer.
5. Perform a cache flush.

Table 12-1 shows the encoding of the two control bits in TR5 for the cache testability functions. Table 12-1 also shows the functionality of the entry and set select bits for each control operation.

The cache tests attempt to use as much of the normal operating circuitry as possible. Therefore when cache tests are being performed, the cache must be disabled (the CE and WT bits in control register must be reset to 0 to disable the cache).

TABLE 12-1
Cache control bit excoding and effect of control bits on entry select and set select functionality

Control Bits		Operation	Entry Select Bits Function	Set Select Bits
Bit 1	Bit 0			
0	0	Enable { Fill Buffer Write / Read Buffer Read	Select 32-bit location in fill/read buffer	—
0	1	Perform Cache Write	Select an entry in set.	Select a set to write to
1	0	Perform Cache Read	Select an entry in set.	Select a set to read from
1	1	Perform Flush Cache	—	—

Reprinted courtesy of Intel Corporation.

Loading the fill buffer is accomplished by first writing to the entry select bits in TR5 and setting the control bits in TR5 to 00. The entry selects bits identify one of four 32-bit locations in the cache fill buffer to put 32 bits of data. Following the write to TR5, TR3 is written with 32 bits of data, which are immediately placed in the cache fill buffer. Writing to TR3 initiates the write to the cache fill buffer. The cache fill buffer is loaded with 128 bits of data by writing to TR5 and TR3 four times using a different entry select location each time.

TR4 must be loaded with the 21-bit tag and valid bit (bit 10 in TR4) before the contents of the fill buffer are written to a cache location.

The contents of the cache fill buffer are written to a cache location by writing TR5 with a control field of 01 along with the set select and entry select fields. The set select and entry select fields indicate the location in the cache to be written. The normal cache LRU update circuitry updates the internal LRU bits for the selected set.

Note that a cache testability write can only be done when the cache is disabled for replaces (the CE bit is control register 0 and is reset to 0). Also note that care must be taken when directly writing to entries in the cache. If the entry is set to overlap an area of memory that is being used in external memory, that cache entry could inadvertently be used instead of the external memory. Of course, this is exactly the type of operation that one would desire if the cache were to be used as a high-speed RAM.

Cache Testability Read

A cache testability read is a two-step process. First the contents of the cache location are read into the cache read buffer. Next the data is examined by reading it out of the read buffer.

Reading the contents of a cache location into the cache read buffer is initiated by writing TR5 with the control bits set to 10 and the desired 7-bit set select and 2-bit entry select. In response to the write to TR5, TR4 is loaded with the 21-bit tag field and the single valid bit from the cache entry read. TR4 is also loaded with the 3 LRU bits and 4 valid bits corresponding to the cache set that was accessed. The cache read buffer is filled with the 128-bit value that was found in the data array at the specified location.

The contents of the read buffer are examined by performing four reads of TR3. Before reading TR3 the entry select bits in TR5 must be loaded to indicate which of the four 32-bit words in the read buffer to transfer into TR3, and the control bits in TR5 must be loaded with 00. The register read of TR3 will initiate the transfer of the 32-bit value from the read buffer to the specified general-purpose register.

Note that it is very important that the entire 128-bit quantity from the read buffer and also the information from TR4 be read before any memory references are allowed to happen, or the contents of the read buffer will be corrupted. This is because the testability operations use hardware that is used in normal memory accesses for the 486 microprocessor whether the cache is enabled or not.

Flush Cache

The control bits in TR5 must be written with 11 to flush the cache. None of the other bits in TR5 have any meaning when 11 is written to the control bits. Flushing the cache will reset the LRU bits and the valid bits to 0 but will not change the cache tag or data arrays.

When the cache is flushed by writing to TR5, the special bus cycle indicating a cache flush to the external system is not run. The cache should be flushed with the INVD (Invalidate Data Cache) instruction or the WBINVD (Write-back and Invalidate Data Cache) instruction.

Translation Lookaside Buffer (TLB) Testing

The 486 microprocessor TLB testability hooks are similar to those in the 386 microprocessor. The testability hooks have been enhanced to provide added test features and to include new features in the 486 microprocessor. The TLB testability hooks are designed to be accessible during the BIST and for assembly-language testing of the TLB.

Translation Lookaside Buffer Organization

The 486 microprocessor TLB is four-way set associative and has space for 32 entries. The TLB is logically split into three blocks, as shown in Figure 12-6.

The data block is physically split into four arrays, each with space for eight entries. An entry in the data block is 22 bits wide containing a 20-bit physical address and 2 bits for the page attributes. The page attributes are the PCD (page cache disable) bit and the PWT (page write-through) bit.

The tag block is also split into four arrays, one for each of the data arrays. A tag entry is 21 bits wide, containing a 17-bit linear address and 4

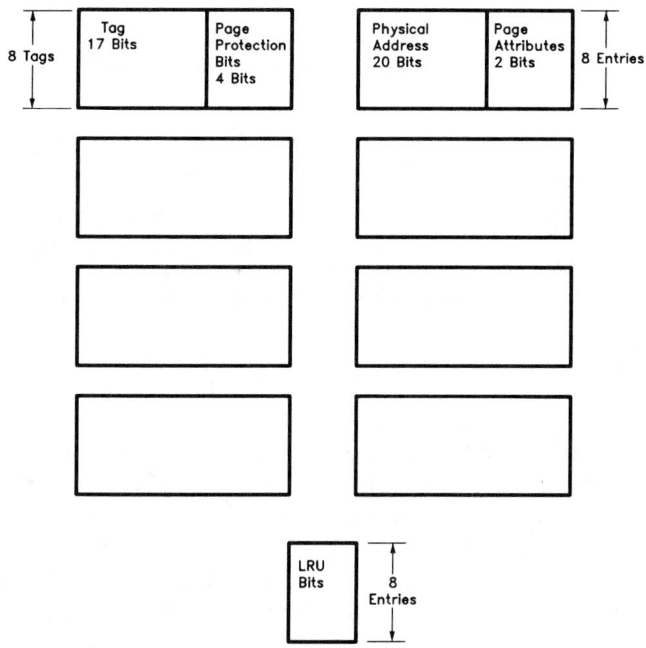

FIGURE 12-6
TLB organization. Reprinted courtesy of Intel Corporation.

protection bits. The protection bits are valid (V), user/supervisor (U/S), read/write (R/W), and dirty (D).

The third block contains eight 3-bit quantities used in the pseudo least recently used (LRU) replacement algorithm. These bits are called the LRU bits. The LRU replacement algorithm used in the TLB is the same as that used by the on-chip cache.

TLB Test Registers TR6 and TR7

The two TLB test registers are shown in Figure 12-7. TR6 is the command test register, and TR7 is the data test register. External access to these registers is provided through MOV reg,TREG and MOV TREG,reg instructions.

Command Test Register: TR6

TR6 contains the tag information and control information used in a TLB test. Loading TR6 with tag and control information initiates a TLB write or lookup test.

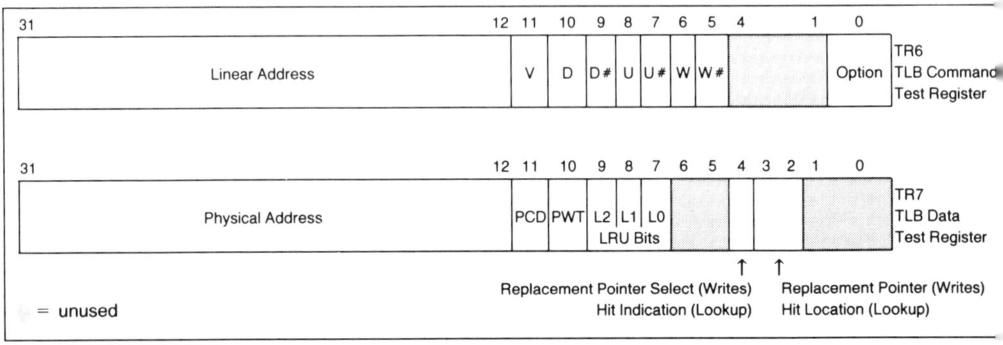

FIGURE 12-7
TLB test registers. Reprinted courtesy of Intel Corporation.

TR6 contains three bit fields, a 20-bit linear address (bits 12–31), 7 bits for the TLB tag protection bits (bits 5–11) and 1 bit (bit 0) to define the type of operation to be performed on the TLB.

The 20-bit linear address forms the tag information used in the TLB access. The lower 3 bits of the linear address select which of the eight sets are accessed. The upper 17 bits of the linear address form the tag stored in the tag array.

The 7 TLB tag protection bits are described as follows:

V: The valid bit for this TLB entry

D,D#: The dirty bit for/from the TLB entry

U,U#: The user/supervisor bit for/from the TLB entry

W,W#: The read/write bit for/from the TLB entry

Two bits are used to represent the D,U/S and R/W bits in the TLB tag to permit the option of a forced miss or hit during a TLB lookup operation. The forced miss or hit will occur regardless of the state of the actual bit in the TLB. The meaning of these pairs of bits is given in Table 12-2.

The operation bit in TR6 determines whether the TLB test operation will be a write or a lookup. The function of the operation bit is given in Table 12-3.

Data Test Register: TR7

TR7 contains the information stored or read from the data block during a TLB test operation. Before a TLB test write, TR7 contains the physical address and the page attribute bits to be stored in the entry. After a TLB test lookup hit, TR7 contains the physical address, page attributes, LRU bits, and entry location from the access.

TABLE 12-2
Meaning of a Pair of TR6 Protection Bits

TR6 Protection Bit (B)	*TR6 Protection Bit # (B#)*	*Meaning on TLB Write Operation*	*Meaning on TLB Lookup Operation*
0	0	Undefined	Miss any TLB TAG Bit B
0	1	Write 0 to TLB TAG Bit B	Match TLB TAG Bit B if 0
1	0	Write 1 to TLB TAG Bit B	Match TLB TAG Bit B if 1
1	1	Undefined	Match any TLB TAG Bit B

TABLE 12-3
TR6 Operation Bit Encoding

TR6 Bit 0	*TLB Operation to Be Performed*
0	TLB Write
1	TLB Lookup

TR7 contains a 20-bit physical address (bits 12–31), 2 bits for PCD (bit 11) and PWT (bit 10), and 3 bits for the LRU bits (bits 7–9). The LRU bits in TR7 are only used during a TLB lookup test. The functionality of TR7 bit 4 differs for TLB writes and lookups. The encoding of bit 4 is defined in Tables 12-4 and 12-5. Finally, TR7 contains 2 bits (bits 2–3) to specify a TLB replacement pointer or the location of a TLB hit.

A replacement pointer is used during a TLB write. The pointer indicates which of the four entries in an accessed set is to be written. The replacement pointer can be specified to be the internal LRU bits or bits 2–3 in TR7. The source of the replacement pointer is specified by TR7 bit 4. The encoding of bit 4 during a write is given by Table 12-4.

Note that both testability writes and lookups affect the state of the internal LRU bits regardless of the replacement pointer used. All TLB write operations (testability or normal operation) cause the written entry to become the most recently used. For example, during a testability write with the replacement pointer specified by TR7 bits 2–3, the indicated entry is written and that entry becomes the most recently used as specified by the internal LRU bits.

There are two TLB testing operations: write entries into the TLB, and perform TLB lookups. One major enhancement over TLB testing in the 386 microprocessor is that paging need not be disabled while executing testability writes or lookups.

TABLE 12-4
Encoding of Bit 4 of TR7 on Writes

TR7 Bit 4	*Replacement Pointer Used on TLB Write*
0	Pseudo-LRU Replacement Pointer
1	Data Test Register Bits 3:2

TABLE 12-5
Encoding of Bit 4 of TR7 on Lookups

TR7 Bit 4	*Meaning after TLB Lookup Operation*
0	TLB Lookup Resulted in a Miss
1	TLB Lookup Resulted in a Hit

Note that any time one TLB set contains the same linear address in more than one of its entries, looking up that linear address will not result in a hit. Therefore a single linear address should not be written to one TLB set more than once.

TLB Write Test

To perform a TLB write, TR7 must be loaded followed by a TR6 load. The register operations must be performed in this order, since the TLB operation is triggered by the write to TR6.

TR7 is loaded with a 20-bit physical address and values for PCD and PWT to be written to the data portion of the TLB. In addition, bit 4 of TR7 must be loaded to indicate whether to use TR7 bits 3–2 or the internal LRU bits as the replacement pointer on the TLB write operation. Note that the LRU bits in TR7 are not used in a write test.

TR6 must be written to initiate the TLB write operation. Bit 0 in TR6 must be reset to zero to indicate a TLB write. The 20-bit linear address and the 7 page protection bits must also be written in TR6 to specify the tag portion of the TLB entry. Note that the 3 least significant bits of the linear address specify which of the eight sets in the data block will be loaded with the physical address data. Thus only 17 of the linear address bits are stored in the tag array.

TLB Lookup Test

To perform a TLB lookup it is only necessary to write the proper tags and control information into TR6. Bit 0 in TR6 must be set to 1 to indicate a TLB lookup. TR6 must be loaded with a 20-bit linear address and the 7 protection bits. To force misses and matches of the individual protection bits on TLB lookups, the 7 protection bits are set as specified in Table 12-2.

A TLB lookup operation is initiated by the write to TR6, and TR7 will indicate the result of the lookup operation following the write to TR6. The hit/miss indication can be found in TR7 bit 4 (see Table 12-5).

TR7 will contain the following information if bit 4 indicated that the lookup test resulted in a hit. Bits 2–3 will indicate in which set the match occurred. The 22 most significant bits in TR7 will contain the physical address and page attributes contained in the entry. Bits 9–7 will contain the LRU bits associated with the accessed set. The state of the LRU bits is previous to their being updated for the current lookup.

If bit 4 in TR7 indicated that the lookup test resulted in a miss, the remaining bits in TR7 are undefined.

Again it should be noted that a TLB testability lookup operation affects the state of the LRU bits. The LRU bits will be updated if a hit occurred. The entry that was hit will become the most recently used.

Tristate Output Test Mode

The 486 microprocessor provides the ability to float all its outputs and bidirectional pins. This includes all pins floated during bus hold as well as pins that are never floated in normal operation of the chip (HLDA, BREQ, FERR#, and PCHK#). When the 486 microprocessor is in the tristate output test mode, external testing can be used to test board connections.

The tristate test mode is invoked by driving FLUSH# low in the clock prior to RESET going low (see Figure 12-8). The 486 microprocessor remains in the tristate test mode until the next RESET.

68332 Microcontroller Test Submodule

The test submodule is a primary tool to support all types of testing, such as production test and user self-test, that are integrated into the MCU. The test submodule supports scan-based testing of the various modules in the MCU. The scan test employed here consists of the test submodule performing the following steps:

1. Serially shifting stimulus data to an idle module under test (MUT)
2. Activating the MUT
3. Serially shifting response data back from the MUT
4. Latching the response data for interrogation by the bus master

Figure 12-9 is a block diagram of the test submodule.

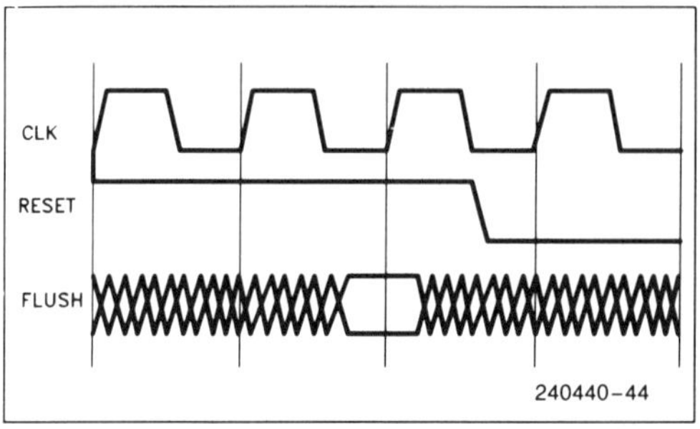

FIGURE 12-8
Tristate test. Reprinted courtesy of Intel Corporattion.

The test submodule serves as a focal point in the test process, applying stimuli, controlling the test sequence, and collecting responses. Communication between the test submodule and the bus master is always 16-bit parallel, reducing the test pattern length to the MCU or external test fixture serving as bus master. Stimulus data can be generated by either the bus master or the test submodule. The response data is always read by the bus master for pass/fail determination. Communication between the test submodule and the module under test is always serial, and responses from the previous stimulus can be shifted concurrently with the next stimulus.

The test submodule can apply two different types of stimuli to the module under test: deterministic and random. Deterministic data is developed either manually or by a test program for application to a specific circuit implementation. This deterministic data is written to the test submodule by the bus master and then serially shifted to the module under test. Pseudorandom stimulus data can be generated directly by the test submodule and shifted to the module under test. This type of pattern is commonly used for exhaustively testing certain sections of the module under test.

The response from either type of stimulus can be collected in either compressed or uncompressed form by the test submodule. Compressed form means the responses from the module are reduced by polynomial division (signature analysis). The bus master then need only periodically interrogate the test submodule to obtain the signature latched there. Uncompressed form means the unmodified response is latched by the test submodule, and the bus master must interrogate the test submodule after each response by the module under test.

Modes of Operation

The test submodule has three modes of operation: manual, automatic, and continuous. All three modes are useful in production test, user test, and system diagnostics.

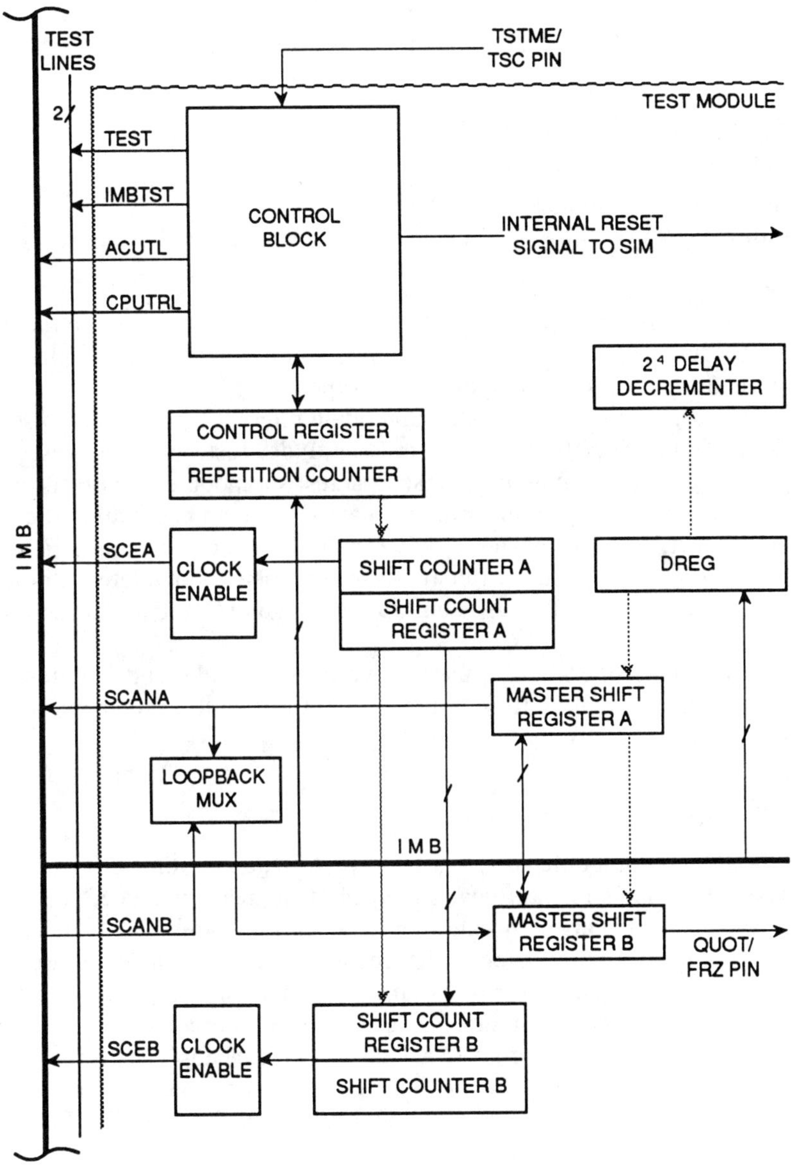

FIGURE 12-9
Block diagram of 68332 test submodule. Reprinted with permission of Motorola, Inc.

In manual mode, a stimulus is applied to the module under test and the response to this stimulus is latched in the test submodule. The bus master must communicate with the test submodule after each incremental step of the test process. The incremental steps consist of the following:

1. Writing the test stimulus into the test submodule
2. Initiating the shifting operation
3. Activating the module under test
4. Interrogating the response data

These steps are repeated until all stimuli have been applied to the module under test and all responses have been interrogated. The manual mode supports all stimulus and response data types.

In automatic mode, the test submodule internally generates a pseudo-random stimulus, shifts it into the module under test, and compresses the response from the module under test. The test submodule also performs the incremental steps of initiating stimulus data shifting and activating the module under test. The bus master is only required for setup, test initiation, and interrogation of test results after the test sequence is complete. Automatic mode is typically used with pseudo-random stimuli and compressed responses.

Continuous operation is used to verify the clocks and counters that have the capability of shifting out the output, by shifting their output into master shift register B and compressing it.

Capabilities

The test submodule supports serial pattern generation, serial signature analysis, and manual or automatic control of the test sequence.

Master shift register A can operate as either a simple shift register or a linear feedback shift register configured as a serial pattern generator. Master shift register B can operate as either a simple shift register or a linear feedback shift register configured as a serial signature analyzer.

Using manual test control, a bus master can apply patterns to any circuit accessible by the scan paths. Figure 12-10 shows the typical steps required for manual operation. Although the flow illustrated is for deterministic stimuli and uncompressed responses, manual control does support the use of all types of stimuli (deterministic and pseudo-random) and responses (compressed and uncompressed). Automatic test sequences can be initiated by either the CPU or an external test system. Figure 12-11 illustrates the typical steps required for automatic test operation. Although the flow illustrated is for pseudo-random stimuli and compressed responses, automatic control can be used for any stimuli and response type, if desired.

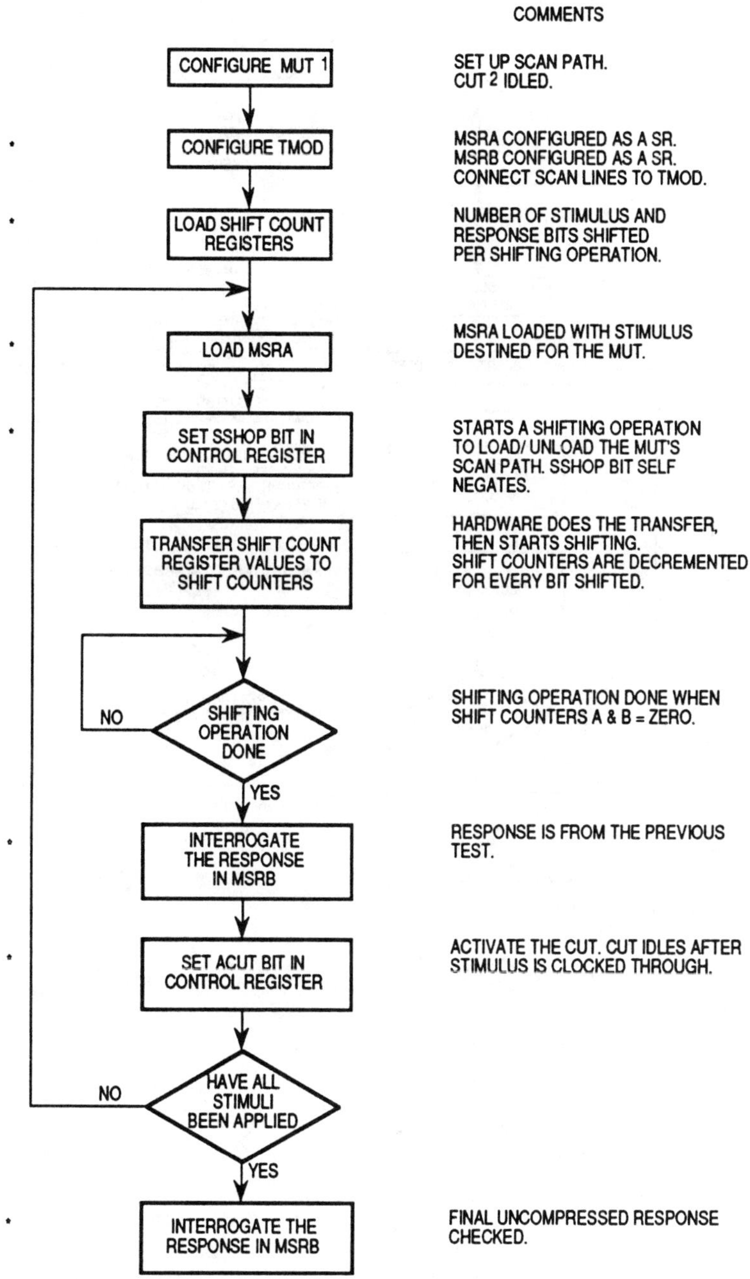

* = STEPS WHERE BUS MASTER MUST READ AND/ OR WRITE THE TMOD.
1 = MODULE UNDER TEST
2 = CIRCUIT UNDER TEST

FIGURE 12-10 68332 manual mode test flow. Reprinted with permission of Motorola, Inc.

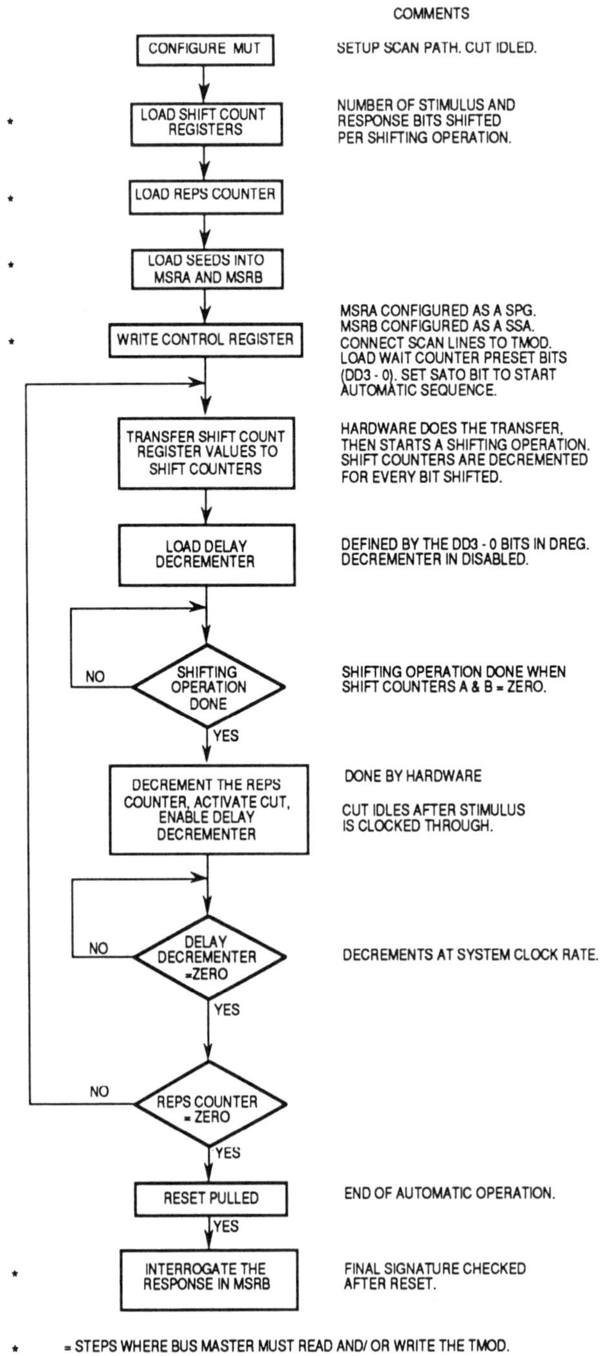

FIGURE 12-11 68332 automatic mode test flow. Reprinted with permission of Motorola, Inc.

Entering Test Mode

Test mode is entered by a combination hardware and software method: a register bit must be set while an external pin is in the correct state. To enter test mode, the following conditions must be met:

The TSTME pin (test mode enabled—active low) must be pulled low.

The ETM bit in the test submodule control register (CREG) must be set. This is a write-once bit, so it must not have been written since reset, and it must be set to one on the first attempt to write it. Writing the bit to zero in initialization software prevents any accidental entry to test mode in normal operation.

The TSTME pin must remain low to stay in test mode. If TSTME goes high, the MCU is reset, exiting test mode.

The TSTME pin has a second function, when driven to 1.4 times V_{DD}, the MCU places all output driver circuits in a high-impedance state, isolating the MCU from the remainder of the system.

Test Submodule Control Register (CREG)

The test submodule control register configures the test submodule for the test operation desired, provides the bus master with the means for controlling the test sequence, and retains status information on the test submodule.

CREG $YFFA38

	15	14	13	12	11	10	9	8	7	6	5	4	3	2	1	0
	BUSY	TMARM	COMP	IMBTST	CPUTR	OBIT	MUXEL					ACUT	SCONT	SSHOP	SATO	ETM
RESET:	1	TSTME	U	0	0	0	0	0	0	0	0	0	0	0	0	0

ETM—Enter Test Mode

1 = Enter test mode.
0 = Stay in normal mode.

This bit can be written only once after reset. It can be set only if the TSTME pin is asserted. This bit can be read anytime.

SATO—Start Automatic Test Operation

1 = Start an automatic test operation.
0 = Stay in normal mode.

The test submodule and module under test must be properly set up before setting this bit. This bit is cleared when the reps counter decrements to zero. Clearing this bit to stop an automatic operation is not recommended. This bit can be read and written at any time.

SSHOP—Start Shifting Operation

1 = Start in a shifting operation.
0 = Stay in normal mode.

The shifting operation ends and SSHOP is cleared when both shift counter A and shift counter B equal zero. If both this bit and the SATO bit are set at the same time, a shifting operation will still start. Clearing this bit does not stop a shifting operation and is not recommended. This bit can be read and written at any time.

SCONT—Start Continuous Operation

1 = Start continuous operation.
0 = Stop continuous operation.

When this bit is set, the test submodule continuously shifts data into master shift register B. Continuous operation ends when the bit is cleared. This bit can be set simultaneously with SSHOP and SATO, and a continuous operation will start. This bit can be read and written at any time.

ACUT—Activate Circuit Under Test

1 = Assert the ACUTL line.
0 = Stay in normal mode.

Setting this bit enables an idle module under test to run for the period defined by the configuration of the module under test. This bit is automatically cleared in less than one bus write cycle so that the same module can run consecutive cycles without this bit being cleared between cycles. This bit can be written at any time, but it always reads zero since it clears in less than a bus cycle.

MUXSEL—Multiplexer Select Bit

1 = Shift in source for MSRB is the external interrupt pin.
0 = Shift in source for MSRB is the internal test line.

The SIM must be correctly configured to shift in external data. This bit can be written only in test mode.

QBIT—Quotient Bit

1 = The least significant bit of master shift register B is available at the QUOT/FRZ pin.
0 = The internal freeze status is available at the QUOT/FRZ pin.

Monitoring the LSB of master shift register B when using it as a signature analyzer reduces the possibility of aliasing. As a shift register, the LSB can be used to monitor uncompressed data on a bit-by-bit basis if development problems prevent a normal read. This bit can be read and written only in test mode.

CPUTR—CPU Test Register

1 = Scan lines connected to the CPU test register.
0 = Scan lines disconnected from the CPU test register.

With this bit set, the test submodule can change the contents of the CPU test register and/or scan the test register contents to the test submodule. This bit can be written only in test mode, but it can be read at any time.

IMBTST—Intermodule Bus Test

1 = Internal interconnect lines are configured as test lines. The output of master shift register A is connected to the SCANA line, and the SCANB line is connected to the input of master shift register B.
0 = Internal interconnect lines have normal function. Master shift register A output is connected to master shift register B input to test the test submodule. This bit can be read and written only in test mode.

COMP—Compare Status Bit

1 = Master shift register B contains the correct answer for the user self-test basic test.
0 = Master shift register B does not contain the correct answer for the user self-test basic test.

This status bit is only useful at the end of the basic test to determine whether the correct signature was generated. This status bit can be read at any time, but it cannot be written.

TMARM—Test Mode Armed Status Bit

1 = TSTME pin is asserted; test mode can be entered by setting the ETM to control it.
0 = TSTME pin is negated; test mode cannot be entered.

This status bit can be read at any time, but it cannot be written.

BUSY—Test Submodule Busy Status Bit

1 = Test submodule is busy.
0 = Test submodule is not busy.

The test submodule is busy when an automatic test operation, shifting operation, or continuous operation is in progress. This status bit can be read at any time, but it cannot be written.

Distributed Register (DREG)

The distributed register configures the master shift registers for either pseudo-random or normal shift register operation. It is also used to access the upper 3 bits of each master shift register and the wait counter preset value.

WAIT3-WAITI—Wait Counter Preset 3–1

These bits program the delay time between automatic test sequences. This delay time allows the circuit under test (CUT) to respond to the ACUTL line. The next shifting operation cannot begin until the delay period is over. These bits can be read or written at any time. The following sections list the value of these bits.

MSRA18-MSRA16—Master Shift Register A Bits 18–16

These bits are the 3 most significant bits of master shift register A that are used only in serial pattern generator mode. These bits can be read or written at any time.

MSRAC—Master Shift Register A Configuration

1 = Master shift register A is configured as a 19-bit serial pattern generator.
0 = Master shift register A is configured as a 16-bit shift register.

This bit can be read or written at any time.

MSRB18-MSRB-16 Master Shift Register B Bits 18–16

These bits are the upper 3 bits of master shift register B that are used only in serial signature analyzer mode. These bits can be read or written at any time.

MSRBC—Master Shift Register B Configuration

1 = Master shift register B is configured as a 19-bit serial signature analyzer.
0 = Master shift register B is configured as a 16-bit shift register.

This bit can be read or written at any time.

Master Shift Register A (MSRA)

Master shift register A is either a 16- or 19-bit multifunction register that can be read and written by the bus master. It contains the stimulus to be transferred from the test submodule to the module under test. Master shift register A can function as either a simple shift register or as a serial pattern generator, determined by the MSRAC bit in the distributed register. Reset selects the shift register function but does not affect the contents of master shift register A.

When MSRA is operating as a shift register, the bus master writes a 16-bit stimulus to it. This data is then shifted into the module under test using manual mode.

When MSRA is operating as a serial pattern generator, it functions as a 19-bit linear feedback shift register with the upper 3 bits accessible through the distributed register. The bus master must write an initial value (seed) into MSRA prior to starting either a manual or automatic operation. The stimulus can be manually shifted by setting the SSHOP bit or automatically shifted by setting the SATO bit.

Whether operating as a shift register or a serial pattern generator, the number of bits shifted into the module under test is controlled by the value that the bus master loads into shift count register A.

Shift Count Register A and Shift Counter A

Shift count register A is an 8-bit shift register that can be read and written by the bus master. Shift counter A is an 8-bit counter that is loaded by shift count register A and not accessible to the bus master. After reset, both shift count register A and shift counter A are initialized to zero.

The value loaded into shift count register A by the bus master determines the number of stimulus bits shifted from master shift register A to the module under test for each shifting operation. When a shifting operation begins, the value in shift count register A is loaded into shift counter A. Shift counter A is decremented for each bit shifted out of master shift register A, which occurs at the system clock rate. While shift counter A has a nonzero value, the scan clocks are enabled and the SCEA line is asserted, allowing the scan paths in the module under test to shift using their own local clocks.

Stimulus data is transferred from master shift register A to the module under test over the SCANA line. When shift counter A reaches zero, the following events occur:

Shift counter A stops decrementing.

Master shift register A stops shifting out stimulus data.

The SCEA line is negated to disable scan clocks in the module under test. The test submodule is notified that the shifting operation is complete.

Master Shift Register B (MSRB)

Master shift register B is either a 16- or 19-bit multifunction register that can be read and written by the bus master. It collects the response data shifted from the module under test to the test submodule. Master shift register B can function either as a 16-bit shift register or as a 19-bit serial signature analyzer. These functions are selected by the MSRBC bit in the DREG. Reset selects the shift register function but does not affect the contents of master shift register B.

As a simple shift register, test responses are manually shifted to master shift register B from the module under test. This uncompressed data can then be read 16 bits at a time by the bus master.

As a serial signature analyzer, master shift register B is a 19-bit linear feedback shift register with the upper 3 bits accessible through the DREG. Responses from the module under test are shifted in and compressed. After the test sequence is complete, the 19-bit result is read by the bus master to determine a pass/fail condition. If the signature is the final result of the user self-test basic test, pass/fail condition can be determined by reading the COMP bit in the CREG. The test sequence can be under manual or automatic control.

Whether operating as a shift register or a serial signature analyzer, the number of bits shifted from the module under test is controlled by the value that the bus master loads into shift count register B.

Shift Count Register B and Shift Counter B

Shift count register B is an 8-bit shift register that can be read and written by the bus master. Shift counter B is an 8-bit counter that is loaded by shift count register B, and it is not accessible to the bus master. After reset, both shift count register B and shift counter B are initialized to zero.

The value loaded into shift count register B by the bus master determines the number of response bits shifted from the module under test to master shift register B for each shifting operation. When a shifting operation begins, the value in shift count register B is loaded into shift counter B. Shift counter B is decremented for each bit shifted into master shift register B, which occurs at the system clock rate. While shift counter B has a nonzero value, the scan clocks are enabled and the SCEB line is asserted, allowing the scan paths in the module under test to shift using their own local clocks. Response data is transferred from the module under test to master shift register B over the SCANB line. When shift counter B reaches zero, the following events occur:

Shift counter B stops decrementing.

Master shift register B stops shifting in response data.

The SCEB line is negated to disable the module under test scan clocks.

The test submodule is notified that the shifting operation is complete.

Reps Counter

The reps counter is a 16-bit down counter that determines the number of pseudo-random vectors that are generated in the automatic mode of operation. In the manual mode of operation, the reps counter is not used. After reset, the reps counter is set to zero.

After loading the shift counter registers, the bus master loads the number of vectors to be generated by master shift register A into the reps counter. The bus master then initiates an automatic test sequence. A shifting operation is performed, and the reps counter is decremented. If the reps counter value is nonzero, the test submodule activates the module under test by asserting ACUT. The automatic sequence, consisting of shifting, decrementing the counter, and activating the module under test, is continued until the reps counter reaches zero. After reset, the bus master reads the compressed test result from master shift register B, which concludes the automatic operation.

Wait Counter

The wait counter generates a delay time between automatic test sequences. This time is set using the WAIT3–WAIT1 bits in the DREG. The wait counter is not used in the manual mode of operation.

The wait counter delay allows a variable amount of time for the module under test to respond to the stimulus before the responses are shifted into the test submodule. The delay starts when the ACUTL line is asserted and ends when the delay count is decremented to zero. The next shifting operation cannot begin until the delay period is over.

Test Lines

Internally there are eight lines that are used for scan testing. Table 12-6 lists the lines and their function.

TABLE 12-6
Test Lines

Test Line	*Test Function*
TEST	Signals test mode status to all modules
IMBTST	Enables the test lines
SCANA	Routes serial stimulus from master shift register A to the module under test
SCANB	Routes serial responses from the module under test to master shift register B
SCEA	Enables shifting of stimulus from master shift register A to the module under test
SCEB	Enables shifting of responses from the module under test to master shift register B
CPUTRL	Permits the test module to scan the CPU test register
ACUTL	Activates an idle module under test

80860(i860) Self-Test Features

The i860 has a boundary-scan mode that may be used in component- or board-level testing to test the signal traces leading to and from the microprocessor. A simple serial interface allows the testing of all signal traces with only seven probe connections. These are CLK, BSCN, SCAN, SHI, BREQ, RESET, and HOLD signals. With this configuration the user can apply test vectors to the system via a serial channel and sample the response.

The CLK input determines the execution rate and timings of the i860 microprocessor. The timings of the other signals are specified relative to the rising edge of this signal. In addition, the clock signal determines the shift-in rate of the test vectors and the shift-out rate of the response. The testability pins are the Boundary Scan Shift Input (SHI), the Boundary Scan Enable (BSCN), the Shift Scan Path (SCAN), and the Boundary Scan Shift Output, which is shared with the Bus Request (BREQ) output. The BREQ pin has two functions. In normal processor operations, the BREQ line is asserted when the i860 microprocessor has a pending memory request, even when the HLDA line is asserted. In the boundary-scan mode it is the serial shift-out pin. The Bus Hold (HOLD) line has a specific function to be described later in this chapter. The RESET line causes initialization of the processor.

When the Boundary Scan Enable (BSCN) signal is asserted, the i860 microprocessor enters the boundary-scan mode on the next rising edge of CLK. When BSCN is deasserted while in the boundary-scan mode, the i860 microprocessor leaves the boundary-scan mode on the next rising edge of CLK. After leaving the boundary-scan mode the internal state is unidentified and RESET must be asserted. The timings to enter and exit the boundary-scan mode are shown in Figure 12-12.

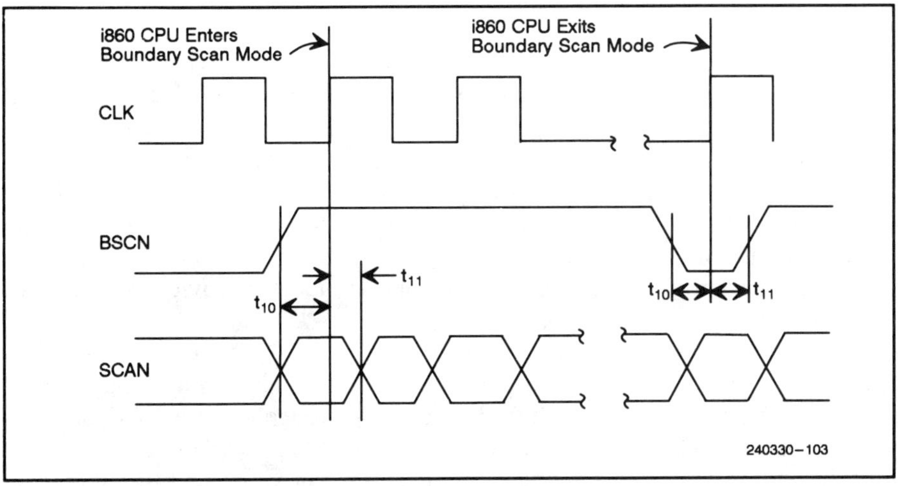

FIGURE 12-12
Entering and exiting the boundary scan mode. Reprinted courtesy of Intel Corporation.

The BSCN signal configures the i860 for the test mode. When in this mode, the processor can operate in a normal mode or a shift mode. The Shift Scan Path (SCAN) causes the processor to assume one or the other. Test mode operations are defined in Table 12-7. The normal mode is entered on the rising clock edge when the SCAN line is deasserted. The shift mode is entered on the rising clock when the SCAN line is asserted. In the normal mode, the output signals are driven, and the inputs are sampled simultaneously. In the shift mode, the test vectors are shifted at one end of the chain, and the response is shifted out simultaneously at the other end. In testing operations the user switches from the shift modc to the normal mode and vice versa until all of the vectors are exercised. At this point the user exits from the boundary-scan mode.

TABLE 12-7
Test Mode Selection

BSCN	*SCAN*	*Testability Mode*
LO	LO	No testability
LO	HI	Intel Reserved
HI	LO	Boundary Scan Mode, Normal
HI	HI	Boundary Scan Mode, Shift SHI as input, BREQ as output

Reprinted courtesy of Intel Corporation.

Shift Mode Operation

To transfer vectors to the system and to read the responses from this system requires a serial path to the microprocessor. This path is provided in shift mode operation. In effect, the shift mode creates a long serial register to and from the microprocessor. Data shifted into the CPU corresponds to new test vectors that will be output to the system. A response from the system can be shifted out while in the shift mode. When switching to and from the normal mode, the i860 microprocessor performs either a serial-to-parallel operation or a parallel-to-serial operation. This provides an effective and simple way to test the system.

In the shift mode, the pins shown in Table 12-8 are organized as a boundary-scan chain. As mentioned above, this chain can be thought of as a long shift register that is shifted on the rising edge of CLK. The order of the boundary-scan chain is shown in Figure 12-13. The SHI pin receives the input on one end of the boundary-scan chain. The other end (the most significant) of the boundary-scan chain is the BREQ pin, shown here in the 127th position. The bits that are shifted in through the SHI pin in the serial mode are in the sequence shown in Figure 12-13. For shift mode operation, BREQ is shifted in the order BE0#, BE1#, ..., BSCN, SHI. A total of 127 cycles is needed. While these values are being shifted in, serial bits are

TABLE 12-8
Test Mode Latches

Input Latch	*Output Latch*	*Associated Control Latch*
SHI		
BSCN		
SCAN		
RESET		
D0-D63	D0-63	DATAt
CC1-CC0		
	A31-A3	ADDRt
	NENE#	NENEt
	PTB#	PTBt
	W/R#	W/Rt
	ADS#	ADSt
	HLDA	
	LOCK#	LOCKt
READY#		
KEN#		
NA#		
INT/CS8		
HOLD		
	BE7#-BE0#	BEt
	BREQ	

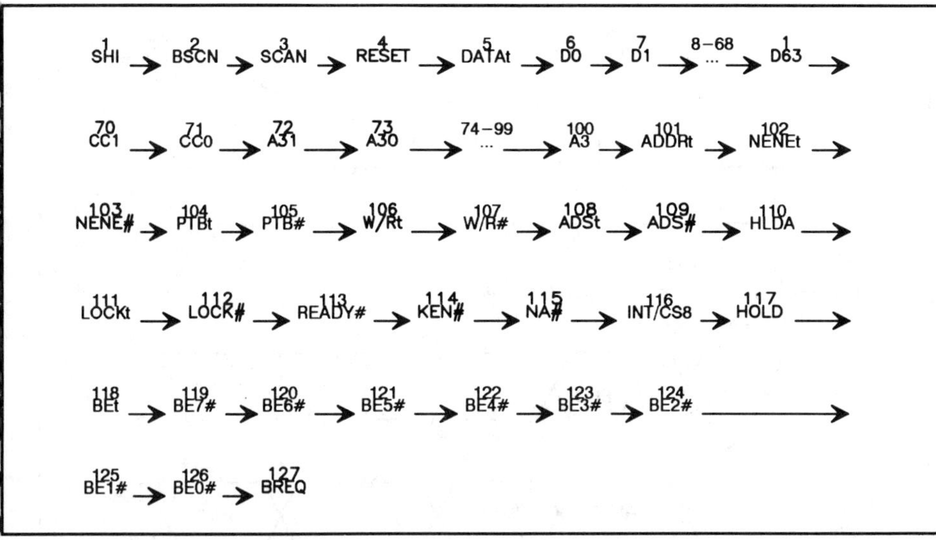

FIGURE 12-13
Order of the boundary scan chain. Reprinted courtesy of Intel Corporation.

coming out of the BREQ pin. These bits are in the order BREQ, BE0#, BE1#,..., BSCN, and SHI. They correspond to the response of a previously loaded vector. To avoid glitches that could occur while the values are being shifted out along the chain, the RESET and HOLD pins must be asserted. All of the tristateable outputs will be disabled when these pins are asserted. Timings related to serially shifting the data into and out of the processor are shown in Figure 12-14.

The shift mode is used to shift in a new test vector, as well as to read the response to the previous test vector. These operations are performed concurrently.

Normal Mode

During normal mode, two operations can occur simultaneously. The first operation is the driving of the output lines with the test vector, while the second operation is the sampling of the response on the input lines. The outputs and inputs are shown in Table 12-8. The tristate output pins A31–A3, BE7#–BE0#, W/R#, NENE#, ADS#, LOCK#, and PTB are enabled by storing a high value in the corresponding control latches ADDRt, BEt, W/Rt, NENEt, ADSt, LOCKt, and PTBt, respectively. If the corresponding output control latch contains a low value, the pins are tristated.

The data pins D63–D0 are I/O pins and are enabled by the control latch, DATAt, which is similar to the other control latches. In addition,

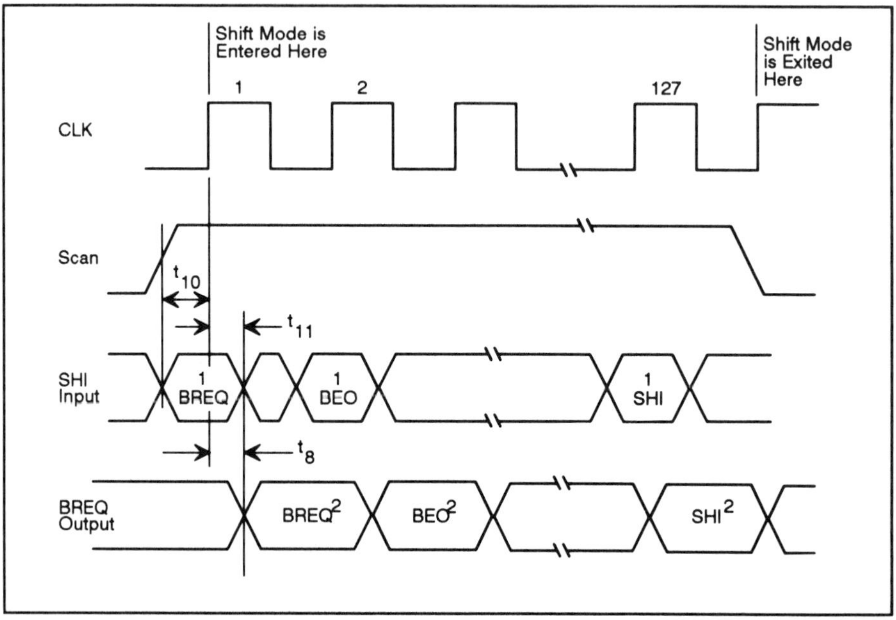

FIGURE 12-14
Shift mode of the boundary scan mode. Reprinted courtesy of Intel Corporation.

when the DATAt contains a low value, the data pins are configured as inputs and the values on the pins are sampled.

For each of the input pins RESET, HOLD, INT/CS8, NA#, READY#, KEN#, SHI, BSCN, SCAN, and CC1-0, the corresponding latch is loaded with the value that is being driven into the pin.

Normal mode is selected when the SCAN is deasserted while BSCN is asserted. All of the signals shown in Table 12-8 are loaded serially in the shift mode (while BSCN and SCAN are high). The timings for normal mode operations are shown in Figure 12-15.

Test Sequence Example

The following steps relate to a typical test session:

a. Enter the boundary-scan mode.
b. Establish a serial link via the shift mode to shift in the test vectors and to sample the response. Data can be shifted into and out of the i860 CPU. Values are assigned to the latches that correspond with the pins, as shown in Table 12-7. The first test vector is then transferred to the i860 microprocessor.

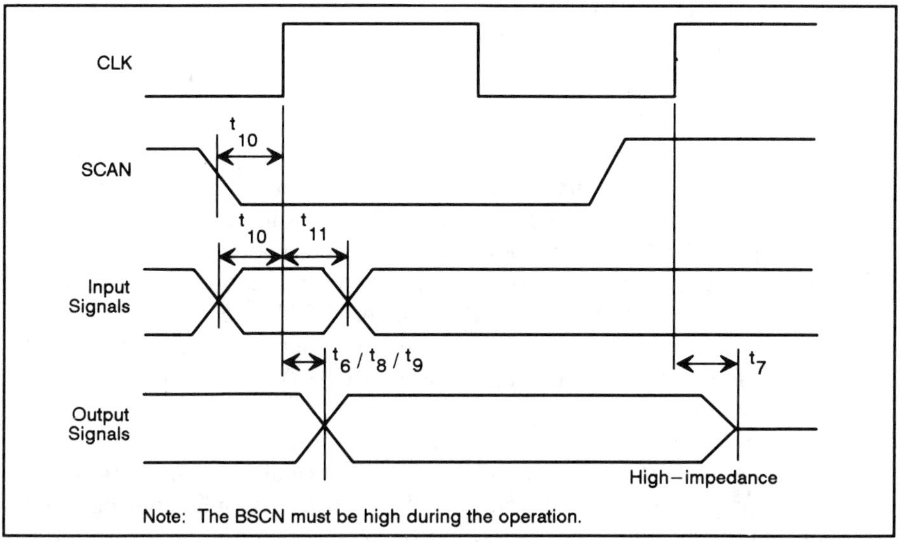

FIGURE 12-15
Normal mode of the boundary scan mode. Reprinted courtesy of Intel Corporation.

c. Enter the normal mode. The processor drives the output and also samples the inputs. This portion of the sequence sends a test vector to the system and also reads the response synchronously to the rising edge of CLK.
d. Reenter shift mode and read the response from the output of the serial chain, while loading new values for the next test vector at the input of thc serial chain. It is possible to loop between steps c and d until finished.
e. Exit the boundary-scan mode and reset the processor.

While running the test sequence, the user must be aware that the response being received during the normal mode may correspond to the test vector presented in the previous cycle. This is true if a scan path also exists in the peripheral circuits.

Using the Boundary-Scan Mode of the i860

As indicated previously, the boundary-scan mode can be useful in both component- and board-level testing.

At the component level in incoming testing, the serial scan device uses a serial channel to provide a means for causing the output pins to toggle. Values can be assigned to the output latches in the shift mode. During normal mode operation, the processor will drive the output pins. The values of the input pins can also be latched in normal mode and read out in the shift mode.

This provides a mechanism for monitoring the inputs and toggling the output via a serial channel and may reduce test program development time.

Board-Level Testing

A microprocessor-based system consisting of peripherals and memory elements is shown in Figure 12-16. When the system is being debugged and brought up, the serial path can set up test vectors to various memory elements or peripherals. The response from the peripherals can be accessed following the propagation delay while in the normal mode. After that, the response is serially shifted out in the shift mode. The received response is compared to the expected response. After the user knows that the peripherals are functioning properly, code from an EPROM can be used for more extensive testing.

The peripherals or the memory elements must observe the setup and hold times of the i860 microprocessor during normal mode operation. This can be accomplished by giving multiple clocks during normal mode operation to meet the delay timing of the longest path. A typical timing diagram for a memory access is shown in Figure 12-17. For example, if the longest delay path is four clock cycles, by asserting the normal mode for five clock cycles, the address and other controls are presented to the memory element, after which the data can be read or written. The latching of the signals in the i860

FIGURE 12-16
A typical i860 system with diagnostics capabilities. Reprinted courtesy of Intel Corporation.

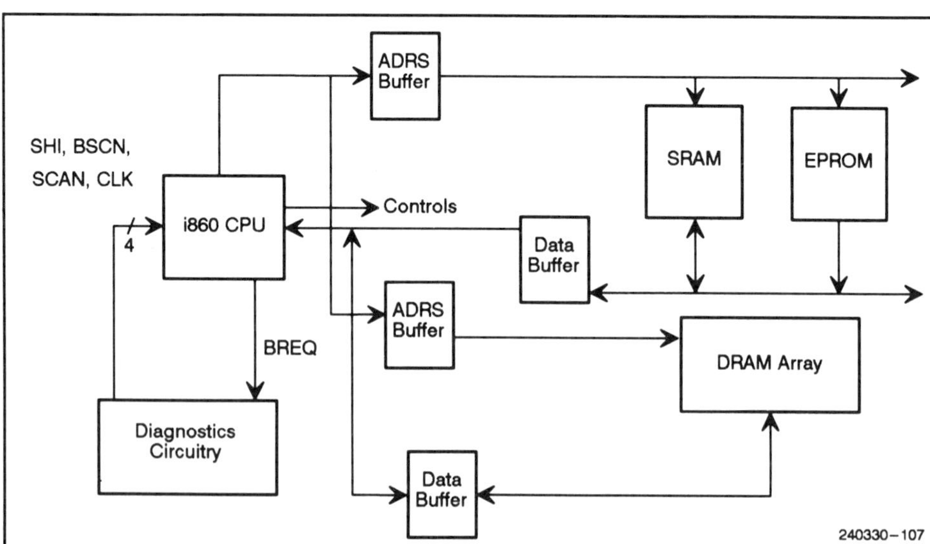

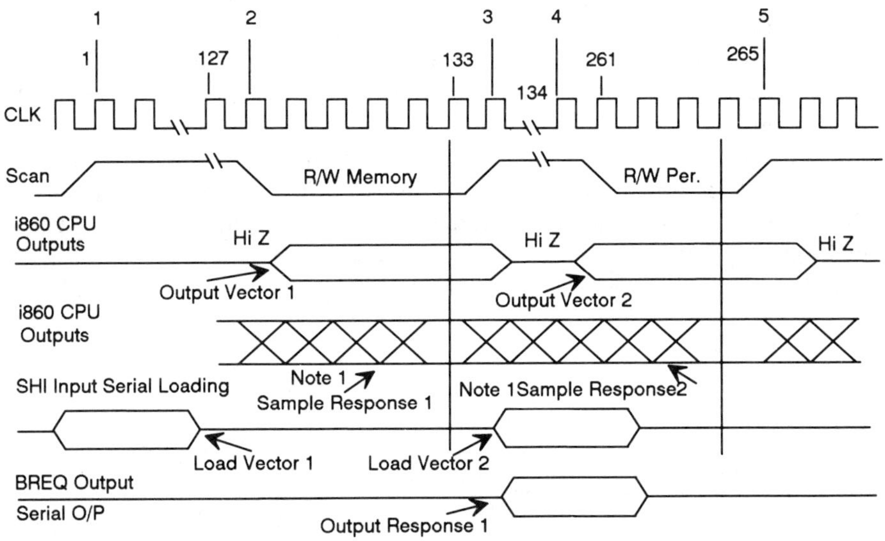

TIMESLOT 1–2 LOAD TEST VECTOR SERIALLY–SHIFT.
2–3 OUTPUT THE TEST VECTOR, AND MONITOR THE INPUTS. THE INPUTS ARE LATCHED ON EVERY CLOCK CYCLE FROM CYCLE 128 TO CYCLE 133. THE EXTERNAL DEVICES MUST HAVE THE DATA SET UP AROUND THE RISING EDGE OF CLOCK NO. 133. THE RESPONSE TO TEST VECTOR 1 IS LATCHED FROM THE INPUT PINS.
3–4 LOAD IN TEST VECTOR 2 SERIALLY, AND SERIALLY SHIFT OUT THE RESPONSE FROM TEST VECTOR 1.
4–5 OUTPUT TEST VECTOR 2, AND MONITOR THE INPUTS. THE NUMBER OF CYCLES THAT SCAN IS IN THE LOW STATE IS SHORTER THAN THAT OF TIMESLOT 2–3 AND THIS IS BECAUSE THE EXTERNAL DEVICE HAS A FASTER ACCESS TIME. THE RESPONSE IS LATCHED ON CLOCK CYCLE 261 TO 265.
NOTE 1 INPUTS ARE SAMPLED IN NORMAL SCAN MODE ON EVERY RISING EDGE OF CLK.

240330–108

FIGURE 12-17
Typical timing for the serial scan mode. Reprinted courtesy of Intel Corporation.

CPU can be carried out reliably on the last clock edge, before the shift mode operations are performed.

RAM and other circuits external to the i860 microprocessor peripherals may consist of state machines. The user must ensure that the state machines are enabled or disabled at the appropriate time during the boundary-scan mode. One way to ensure this is to have separate clocks for the on-board state machines.

If the peripheral or memory elements have a serial scanning scheme as well, the clock can be asserted for as briefly as one cycle during normal mode. This implies that the system will be responding to the test vectors of the previous cycle. The timings should be similar to those of Figure 12-17 with a smaller number of clock pulses for time slots 2–3 and 4–5. Again, this assumes that the peripherals have a serial scan chain as well. This method reduces the number of clock pulses for the normal mode. However, the overhead in the number of external components is larger than in the first method.

Glossary

Analysis Pod: Unit containing emulation components for application to microprocessor or memory analysis.

Application Binary Interface: A collection of object or binary code standards.

Assembler: A program that translates assembly-language source code into object code.

Background Monitor: Monitor present in a background or shadow memory to control microprocessor emulation in the target system.

Bed of Nails: A test technique in which the PCB to be tested is pressed on "elastic" nails arranged in front of the testing points of the PCB.

Big Endian: Byte ordering or addressing scheme to set the format for sending data to a microcomputer. The most significant byte is sent first.

Boundary Scan: A testing technique that allows a circuit to be tested at the board edge connector. A shift register, included on every chip, is used to shift, apply, or capture test data. Thus both the individual chips and the board interconnect can be tested.

Bug: An error in a program.

Bus: The physical channel over which electrical signals are transferred between the components of a microcomputer or a system of computers. Protocol rules govern the transfer.

Cache: A small buffer memory located close to the processor to give it speedy access to the information (copied from a much larger main memory) that the processor is likely to request next.

Clock Stretcher: A circuit to provide extra access time. The low portion of the clock is extended as required to complete access.

Compiler: A program that translates high-level language source code into object code.

Context: The internal state of the processor that defines the current execution. It includes the contents of the various execution pipelines and the state of the register file.

Control Register: The register that controls processor execution, provides processor status, or holds context information during an exception.

Coprocessor: A device that performs specialized processing in conjunction with the main microprocessor of the system.

Cross-Assembler: A program that makes possible the development of an assembly-language program on a host computer having a different architecture than the microprocessor or microcontroller for which it is being developed.

Cross-Compiler: A program that makes possible the development of a high-level language program on a host computer having a different architecture than that of the microprocessor or microcontroller for which it is being developed.

Data Bus: A bus that carries data from memory sections and distributes it to other areas of the computer system.

Debug Register: A register that provides on-chip support for debugging. A debug control register is used to set the breakpoints, and a debug status register displays the current state of the breakpoints.

Debugger: A program, generally interactive, that aids a programmer in debugging programs by providing facilities such as breakpoints, single-stepping, and access to the contents of variables.

Design for Testability: Techniques that involve designing chips with self-testable circuits or providing for testing with scan techniques.

Emulation: A technique by which a microprocessor or memory in a target system is replaced by a device that links the target system to an external controller, ranging from a PC to a mainframe.

Exception Handler: A software routine that executes when an exception occurs. Normally, the exception handler will correct the condition that caused the exception or perform some other meaningful task, such as aborting the program that caused the exception. Exception handlers are defined by a two-word exception vector, which is branched to automatically when an exception occurs.

Fault: A logical memory error that causes an exception. Faults are caused by invalid memory descriptors or protection violations. Both situations can be corrected by software.

Ferret: A symbolic debugger for C language programs written for the target systems of certain AT&T microprocessors.

File: A collection of data that may be read or written by an operating system command, assembler, compiler, or other programs.

Floating Point: Description of a data type used in computers in which an exponent specifies the location of a floating radix point.

Foreground Monitor: Monitor residing in the target processor's directly addressable memory space along with the user program to control microprocessor emulation in the target system.

Forth: An application-oriented computer language.

General-Purpose Register: As distinguished from control and internal registers, a general-purpose register provides the source operands and destination results for data manipulation instructions. Load instructions move data from memory to registers, and store instructions move data from registers to memory.

Glitch: Any even number of signal transitions between two of the logic analyzer's sample points.

Hardware Breakpoint: A method of halting execution when data is read, written, or executed from a particular address.

Harvard Architecture: A computer design featuring separate memory access points for instructions and data.

Host: The computer that runs the development system.

In-Circuit Debugger: An in-circuit monitor that, coupled with a logic analyzer, provides for hardware and software debugging.

In-Circuit Emulator: A module, composed of hardware and firmware, which replaces a microprocessor or memory to control either function in a target system.

Integer Unit: The functional unit in a processor for executing all instructions except floating-point, integer multiply and divide, and memory access instructions.

Interactive Monitor Program (IMP): An assembly-level, hex-oriented debugger contained in EPROM on the analysis pod.

Interrupt: A signal triggered by an external event that causes a microprocessor CPU to stop what it is doing, perform a special routine to handle the interrupt, and resume the interrupted processing where it left off.

Interrupt Handling: A routine that is invoked when an interrupt occurs. It performs whatever processing is necessary to take care of the event that caused the interrupt.

Library: A file that contains object modules and is created and maintained by a library utility.

Linker: A program that combines object modules.

Little Endian: Byte-ordering or addressing scheme setting the format for sending data to a microcomputer. The least significant byte is sent first.

Loader: A program that loads object modules into memory so that they are ready to run.

Memory Management Unit: A hardware unit or component that translates virtual addresses as seen by software into physical addresses as seen by hardware.

Memory Mapping: The movement of emulation or overlay memory to target memory space, which has no memory present.

Microcontroller: A single-chip device containing a core CPU plus the RAM and ROM required to accomplish a desired result in a product or system. Clock generation and I/O circuits are also on board the same chip.

Microprogram: The program in some microprocessors that defines their instruction set.

Multitasking: The capability of a computer system to run more than one task at the same time.

Operating System: A collection of programs that provide the functional environment in which other programs work. It is the functional environment in which assemblers, compilers, locators, linkers, in-circuit emulation software, and other user programs run. Examples are DOS and UNIX.

Optimized Compiler: A high-level language translator, which analyzes the input program and uses the analysis results to devise optimal code production strategies. Optimized compilers are especially important for the development of programs for RISC microprocessors.

Page: The smallest managed unit of a virtual memory scheme.

Pipelining: A technique that breaks instruction execution into distinct steps so that multiple steps can be performed at the same time. While one instruction is executing, the next instruction is being decoded and the next is being fetched from the memory.

Real-Time Trace Buffer: Special high-speed memory into which instructions (and, possibly, addresses and data) can be stored in real time, as they are executed. When execution is halted, a history of how execution got there is available.

Register Window: An arrangement that permits parameter passing from the calling to the called procedure, without actually moving these parameters from location to location.

Remote Debugger: A software debugger that runs on the host system but debugs code running on the target system.

Simulation: As applied to the development of microprocessor-based systems, a software approach to coding and debugging the program before the hardware is built. It is also used by hardware and software designers to download and debug the software on a bare machine in order to bring up a totally functioning system immediately after the first hardware is built.

Source Code: A program written in assembly language or a high-level language before it is assembled or compiled into object code.

State Analysis: Analysis of the execution of instructions in a processor system as displayed by a logic analyzer. Data, address, and status codes are captured and displayed as they occur.

Symbolic Debugger: A development tool that permits the user to interactively examine and modify the execution of an application program at the level of a high-level language, such as the C language.

TAP Controller: In boundary scan, a synchronous finite state machine whereby sequencing through the various operations of the JTAG circuitry occurs under control of the test mode select signal.

Target System: The circuit of the product or system under development.

Text Editor: A program, generally interactive, that allows a user to create and modify files that contain text.

Time Processor Unit: An intelligent, semiautonomous microcontroller dedicated to timing control.

Timing Analysis: Analysis of a waveform display provided by a logic analyzer.

Translation Lookaside Buffer: The memory cache of the most recently used page table entries within the memory management unit.

Translator: A program that translates program source code into object code—that is, an assembler, compiler, or interpreter.

Transparency: A feature desired in emulators to make them as transparent as possible to the target system.

Utility: A software tool (program) that aids in the development of programs.

Abbreviations

ABI–Application Binary Interface
AP–Analysis Pod
ASIC–Application-Specific Integrated Circuit
BCD–Binary Code
BIDS–Built-in Development Support
BIST–Built-in Self-Test
BSA–Bus State Analyzer
CAD–Computer-Aided Design
CAE–Computer-Aided Engineering
CAM–Content-Addressable Memory
CASE–Computer-Aided Software Engineering
CISC–Complex Instruction Set Computer
CMMU–Cache Memory Management Unit
CMOS–Complementary Metal Oxide Semiconductor
COMP–Compare Status Bit

CPI–Clocks per Instruction
CPO–Coprocessor
CPU–Central Processing Unit
CPUTR–CPU Test Register
CREG–Test Submodule Control Register
DFT–Design for Testability
DOS–Digital Operating System
DRAM–Dynamic Random Access Memory
DREG–Distributed Register
ECL–Emitter Coupled Logic
EPROM–Electrically Programmable Read Only Memory
EVM–Evaluation Model
EVS–Evaluation System
FE–Falling Edge
FPU–Floating-Point Unit
H/W–Hardware
ICD–In-Circuit Debugger
ICE–In-Circuit Emulator
IEEE–Institute of Electrical and Electronic Engineers
I^2ICE–Integrated Instrumentation and In-Circuit Emulation
IMB–Intermodule Bus
IMP–Interactive Monitor Program
I/O–Input/Output
IPIPE–Interactive Pipeline
IU–Integer Unit
JEDEC–Joint Electron Device Engineering Council
JTAG–Joint Test Action Group
LENi–Breakpoint Length Specification Bits
LRU–Least Recently Used (referring to line replacement)
MAU–Math Acceleration Unit
MBTST–Intermodule Bus Test
MCU–Microcontroller Unit
MDS–Microprocessor Development System
MIPS–Microprocessor without Interlocking Pipeline Stages
MSR–Master Shift Register

MUT–Module Under Test
PCB–Printed Circuit Board
PDT–Prototype Debugging Tool
QSM–Queued Serial Module
QSPI–Queued Serial Peripheral Interface
RAM–Random Access Memory
RE–Rising Edge
RF–Resume Flag
RISC–Reduced Instruction Set Computer
ROM–Read Only Memory
RWi–Memory Access Qualifier Bits
SCI–Serial Communications Interface
SFU–Special Function Unit
SIM–System Integration Module
SLD–Source-Level Debugger
SPARC–Scalable Processor Architecture
SPP–System Programmers Package
SRAM–Static Random Access Memory
SVID–System V Interface Definition
S/W–Software
TAP–Test Access Port
TCK–Test Clock
TDI–Test Data Input
TDO–Test Data Output
TLB–Translation Lookaside Buffer
TMARM–Test Mode Armed Status Bit
TMS–Test Mode Select
TPU–Time Processor Unit
UART–Universal Asynchronous Receiver Transmitter
UUT–Unit Under Test

References

Articles and Papers

1. Baron, M., Ingram, K., Kitagawa, K., 1989, "Development Tools for RISC: An Array of Tools for SPARC Designs," Proceedings of BUSCON/89-West, 243–246.
2. Berger, A. S., 1988, "Host Independent Microprocessor Development Systems," *Hewlett-Packard Journal*, 45–51, December.
3. Berger, A. S., 1988, "Foreground/Background Monitor Increases Development System Transparency," *Personal Engineering & Instrumentation News*, 39–46, February.
4. Cates, R., and Farrell, J., 1989, "VL86C020(ARM III)-RISC Enters the Second Generation," MIDCON/89, Conference Record, 60–63.
5. Clarke, D., 1988, "System Programmer's Package," IEEE Program Session Record 36/4, Wescon, 1–8.
6. Davidow, W., 1975, "The Coming Merger of Hardware and Software Design," *Electronics*, 91–94, May 29.
7. Garrow, R., Hou, S., Lally, J., and Walker, H., 1975, "Microcomputer-Development System Achieves Hardware-Software Harmony," *Electronics*, 95–102, May 29.

8. Gelsinger, P. P., Gargini, P. A., Parker, G. H., Yu, A. Y. C., 1989, "Microprocessors Circa 2000," *IEEE Spectrum*, 43–47, October.
9. Goering, R., 1988, "Emulators Offer Improved Debugging and Increased Processor Coverage," *Personal Engineering & Instrumentation News*, 39–45, November.
10. Goss, R., 1989, "Motorola's 88000: Integration, Performance and Applications," Spring COMCON 89, 20–26.
11. Grimes, J., and Crawford, J., 1988, "What We Learned from RISC," IEEE Program Session Record 23/4, Wescon, 1–6.
12. Heath, P. N., 1988, "Advanced 16/32 Bit Microcontrollers," IEEE Program Session Record 18/4, Wescon, 1–8.
13. Hennessey, J. L., 1984, "VLSI Processor Architecture," *IEEE Transactions on Computers*, Vol. C-33(12), 1221–1246.
14. Izumikawa, S., 1986, "Analyzing Recent Logic Analyzers (1)," AEU, 39–43, May.
15. Izumikawa, S., 1986, "Analyzing Recent Logic Analyzers (2)," AEU 50–53, June.
16. Kohn, L., 1989, "The 64-bit i860 Microprocessor," MIDCON/89, Conference Record, 68–74.
17. Patterson, D. A., 1985, "Reduced Instruction Set Computers," *Communications of the ACM*, Vol. 28(1), 8–21.
18. Piepho, R. S., and Wu, W. S., 1989, "A Comparison of RISC Architecture," *IEEE MICRO*, Vol. 9(4), 51–52.
19. Puerifoy, S. W., and Berger, A. S., 1988, "Cabling Technology Vital to High-end Emulator Performance," *Personal Engineering & Instrumentation News*, 47–54, February.
20. Radin, G., 1983, "The 801 Minicomputer," *IBM Journal for R&D*, Vol. 27(3), 237–246.
21. Slater, M., and Wharton, J. H., 1989, "Revenge of the CISCs," *Byte*, 323–340, November.
22. Weaver, D., 1989, "Debugging Software for SPARC-based Systems," Proceedings of BUSCON/89-West, 451–461.

Manuals and Data Publications

Advanced Micro Devices

1. AM29000 32-Bit Streamlined Instruction Processor Users Manual
2. AM29000 32-Bit Streamlined Instruction Processor Memory Design Handbook.

American Telephone and Telegraph

1. WE321DS Microprocessor Development System User Manual
2. WE321DM CPU & MMU Device Monitor User Manual
3. WE321EB Microprocessor Evaluation Board (Data Sheet)

Cypress Semiconductor

1. Cypress RISC Seminar Notebook

Hewlett-Packard

1. From Scopes to Basic Logic Analyzers (Seminar Handbook)
2. Getting More from Your Logic Analyzer (Seminar Handbook)

Intel

1. 80386 Programmer's Reference Manual
2. i486 Microprocessor
3. i860 64-Bit Microprocessor
4. i860 64-Bit Microprocessor Programmer's Reference Manual
5. 80960 CA Product Overview
6. 80960CA-33, -25, -16, 32-Bit High-Performance Embedded Processor
7. 80960KB Programmer's Reference Manual
8. 80960KB Hardware Designer's Reference Manual

Joint Test Action Group (JTAG)

JTAG Boundary-Scan Architecture Standard Proposal Version 2.0

Motorola

1. MC88100 RISC Microprocessor User's Manual
2. MC88200 Cache/Memory Management User's Manual
3. MC68332 32-Bit Microcontroller, MC68300 Embedded Control Family Overview, Preliminary Edition, 1989.
4. TPU Time Processor Unit, MC68300 Embedded Control Family Reference Manual, Preliminary Edition, 1989

Prentice Hall

1. MIPS RISC Architecture, Gerry Kane, MIPS Computer Systems, 1988.
2. VL86C010 32-Bit RISC MPU and Peripherals Users Manual, VLSI Technology, 1989.

Appendix A

SPARC Uni-Module Board

Ross Technology, a subsidiary of Cypress Semiconductor, has developed several boards to assist in the application of its RISC microprocessors. An example is the SPARC Uni-Module Board running at 40 MHz, utilizing the CY7C604 Cache Controller and Memory Management Unit. This design presents a somewhat different configuration than that of the SPARC Evaluation Board described in Chapter 7.

The SPARC Uni-Module Board is a single board computer designed to show the complete SPARC chip set (Integer Unit, Floating Point Unit, Cache Memory Management Unit and Cache RAMs) operating at maximum speed. It contains a 64K-byte, direct mapped, *virtual* cache (1 CMU and 2 CRAMs). The PC board size is 3.3″ × 7.25″ and it has an Mbus interface via a 100 pin connector (see Figure A-1).

It is intended to allow the running of benchmark programs as well as providing a system environment for running/de-bugging the CMU. It also shows their customers how to interconnect their chip set via the *virtual* bus and, also, memory and I/O via the Mbus.

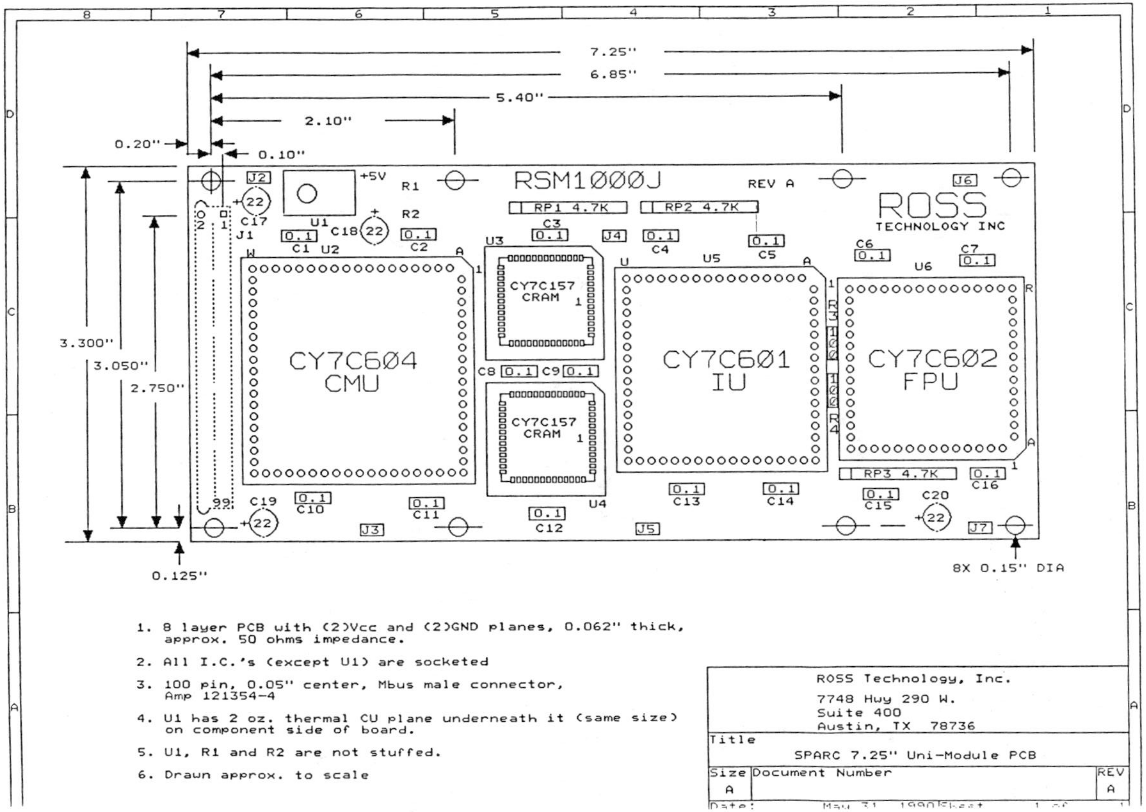

FIGURE A-1 SPARC Uni-Module Board developed by Ross Technology.

Features

1. CY7C601 SPARC Integer Unit
2. CY7C602 Floating Point Unit
3. CY7C604 Cache Controller and Memory Management Unit
4. 64K-bytes of direct mapped cache using 2, CY7C157 CRAM's (0 Wait States on *Virtual* bus)
5. Operates over a frequency range of 10–40 MHz at ambient temperature and nominal +5V
6. The board requires approximately 2A @ +5V via the Mbus connector
7. Onboard voltage regulator that converts +5V to +3.3V to drive all output buffers on all chips (including all outputs to the Mbus)

Basic Mbus Operation and Timing

The Mbus is a fully synchronous (same clock as IU and CMU), multiplexed (address and data), 64-bit bus. A cycle is started when MAS/ (Mbus Address Strobe) is asserted via the CMU and is completed successfully upon the assertion of MRDY/ alone or unsuccessfully with the assertion of various combinations of MERR/ (Mbus ERRor), MRDY/ (Mbus ReaDY) or MRTY/ (Mbus ReTrY) from the Mbus.

It allows multiple masters via an external arbiter on the Mbus. An Mbus master can request the Mbus by asserting his MBR/ (Mbus Bus Request) line to the arbiter and the arbiter can grant the bus by asserting the MBG/ (Mbus Bus Grant) line to the requester. Each potential bus master monitors the MBB/ (Mbus Bus Busy) line and, after receiving his bus grant and observing that MBB/ has been de-asserted, will synchronously assert MBB/ on the next clock and keep it asserted until his access is finished.

A Power On Reset (POR/) signal is generated to the CMU from the Mbus.

Level sensitive interrupts (15 max.) are generated to the CY7C601 IU via the IRL[3:0] lines from the Mbus. A value of 0000b means that there is no interrupt while a value of 1111b means an NMI is being asserted.

Basic Mbus timing is as follows:

1. The CY7C604, running at 40 MHz, makes address, data and status signals available 18 nS after the clock rising edge (RE) which gives 7 nS setup time before the following clock RE to latch them. It also holds these signals until 4 nS after the next clock RE. For control signals the respective timings are 16 nS after the clock RE (gives 9 nS setup time) and holds the signals for 4 nS after the clock RE.

2. The CMU requires that data, control and status signals be valid no later than 3 nS before the clock RE and that it be held for 2 nS after the clock RE.

Board Detailed Description

Fig. A-2 is a block diagram of the SPARC Uni-Module Board. For a more complete understanding, consult the Uni-Module Board schematic diagrams. The logic can be broken down into the following sections:

IU, CMU, CRAMs and FPU

The socketed chip set consists of the CY7C601 Integer Unit in a 207 pin PGA package, the CY7C602 Floating Point Unit in a 143 pin PGA package, the C67C604 Cache Controller and Memory Management Unit in a 243 pin PGA package and two CY7C157 Cache RAM's in 52 pin PLCC packages. These chips are connected together in a tightly coupled configuration to provide integer, floating point and memory management capabilities as well as 64 K-bytes of direct mapped, virtual, cache. The board/Mbus does not make use of any coprocessor signals nor the MHOLDB/, BHOLD/ or MAO signals.

It should be noted that the Uni-Module board is laid out to accept either the CY7C604 or CY7C605 and will operate properly with either chip. The extra signals (MIH/ and MSH/) are connected to the Mbus connector.

+3.3V Regulator

This circuitry consists of a 1.5A, 3-terminal, voltage regulator, a voltage divider consisting of a pair of 1% resistors to set the feedback voltage and a 22uF tantalum capacitor located close to the output terminals to stabilize the output voltage.

There are 6 jumpers (J2–J7) on the board which are connected between the +5V plane and the +3.3V plane. For test purposes the voltage regulator and voltage divider resistors can be removed and wire jumpers soldered onto the board. This allows testing speed and noise improvements when all outputs are operating at +3.3V vs +5V.

Board De-coupling and Signal Termination

There are various pull-up andpull-down resistors on the Uni-Module board to improve operation, testability and to allow the removal of the FPU. Multiple 0.1uF ceramic de-coupling capacitors are placed around each chip to provide power for instantaneous, high frequency, current requirements. Multiple 22uF tantalum de-coupling capacitors are placed near the Mbus connector and at the board edges to help provide a stable, low frequency, low impedance, power source.

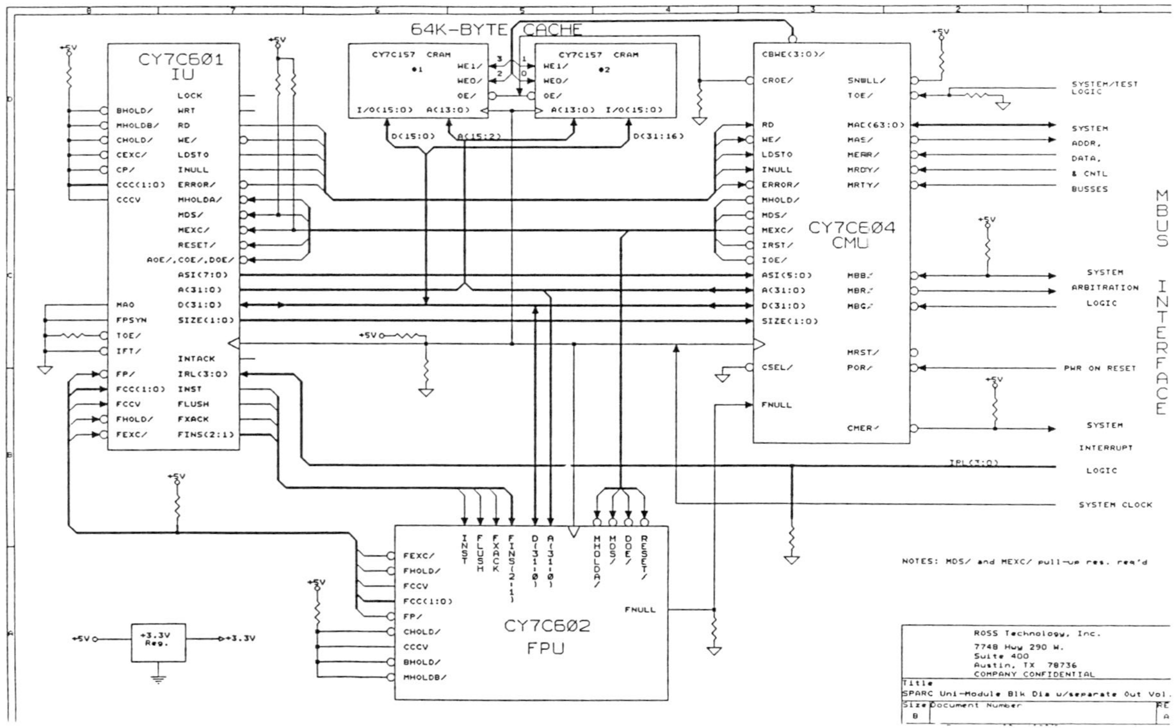

FIGURE A-2 Block diagram of SPARC Uni-Module Board. Courtesy of Ross Technology.

MDS/ and MEXC/ lines have pull-up resistors on them (required) since these are 3-state lines driven by the CMU. The MBB/line (OC) from the Mbus has a pull-up resistor as protection in case the connection to the Mbus fails in which case the CMU would never generate an Mbus request. There are similar pull-down resistors on the Mbus IRL3:0 lines so that the interrupt generation logic on the Mother board can be removed as an aid in de-bugging.

The CLK line from the Mbus connector has a parallel termination at its end (FPU). This is normally meant to be terminated by a Thevinin equivalent 50 ohms since this is the design impedance of the board but various types of terminations will be tesed for effectiveness.

There is a pull-down resistor on the TOE/ pin of the IU (DOE/ pin of the FPU, OE/ pin of the CRAM's) so that the outputs are always enabled except when 3-stated by the CMU via the IOE/ (CROE/) signal. The FNULL line from the FPU has a pull-down resistor on it so that if an FPU is not present, the IU and CMU will still operate correctly. The IFT/ pin of the IU is connected to GND so that the execution of an IFLUSH instruction will cause an illegal instruction trap. There are pull-up resistors on all of the coprocessor lines, MHOLDB/, BHOLD/, SNULL/, the floating point lines from the FPU to the IU and on the CMER/, MERR/ and MRTY/ lines of the CMU.

There is a pull-down resistor on the TOE/ pin of the CMU and the line is also brought out to the Mbus so that in a multi-master environment the CMU can be 3-stated off of the Mbus.

The CMU is always selected since its chip select pin is tied to GND. The MAO pin of the IU is similarly grounded to prevent falsely switching the internal source address mux of the IU.

Mbus Connector

The SPARC Uni-Module Board interface is via a 100 pin Mbus connector which has two signal rows plus center +5V and GND blades, 0.050″ spacing PCB pins, male, connector. It is a high density controlled impedance (50 +/− 10%) connector mfg. by Amp.

Each signal line within the mated connector is located at a specific distance from an integral, separable, bus bar serving as a ground plane in a microstrip configuration. The microstrip dimensions provide the controlled impedance and configuration reduces crosstalk and ground inductance.

Each inch length of connector houses two electrically isolated, high current (10A), contacts (blades). When soldered to the PCB GND (PWR) plane (via 4 pins), that plane is extended through the mated connctor. Resistance is minimized by a large contact area and short electrical length which provides signal return paths with negligible ground loop voltages.

Appendix B

In Tables 1 and 2, representative in-circuit emulators and logic analyzers for use in the development of products and systems, utilizing RISC microprocessors, are listed.

Table 1—Representative RISC ICEs

Vendor	Model	Supported processor	Price[1]	Comments
Embedded Performance	SYS29K	Am29000 Am29005	$15,000	Other supported products include R3000, R3000A, and R3001. Host can be IBM PC/AT or workstation from Sun, DEC or HP.
Hewlett-Packard Co	64774B	Am29000	$38,000	Has 512k bytes of emulation memory; $33,000 without memory. Use HP9000 Series 300 as host.
Intel	ICE960KB	i960KA/KB	$16,495	Requires IBM PC or compatible computer with 640k bytes of RAM, 1M byte of expanded RAM, 20M-byte hard disk, and MS-DOS V3.3.
Step Engineering	Adapt II	Am29000 Am29027	$13,500	Step offers an extensive set of integrated hardware and software development tools for these ICs. Use PC or ASCII terminal as host.
	Express960	i960CA	$13,500	

Notes: 1. Approximate US list price of the minumum usable configuration. Vendor-supplied software required for ICE operation is included; cost of the host program-development system is not included.

Table 2—Representative logic analyzers that support RISCs

Vendor	Model[1]	Max state-analysis speed	Supported processor	Price[2]	Comments
Arium	ML4400S	50 MHz	88100	$16,000	256k-byte ROM emulator $1995 extra.
		25 MHz	R3000	$9600	
Biomation	CLAS 4000	50 MHz	CY7C601 (SPARC)	$47,000	Packages support R3000, Am29000, 88100, 88200, i860, i960, and CY7C611.
Fluke	PM 3655/R	100 MHz	I960CA R2000	$13,200	400-MHz timing analysis $5500 extra.
	PM3585/90	50 MHz	See note 3.	$10,800	200-MHz timing analysis standard.
Hewlett-Packard Co	1650B[4]	35 MHz	88200	$9000	Preprocessors for i860 (PGA), 88100, 88200, and R3000[5]. Bus interfaces for Am29000, 88100, 88200, and i960CA.
Tektronix	DAS 9200	>33 MHz	88100	$40,000	Monitors M and P buses.
			I860		
VMEtro	PMA-030	25 MHz	88100 and 88200	$9000	Monitors M or P bus; for $18,000, monitors both. Require terminal or PC.

Notes: 1. In general, to specify a functioning logic-analysis system, your purchase order must list model numbers for a mainframe and a group of system components. This column shows only the mainframe model number.

2. These are approximate US list prices of complete systems equipped to work with the listed processor.
3. Introduction date of this product is October 1, 1990. Analyzer currently provides general-purpose RISC support.
4. In general, H-P's RISC accessories also work with the 1652B logic analyzer/scope and 16500A modular logic-analysis system.
5. Preprocessor for R3000 is offered by a third party. With R3000, analyzers support processor operation to 17.5 MHz.

Index

A

B

C

D

E

F

G

H

I

L

M

T

U

V

W

X

Z